Portable X-ray Fluorescence Spectrometry

Capabilities for *In Situ* Analysis

Portable X-ray Fluorescence Spectrometry

Capabilities for In Situ Analysis

Edited by

Philip J Potts
The Open University, Milton Keynes, UK

Margaret West
West X-ray Solutions Ltd, Sheffield, UK

RSC Publishing

ISBN: 978-0-85404-552-5

A catalogue record for this book is available from the British Library

Published by The Royal Society of Chemistry,
Thomas Graham House, Science Park, Milton Road,
Cambridge CB4 0WF, UK

Registered Charity Number 207890

For further information see our web site at www.rsc.org

Preface

Over the last 50 years, X-ray fluorescence (XRF) spectrometry has evolved from a manual wavelength dispersive analytical technique used by academics and a few adventurous industrial chemists to an automated, major tool recognised for its ability to provide comprehensive, quantitative analytical data for scientists and industrialists working in a laboratory, plant or field environment. Developments in excitation sources, detectors and microprocessor technology have facilitated the design of energy dispersive configurations capable of analysing most elements in the periodic table with detection limits that, under optimum conditions, fall below the $mg\,kg^{-1}$ level. In the last 20 years, portable XRF systems have also been developed and are now commercially available, making it possible to take the spectrometer to the sample rather than the other way round. These hand-held devices are particularly appropriate for the inorganic analysis of a variety of sample types *in situ* in the field to satisfy many needs of contemporary scientific interests.

This book brings together the knowledge and expertise of internationally recognised scientists with practical experience of *in situ* analysis using portable X-ray fluorescence technology. It offers a general introduction to the technique and its applications, including the assessment of contaminated land, surfaces, coatings and paints, workplace monitoring, metal and alloy sorting, geochemical prospecting, archaeological investigations, museum samples and works of art and extraterrestrial analysis. In addition, information is given on the analytical capabilities, instrumentation, quantification, correction procedures and sampling considerations specific to hand-held systems. Whilst many authors use commercially available instrumentation, applications such as the analysis of works of art, geochemical prospecting and extraterrestrial studies demonstrate the ingenuity of the authors to develop and build equipment for specifically demanding studies.

The chapters in this book are designed to enable scientists and students alike to appreciate the advantages offered by portable X-ray fluorescence spectrometry to support the concept of interactive sampling and analysis. *In situ* analysis entails placing the analyser in contact with the surface of the material of interest. No sample preparation is required – therefore the technique is particularly appropriate for situations where it is not possible or desirable to

Portable X-ray Fluorescence Spectrometry: Capabilities for *In Situ* Analysis
Edited by Philip J Potts and Margaret West

move the test piece. In many cases quantification for *in situ* XRF analysis may be complex, requiring careful consideration of physical matrix effects (related to particle size, surface irregularity, sample mineralogy and moisture), chemical matrix effects (including absorption and enhancement), and spectral interferences. All the interfering effects, together with the relevant correction procedures, are introduced and critically reviewed. In addition, parameters related to the applications of portable XRF spectrometers are defined, such as critical penetration depth, detection limit, precision, accuracy, role of blank samples, and total uncertainty of the *in situ* XRF measurements.

Several chapters demonstrate the ability of *in situ* measurements to provide supplementary information such as the determination of coating thickness and the identification of a manufactured alloy. Hand-held systems are particularly suited to situations where conformance to legislation is a priority, as demonstrated in the chapters on surveying contaminated land and workplace monitoring.

Future improvements in portable XRF systems are considered where the analyser will be expected to be a tool to solve problems and enable decisions to be made rather than simply produce data. Devices today may be equipped with wireless technology to link the spectrometer to a larger computer or with GPS capability for linking the measurement results with the geographical location. Given the multitude of possible situations and objects that may require testing there is still much untapped potential for the use of *in situ* hand-held XRF spectrometry. The way this potential will be exploited in the future, especially in the area of environmental analysis, will depend on changes in regulatory requirements, recognizing the continuing societal concern about minimizing the impact of human activities. Details in this book cover the contemporary standing of the technique, but readers will need to check and monitor changes that will influence the applicability of PXRF in the future.

Acknowledgements

The editors and authors thank the many scientists and practitioners who generously shared the information and data used in this book.

Philip J. Potts
Margaret West

Contents

Portable X-ray Fluorescence Spectrometry: Capabilities for *In Situ* Analysis
Edited by Philip J Potts and Margaret West

CHAPTER 1

Introduction, Analytical Instrumentation and Application Overview

PHILIP J. POTTS

Faculty of Science, The Open University, Walton Hall, Milton Keynes, MK7 6AA, UK

1.1 Portable X-ray Fluorescence

Portable X-ray fluorescence analysis offers some unique advantages that cannot be rivalled by any other analytical technique. These advantages arise not just from the multi-element capability of the technique, nor just from its non-destructive nature, but most importantly from the immediate availability to the operator of information on the chemical composition of a sample in the field. Thus, the operator has information in real time that can contribute directly to solving the problem for which analytical results are required, and indeed far more quickly than conventional sampling and laboratory analysis. These advantages of direct *in situ* analysis also give rise to several limitations to the technique that are summarized later in this chapter. However, before exploring these characteristics further, it is important to expand on the meaning of "portable" and, in particular, the capability of using PXRF for *in situ* determinations.

Many analytical techniques can be considered "portable" in the sense that they can be operated in a mobile laboratory. In this mode of operation, the mobile laboratory can be transported to a field site to provide an immediate laboratory analysis facility. The advantages of this mode of operation are that sample submission times and the reporting of analytical results can occur rapidly. Furthermore, investigators collecting and submitting samples

Portable X-ray Fluorescence Spectrometry: Capabilities for *In Situ* Analysis
Edited by Philip J Potts and Margaret West

for analysis also have immediate access to the chemical analyst operating the laboratory to provide expertise on data quality and the resolution of any analytical difficulties. Many techniques can be used in this way just as effectively as PXRF, and the analytical procedures used are likely to be the same as those adopted in a permanent laboratory. However, the particular advantage of PXRF results from the capability of the technique to undertake *in situ* analytical measurements. In this context, "*in situ*" is used to mean that the analytical instrument is taken to, and placed in contact with, the sample. An analytical measurement is undertaken and the result is immediately available to the operator. The operator can then make use of this information in deciding what to analyse next, giving rise to the concept of an "interactive sampling and analysis" capability. PXRF belongs, therefore, to a special category of "hand held" instrumentation.

The importance of PXRF in the field of *in situ* analysis is that few other techniques are capable of this mode of operation. Considering only the techniques used to determine inorganic elements (other portable techniques can measure organic species), the capabilities of the principle techniques that can be operated in this mode are shown in Table 1.1. The analytical characteristics of these techniques are outlined briefly below.

1.2 Techniques Capable of *In Situ* Portable Analytical Measurements

1.2.1 Spark Source Optical Emission Spectrometry

This technique uses the energy of an electrical discharge to ablate and excite a small mass of sample material. The energy of the spark causes atoms (and ions) of the sample to become excited and to emit characteristic optical lines from which elemental abundances can be determined. The technique offers high sensitivity to many low- and middle-order atomic number elements, with lower sensitivity for higher atomic number elements. The technique is very effective in the analysis of electrical conducting samples, and for this reason is widely used in applications such as alloy sorting. Non-conducting samples can only be analysed effectively if crushed and mixed with a conducting binder as, otherwise, the spark emission is highly erratic. The technique is not, therefore, suitable for the *in situ* analysis of non-conducting material where sample preparation is not possible.

1.2.2 Gamma Spectrometry

When applied to the analysis of natural samples, gamma spectrometry is used to measure gamma emissions associated with the radioactive decay of three elements, potassium, thorium and uranium. Potassium has a naturally occurring radioactive isotope, ^{40}K and the naturally occurring parent isotopes ^{235}U and ^{232}Th form a series of radioactive progeny isotopes. Because the half-life of these isotopes is in excess of 10^8 years, all occur naturally in rock samples. Gamma rays associated with each of these elements may be detected with a large volume

Table 1.1 Techniques capable of *in situ* portable analytical measurements of inorganic materials.

Technique	*Element range*	*Non-destructive?*	*Typical sample mass*	*Comments*
Portable X-ray fluorescence	Most of elements in periodic table above Si	Yes	mg to g, depending in energy of characteristic X-ray	Used in a wide range of applications involving the analysis of solid samples
Spark-source optical emission spectrometry	Many elements in periodic table, reduced sensitivity for high atomic elements	Small mass of material ablated	ng to μg range	Effective for analysing conducting samples, particularly metals and widely used for alloy sorting in scrap metal recycling
Gamma spectrometry	K, Th, U	Yes	tonne range	Normally used for field testing of rocks and soils, detection of mineralization and surveying for artificial radioactive contamination in the environment
Chemical test kits, anodic stripping voltammetry	Mainly trace elements	No	mL or g of sample	Mainly used to analyse waters, sometimes soils
Laser-induced breakdown spectrometry	Similar to SS-OES[a]	Small mass of material ablated	ng to μg range	Technique is rapidly developing and is not just limited to conducting samples

[a]SS-OES = spark source optical emission spectrometry.

sodium iodide scintillation detector. The technique has been used in geochemical exploration in the search for U/Th mineralization and also for the non-destructive provenancing of archaeological rock samples. However, the gamma emissions detected by this technique can penetrate through a significant mass of sample, so the analysed volume is in the cubic metre range, and the analysed mass is several tonne of material. The technique has $\mu g\ kg^{-1}$ detection limits for Th and U and is effective in measuring the average composition when large masses of material are present, a factor that could be an advantage or a limitation, depending on the application.

1.2.3 Chemical Test Kits

Some chemical test kits have been developed, particularly for geochemical exploration and lead in paint applications. A gel containing appropriate chemical reagents is painted onto a sample and changes colour if the element of interest is present. Normally the procedure is only capable of detecting a single designated element and this mode of operation would only be suitable for certain applications – particularly in geochemical exploration programmes in the search for mineralization.

1.2.4 Laser-Induced Breakdown Spectroscopy (LIBS)

This is a developing technique that is capable of the *in situ* analysis of samples. Essentially, the technique is the laser equivalent of spark source optical emission. In LIBS, the sample surface is excited by a laser. Interactions between sample and laser cause the ablation of a small mass, forming a highly energetic plasma. Atomic and ionic emission lines from this plasma may then be detected by an appropriate spectrometer, giving the technique a multi-element capability. The technique does not suffer the restrictions in requiring a conducting sample of the spark source technique, although matrix effects occur in relation to the amount of material ablated and influences on atomic and ionic emission intensities. Hand held instrumentation and quantification techniques are an area of significant activity.

1.2.5 Portable X-ray Fluorescence

The technique that is the subject of this monograph uses either a miniature X-ray tube or a sealed radioactive source to excite the sample with X-ray photons. These primary X-ray photons can excite secondary X-ray photons characteristic of the atoms present in the sample; the resultant X-ray spectrum is recorded with a suitable detector. In theory, the technique can measure almost all the elements in the periodic table. However, as PXRF measurements are normally undertaken in air, severe attenuation of the low-energy fluorescence X-rays occurs (unless special precautions are taken) so that elements below about Si (depending on instrument design) cannot be effectively detected. Elements that can be detected with highest sensitivity are those measured from the K-line series with absorption edges just below the energy of the characteristic emission lines from the excitation source. These are normally elements up to about Mo in the periodic table, but depend on the excitation source selected. The K-lines of higher atomic elements cannot always be excited with adequate sensitivity, but may be determined at a lower sensitivity and with some risk of additional spectrum overlap interference from their corresponding L-series lines. The mass of sample analysed depends on the energy of the characteristic fluorescent X-ray and its associated critical penetration depth within the sample. The critical penetration depth is the depth below the surface of the sample beyond which over 99% of the X-ray line emission of an element is absorbed within the sample and is not available for detection. Because the absorption characteristics of X-rays vary with energy, for the lower atomic number elements (which emit low energy fluorescence lines), critical penetration depths are in the μm range. For the K-lines of higher atomic elements, critical penetration depths are in the 1–10 mm range (noting the restrictions on exciting the higher atomic number K-lines mentioned above). The best estimate of analysed mass for elements routinely determined by PXRF is, therefore, in the tens of μg to hundreds of mg range. These concepts are described in more detail in the following chapters.

When evaluating the analytical characteristics of these techniques, it is apparent that PXRF has several advantages related to the range of elements that can be determined and the lack of restriction on the sample types to which the technique can be applied. However, several considerations additional to those applicable to conventional laboratory techniques must be taken into account.

1.3 PXRF Instrumentation

Portable XRF is one of the instrumental techniques that has developed rapidly in recent years, largely because of advances in miniaturization and semiconductor detector technology. Like other XRF instrumentation, PXRF consists of an excitation source, sample positioning facility, detector and pulse processing and analysis facility. However, portability means that in the selection and integration of these components, there is an emphasis on minimizing both mass and power consumption in an ergonomically designed instrument. The characteristics of each of these components are as follows.

1.3.1 Excitation Sources

1.3.1.1 Sealed Radioactive Excitation Sources

For earlier generations of instrument, sealed sources were the only practicable option and certainly meet the criteria of minimum mass with no inherent power consumption. However, only a limited number of sources have decay characteristics suitable for PXRF applications and relevant details; Table 1.2 lists the principal applications in the range of K-lines that can be excited.

One way of assessing the capabilities and applications of these sources is to compare them with the source excitation characteristics of the more familiar

Table 1.2 Sealed radioactive sources commonly used in PXRF instrumentation.

Source	*Half-life*	*Decay mode*	*Principal emission lines*	*Nearest equivalent X-ray tube*	*Range of element K-lines that can be effectively excited*
^{55}Fe	2.7 years	EC[a]	Mn Kα/Kβ	Cr anode	Na–Ti
^{238}Pu	86.4 years	Alpha[b]	U Lines	Mo anode	Ca–As
^{109}Cd	453 days	EC	Ag Kα/Kβ	Ag or Rh anode	Ca–Mo
^{241}Am	432.7 years	Alpha	59.5 keV	None	Fe–Gd

[a]EC = electron capture
[b]alpha = alpha particle decay

conventional X-ray tube. As can be seen from Table 1.2, the closest X-ray tube equivalents to these sources are ^{55}Fe – Cr anode, ^{109}Cd – Rh or Ag anode and ^{238}Pu – Mo anode. The 59.5 keV gamma emission from ^{241}Am has no direct X-ray tube equivalent, mainly because tube generators are often restricted to a maximum operating potential of 60 or 75 kV and so are not capable of efficiently exciting characteristic X-rays of this energy (as a rule of thumb, a potential of 3 to 4 times the X-ray emission energy is required for effective excitation). As with all XRF applications, choice of source is dictated by the intended application, with a combination of ^{55}Fe, ^{109}Cd and ^{241}Am being required for a comprehensive multi-element analysis capability. PXRF instruments that incorporate sources are generally simple in design, with the source offering stable and essentially monochromatic excitation characteristics. For elements with absorption edges just below the source emission energies, which are particularly well excited, the absence of a continuum component to the excitation spectrum offers the advantage of minimizing scatter under the fluorescence lines of interest, so avoiding a degradation in detection limits. The predictable decay characteristics (characterized by source half-life) permits a simple calculation to be made of the progressive reduction in source intensity based on the known half-life of the source.

However, there are several specific disadvantages to the use of radioactive source excitation especially in comparison with miniature X-ray tubes, as follows:

1. Instruments containing radioactive materials are covered by specific regularity requirements – in the UK the Radioactive Substances Act places a legal obligation on users to register instrumentation as "mobile radioactive apparatus" and places specific limitations on the countries covered by a particular operating licence. There are also additional duties of care placed on operators to ensure the security and safe operation of instrumentation. Because miniature X-ray tube instruments can be turned off when not in use, restrictions only apply when the instrument is in use.
2. Because of the need to provide radiation shielding, a restriction on the maximum activity of sources that can be incorporated in hand-held instruments means that sources used in PXRF are not as bright as is the emission available from miniature X-ray tubes.

3. Source excitation spectra cannot be optimized as flexibly as can miniature X-ray tubes (see below). In addition, the absence of a continuum component in the source emission spectrum means that sources may not be as efficient at exciting the lower atomic number elements.
4. The finite half-life of sources, in particular ^{55}Fe and ^{109}Cd (see Table 1.2), limits their useful life and dictates periodic replacement.
5. The permanent nature of source emissions means that special consideration is required to provide effective shielding and interlocks, even when the instrument is not in use.

1.3.1.2 Miniature X-ray Tubes

The development of miniaturization technology has led to the introduction of miniature X-ray tubes with power requirements that are compatible with battery operated instrumentation. As with all XRF instrumentation, in considering a specific application, a significant issue is the choice of anode in relation to the range of elements to be excited most efficiently (Table 1.2). To allow further optimization of excitation conditions, the most versatile instruments allow users to select tube kV and mA to optimize excitation conditions and, in the most adaptable instruments, by the selection of a primary beam filter. The latter normally consists of a thin metal foil placed between the X-ray tube and sample and is designed to modify the tube spectrum available to excite the sample. This can have the beneficial effect of reducing the intensity of the tube continuum in comparison with the characteristic tube lines, and so reducing the proportion of tube continuum available for scatter off the sample that would otherwise contribute to the detected background, so degrading detection limits. For highly specific applications (*e.g.* instruments optimized for the determination of single elements), balanced filters may be used but, in this instance, placed between the sample and the detector to attenuate the fluorescence spectrum available for detection. Balanced filters consist of a pair of foil filters of metals a few atomic numbers apart that provide a narrow energy window of high transmission centred on the emission line of the element of interest. Selectivity is provided by the adjacent absorption edges of the metal foils, such that higher energy radiation is absorbed by the higher atomic foil and *vice versa*. In this way, high transmission of the element line of interest is achieved in a manner that can be used to compensate for the poor resolution of proportional counters, for example, when used as energy dispersive detectors. Primary beam filters may be selectable by the operator, or fixed, depending on the design flexibility.

1.3.2 Sampling Positioning

Many laboratory XRF instruments possess elaborate sample exchange devices, often with samples held within an evacuated sample chamber. Whereas the same issues in ensuring accurate positioning of the sample in relation to excitation source and detector are as relevant to PXRF as to laboratory instrumentation, PXRF sample exchange is down to the operator! However, PXRF instruments must be provided with a clean surface for sample registration if results of the highest accuracy are required. The shape of the analyser snout may also be an important feature. For instruments designed for the analysis of welds, for example, a narrow pointed "snout" is more likely to allow the instrument to be positioned on the surface of a weld, if access is restricted. For instruments that are to be used for the analysis of soils in the field, the ability to set the instrument on the surface of the soil, without the risk of it falling over, is likely to be a significant consideration. Apart from these issues, the ideal sample surface is one that is perfectly flat and, therefore, exactly aligned with the analytical plane of the instrument. This is an ideal that is rarely achieved and bias in results can then be expected – associated with the inverse square law variation in detected intensities linked to deviations of the surface from the plane of registration. Several correction procedures have been developed to compensate for this effect (see Chapter 2).

A further issue is attenuation in air. Significant attenuation occurs for elements of atomic number below about Ca in the periodic table. This attenuation is severe for the lowest atomic number elements of interest such as Na. In laboratory XRF applications, this problem is avoided since many instruments incorporate a specimen chamber that can be evacuated. In many PXRF applications, this attenuation is either irrelevant, because the elements affected are not of interest to the application or alternatively can be tolerated, if the effect is not too severe. If, however, it is important to extend the range of detectable low atomic number elements, instruments have been developed that incorporate a small local vacuum pump or a helium flush facility. This facility is simple to apply by displacing the air path within the instrument, recognizing that the excitation source and detector assembly must be protected from the sample and external environment by a thin polymer membrane (to exclude dust and moisture). With the benefit of this facility, air attenuation will remain between the surface of the sample and analyser window, the magnitude of which will depend on the air path length.

1.3.3 Detectors

The single development that has had most influence on the development of hand-held XRF is the development of non-cryogenic semiconductor detectors. Energy dispersive XRF came of age in the 1970s with the introduction of Si(Li) detectors with their integral liquid nitrogen cryostats. Although semi-portable instruments were developed based on these devices, the additional size and weight of the cryostat hardly contributed to a practical portable device. The development of non-cryogenic detector technology swept away this limitation, the most influential of which has been the Si(PIN) device.

Si(PIN) Detector. This typically consists of a 300 μm thick layer of silicon with an active area 7 to 25 mm^2 diameter. Si(PIN) detectors are compact and offer good performance characteristics, without the need for cryogenic cooling (a small degree of Peltier cooling is sufficient). Best resolutions are of the order of 180 eV FWHM at 5.9 keV with detector effective energy range of 1–20 keV. Si(PIN) detectors have essentially replaced the Si(Li) detectors used in the earliest generation of instrumentation and are currently the detector device used in standard PXRF instrumentation.

Silicon drift detectors (SDDs). Silicon drift detectors represent a significant development in silicon wafer technology in which the capacitance of the detector has been substantially reduced to allow performance at substantially higher count rates than is possible with a conventional Si(PIN) device. This is achieved by designing the detector such that an electric field is applied, parallel to the surface such that electrons ionised by the detection of an X-ray event are caused to drift towards an anode at the centre. This field is created by many concentric ring electrodes etched into the surface, substantially increasing manufacturing costs. However, these costs are now sufficiently reduced to allow the more widespread use of SDDs in PXRF instruments, allowing detection at much higher count rates, whist retaining the advantage of performance with only Peltier cooling. This is an active area of development.

HgI_2. This is an energy dispersive X-ray detector based on a high purity mercury(II) iodide semiconductor crystal. Mercuric iodide detectors represent an evolving technology in which further improvements are likely to occur. The principal property of these devices is that they offer a reasonably good resolution response (about 250 eV at 5.9 keV) with a small degree of Peltier cooling. Furthermore, detection efficiency extends to significantly higher energies than silicon devices, allowing their use in measuring the higher energy fluorescence spectrum. Escape peaks from Hg (L lines) and I (K lines) may cause spectral overlap interferences in some applications.

CZT. Cadmium-zinc-telluride (CZT) represents one of several semiconductor materials that are being investigated for their X-ray detection properties. New detector materials will need to show clear advantages in detection characteristics, longevity and/or robustness compared with the detector types described in the above sections before they are likely to find application in commercial instrumentation.

Proportional Counters. Proportional counters do not represent semiconductor technology, but an older design that was used in earlier generations of PXRF instrumentation. Although the design is robust and energy efficient, the best resolution is of the order of 900 eV FWHM at 5.9 keV and it is not suitable for most multi-element applications because of the serious overlap interferences encountered in fluorescence spectra.

1.3.4 Pulse Processing and Data Analysis

The function of pulse processing electronics is to integrate the electronic charge created in the detector crystal each time an X-ray is detected and then convert it into a signal that can contribute to the acquisition of the fluorescence spectrum in a multichannel analyser. The latter device stores the accumulating spectrum as channel number (essentially linearly related to X-ray photon energy) *versus* counts (X-ray intensity) in a manner whereby one detected event contributes one additional count in the channel corresponding to the energy of the original detected event. Intensities are quantified in detected spectra, often using a simple region of interest integration and most instruments are supplied with matrix correction and quantification packages suitable for the application in question. Details of these procedures are covered in Chapter 2. Of interest to the user is likely to be the ease with which analytical results can be reviewed during an operating session, the number of results (and spectra) that can be stored in the instrument and the efficiency with which all data can be transferred to a computer for subsequent data processing. In some applications, the availability of integral GPS data will be important – especially where applications require the reconstruction of data for mapping.

1.4 Modes of Operation

In situ analysis by PXRF involves placing the analyser in contact with the surface of the sample to be analysed. No sample preparation is involved, the only flexibility available to the operator being sample *selection*. In many applications, the desire is to estimate the bulk composition of the sample from what is essentially a surface measurement. This gives rise to several issues concerning the representativeness of the measurement and the extent to which it can be used to estimate bulk composition. Issues include weathering or corrosion or compositional segregation at the sample surface, grain and mineral-size effects that may contribute to a significant sampling uncertainty affecting individual measurements, and surface texture effects that may require correction. These factors are discussed in more detail in later chapters, but are important if the full potential for quantitative analysis by PXRF (as opposed to qualitative identification of the elements present in a sample) is to be attained. Indeed, for the successful use of PXRF in ensuring that analytical results are interpreted correctly, it is essential that operators have the knowledge to bridge the gap that often blights other techniques between those with field experience and knowledge of the sample (who do not always have a full appreciation of the quality of analytical data and limitations in its interpretation) and those with laboratory expertise and knowledge of the analytical technique (who do not always understand sampling, the concept of sampling design and uncertainty and limitations on the interpretation of data associated with the nature of the sample). In PXRF, these activities are normally rolled into one.

A further issue that requires some consideration is "portability". Earlier PXRF instruments might be regarded as "transportable" as miniaturization had not developed, to the extent that instruments remained bulky and heavy for use over extended periods. This is not the case with the modern generation of instruments, which can truly be regarded as hand-held devices. The apotheosis of such instrumentation are systems designed for extraterrestrial measurements, exemplified by the Beagle 2 PXRF, of mass 280 g, designed for measurements of the rock and soil on the surface of Mars (see Chapter 10).

No discussion about PXRF can be completed without reference to health and safety. In one sense, "XRF" and "hand-held" operation are incompatible concepts, because of the risk associated with the exposure of the operator or others to ionizing radiations. However, modern instruments incorporate a range of safety features, including key operation, relevant interlocks and contact sensors all designed to minimize the risk of significant exposure. It is essential that any operator is fully trained in these features.

1.5 Applications

The principal advantages of PXRF – its non-destructive characteristics, multi-element capability, coupled to the immediate availability of analytical results on completion of a measurement – give the technique significant advantages in a range of applications, as highlighted in following chapters. However, the concepts into which these applications fall can be summarized as follows.

1.5.1 Interactive Sampling and Analysis

As has already been alluded to, the way in which analytical results are available to the operator immediately after (or, in the case of some instrument designs, during) a measurement permits the use of instrumentation in an interactive sampling and analysis mode. As a result, decisions about what to analyse next can be made on the basis of measurements just made. One example is the analysis of contaminated soil for lead (Pb) (see Chapter 3). Systematic sampling of a site might reveal an unexpected hotspot that could then be characterized in detail by supplementary *in situ* measurements. In this way, a significant reduction in time may be achieved in resolving the task to hand and expenses might be confined to a single field visit. There are, of course, dangers in discarding a systematic sampling approach to avoid a preliminary survey being undertaken on the basis of unsubstantiated assumptions but these are likely to be more than outweighed by the advantages above.

1.5.2 Judgemental Sampling and Analysis

A second area where PXRF offers significant advantages is in applications where a systematic sampling design is not possible. In this mode of operation, PXRF is likely to be used in a problem solving or trouble shooting mode and/or to answer questions such as "What is it?" or "Where is it?". One example was a project in which PXRF was used to identify residual hazards associated with arsenic contamination of an abandoned copper/tin ore calciner of industrial heritage interest in Cornwall UK.[1] Assumptions about the major source of contamination were not supported by field measurements, and on site interpretation of field results led to subsequent work that demonstrated that brick work (but not non-porous stone) used to construct flues formed a significant reservoir of arsenic, which was continually being leached out by the action of rain water. It would have been almost impossible to test this hypothesis by a pre-planned programme of sampling.

1.5.3 Conventional Sampling and Analysis Not Possible

The final area highlighted here where PXRF shows substantial advantages is in applications, for example, involving unique and/or valuable archaeological artefacts, museum samples and works of art (see Chapter 9), where it is not acceptable to remove a sample for conventional laboratory analysis. The only acceptable technique must be both portable and non-destructive, properties for which it is hard to rival PXRF. In addition, it is sometimes not possible to remove the complete artefact to the laboratory for analysis (*e.g.* samples that are integral to either monumental stone work or with a building). In other circumstances (notably in the case of museum artefacts or any particularly valuable object) curators or custodians are not willing to relinquish custody, because of their duty of care or security issues. In these circumstances, PXRF offers many advantages and in some circumstances is the only technique capable of making the measurements.

This monograph has, therefore, been designed to offer an overview into PXRF sampling design, sample selection, optimization of the analytical methodology, correction of results and the interpretation of data, taking into account the integrated nature of PXRF sampling and analysis. Details are also given of a range of applications to which PXRF can make a unique and essential contribution.

Reference

1. P.J. Potts, M.H. Ramsey and J. Carlisle, *Journal of Environmental Monitoring*, 2002, **4**, 1017–1024.

CHAPTER 2

Quantification and Correction Procedures

ANDRZEJ A. MARKOWICZ

IAEA Laboratories, A-2444 Seibersdorf, Austria

2.1 Overview

This chapter gives an introduction to quantitative X-ray fluorescence analysis, and includes basic expressions for the intensities of the characteristic X-rays and scattered primary radiation. Major parameters related to the applications of portable XRF spectrometers are defined, such as critical penetration depth, detection limit, accuracy, precision, role of blank samples and total uncertainty of the *in situ* XRF measurements. In most cases, quantification for *in situ* XRF analysis is complex and requires a careful consideration of (a) physical matrix effects related to particle size, surface irregularity, sample mineralogy and moisture, (b) chemical matrix effects, including absorption and enhancement and (c) spectral interferences. All the interfering effects, together with the relevant correction procedures, are introduced and critically reviewed. Overall, the chapter provides the basic foundation required to understand quantitative XRF analysis and gives a necessary review of the practical applications of portable XRF instruments for *in situ* analysis described in other chapters of this book.

2.2 Introduction

Quantitative X-ray fluorescence analysis is based on the relationship between the measured characteristic X-rays of an analyte element and excitation source

Portable X-ray Fluorescence Spectrometry: Capabilities for *In Situ* Analysis
Edited by Philip J Potts and Margaret West

intensity. Derivation of the expression for the measured characteristic X-rays is available in relevant books.[1,2]

When continuous (polychromatic) radiation is used for excitation, the intensity of the characteristic X-rays from a homogeneous and flat sample is described by:[3]

$$I_i \mathrm{d}\Omega_1 \mathrm{d}\Omega_2 = \frac{\mathrm{d}\Omega_1\, \mathrm{d}\Omega_2}{4\pi} \frac{\varepsilon(E_i)}{\sin \Psi_1} \times \int_{E_{c,i}}^{E_{\max}} a_i(E_0) \frac{1 - \exp[-\rho\, T(\mu(E_0) \csc \Psi_1 + \mu(E_i) \csc \Psi_2)]}{\mu(E_0) \csc \Psi_1 + \mu(E_i) \csc \Psi_2} I_0(E_0) \mathrm{d}E_0 \tag{2.1}$$

with:

$$a_i(E_0) = C_i \tau_i(E_0) \omega_i p_i \left(1 - \frac{1}{j_i}\right) \tag{2.2}$$

where $\mathrm{d}\Omega_1$ and $\mathrm{d}\Omega_2$ are the differential solid angles for the primary and characteristic radiation, respectively, $\varepsilon(E_i)$ is the intrinsic detector efficiency for the characteristic radiation of energy E_i, $E_{c,i}$ and $E_{\max}$ are the critical absorption energy for the analyte i and the maximum energy of the primary spectrum, T is the thickness of the analyzed sample, Ψ_1 and Ψ_2 are the effective incident and take-off angles, respectively, ρ is density of the analyzed sample, $\mu(E_0)$ and $\mu(E_i)$ are the total mass attenuation coefficients of the primary and characteristic radiation in the whole sample, respectively, $I_0(E_0)\mathrm{d}E_0$ is the number of the primary photons per second per steradian in the energy interval E_0 to $E_0 + \mathrm{d}E_0$, C_i is the concentration (weight fraction) of the analyte i in the analyzed sample, $\tau_i(E_0)$ is the photoelectric mass absorption coefficient for the analyte i at the energy E_0, ω_i is the fluorescence yield for the analyte i, p_i is the transition probability for a given characteristic X-ray line of the analyte i and j_i is the jump factor at the absorption edge of the analyte i.

In case of monochromatic excitation and fixed measurement geometry, Equation (2.1) can be simplified to:[3]

$$I_i = G \frac{\varepsilon(E_i) a_i(E_0) I_0(E_0)}{\sin \Psi_1} \frac{1 - \exp[-\rho\, T(\mu(E_0) \csc \Psi_1 + \mu(E_i) \csc \Psi_2)]}{\mu(E_0) \csc \Psi_1 + \mu(E_i) \csc \Psi_2} \tag{2.3}$$

where G is the geometry factor (constant for the fixed measurement geometry).

Equation (2.1) or (2.3) shows that the intensity of the characteristic X-rays of the ith element present in a multi-element sample depends not only on the concentration of the ith element but also on the overall composition of the sample (through the total mass attenuation coefficients of the primary and characteristic radiation in the whole sample); this dependence is the source of the so-called matrix absorption effects.

In the case that the analyte, element i, is accompanied by matrix elements that are excited by the primary radiation and emit the characteristic X-rays of energy higher than the critical absorption energy for the ith element, an additional excitation of the characteristic X-rays of the *i*th element can be observed; this extra excitation by the characteristic X-rays of matrix (interfering) elements is responsible for an enhancement effect. For monochromatic excitation the enhancement effects will modify Equation (2.3) by a factor $(1 + H_i)$, where H_i is the enhancement term given by:[3]

$$H_i = \frac{1}{2\mu_i(E_0)} \sum_{k=1}^{m} C_k \omega_k \left(1 - \frac{1}{j_k}\right) \mu_i(E_k) \mu_k(E_0) \times \left[\frac{\ln(1 + \mu(E_0)/[\mu(E_k)\sin\Psi_1])}{\mu(E_0)/\sin\Psi_1} + \frac{\ln(1 + \mu(E_i)/[\mu(E_k)\sin\Psi_2])}{\mu(E_i)/\sin\Psi_2} \right] \tag{2.4}$$

where $\mu_i(E_0)$ and $\mu_i(E_k)$ are the total mass attenuation coefficients for the *i*th element at the energy of primary radiation (E_0) and characteristic X-rays of the element k (E_k), respectively, $\mu_k(E_0)$ is the total mass attenuation coefficient for the element k at the energy E_0, $\mu(E_k)$ is the total mass attenuation coefficient for the whole sample at the energy E_k, m is the number of the enhancing elements in the analyzed sample.

2.2.1 Types of X-ray Fluorescence Technique

Portable X-ray fluorescence spectrometers are applied for the analysis of various samples of different thickness (or total mass per unit area m). Taking into account the thickness of analyzed samples, XRF can be classified as follows (for simplicity, considerations are limited to excitation using monochromatic radiation).

2.2.1.1 Thin Sample Technique

In this case, the mass per unit area of a sample m must be smaller than a limiting value m_{thin} defined by:

$$m_{\text{thin}} \leq \frac{0.1}{\mu(E_0)\csc\Psi_1 + \mu(E_i)\csc\Psi_2} \tag{2.5}$$

the intensity of the characteristic X-rays is given by:

$$(I_i)_{\text{thin}} = K_i m_i \tag{2.6}$$

where K_i is the calibration factor (also called the sensitivity factor), which can be calculated theoretically or evaluated experimentally from the measurements of thin homogeneous calibration samples; and m_i is the mass per unit area of the *i*th element in the homogeneous thin sample. Since the mass per unit area of the *i*th element (or its total mass on a sample) is a linear function of the intensity of the characteristic X-rays, the matrix effects (both absorption and enhancement) can be neglected in the thin sample technique.

2.2.1.2 Thick Sample Technique

If the mass per unit area (thickness) of a sample is sufficiently large, and exceeds the so-called saturation mass m_{thick} defined as:

$$m_{\text{thick}} \geq \frac{4.61}{\mu(E_0)\csc\Psi_1 + \mu(E_i)\csc\Psi_2} \tag{2.7}$$

the intensity of the characteristic X-rays of the ith element $(I_i)_{\text{thick}}$ is described by:

$$(I_i)_{\text{thick}} = \frac{G\varepsilon(E_i)a_i(E_0)I_0(E_0)}{\mu(E_0) + (\sin\Psi_1/\sin\Psi_2)\mu(E_i)} \tag{2.8}$$

For samples of thickness greater than m_{thick}, practically no further increase in the intensity of the characteristic X-rays is observed.

2.2.1.3 Intermediate Thickness Sample Technique

To cover the whole range of sample thicknesses, one has to consider samples whose masses per unit area m fall into the range:

$$m_{\text{thin}} < m < m_{\text{thick}} \tag{2.9}$$

For these samples (called intermediate thickness samples), the intensity of the characteristic X-rays of the ith element is given by Equation (2.3). One can see that, in this case, the intensity depends not only on the overall composition of the sample but also on its thickness (mass per unit area) and this factor introduces an additional complication in quantitative XRF analysis.

2.2.2 Scattering of Primary Radiation

Some quantitative procedures in XRF analysis are based on scattered primary radiation, which can be treated as a "fluorescent peak" from an internal standard, because it suffers matrix absorption effects similar to that of the characteristic X-rays of the elements to be determined. Moreover, the scattered primary radiation behaves in a similar manner to fluorescent radiation with variations in instrumental design as well as with the interfering effects often present in *in situ* XRF measurements. Two types of the scattering effects exist.

2.2.2.1 Compton (Incoherent) Scattering

In this interaction an incoming photon collides with a free electron, loses some of its energy and is deflected from its original direction.[3,4] The energy of the Compton scattered photon (E_{Com}) depends on the scattering angle, and the intensity of the Compton scattered radiation I_{Com} for monochromatic excitation is given by:[3]

$$I_{Com} = \frac{G' I_0(E_0)\sigma_{Com}(E_0)\{1 - \exp[-\rho T(\mu(E_0)\csc\Psi_1 + \mu(E_{Com})\csc\Psi_2)]\}}{\mu(E_0)\csc\Psi_1 + \mu(E_{Com})\csc\Psi_2} \tag{2.10}$$

where G' is a constant for a given measurement geometry and detection efficiency, $\sigma_{Com}(E_0)$ is the Compton mass scattering coefficient of the analyzed sample for the primary radiation of energy E_0, and $\mu(E_{Com})$ is the total mass attenuation coefficient of the Compton scattered primary radiation of energy E_{Com} in the analyzed sample.

Assuming that the ratio of the atomic number to the mass number (Z/A) is the same (or similar) for all elements present in the sample and that $\mu(E_o) \cong \mu(E_{Com})$, the following simplified formula for the intensity of Compton scattered radiation can be used:[3]

$$I_{Com} = \frac{G'_1 I_0(E_0)\{1 - \exp[-\rho T\mu(E_0)\csc\Psi_1(1 + \csc\Psi_2/\csc\Psi_1)]\}}{\mu(E_0)\csc\Psi_1(1 + \csc\Psi_2/\csc\Psi_1)} \tag{2.11}$$

where G'_1 is a constant factor. Based on the measurement of the intensity of the Compton scattered radiation I_{Com} one can assess from Equation (2.11) the mass attenuation coefficient of the primary radiation in the whole sample $\mu(E_o)$.

2.2.2.2 Rayleigh (Coherent) Scattering

In this interaction, the incoming photons are scattered by bound atomic electrons, with the atoms of the target being neither ionized nor excited. The coherent scattering has the highest probability for low energies and high Z materials, and is confined to small angles in the forward direction. The intensity of the coherent scattered radiation I_{coh} can be calculated from:[3]

$$I_{coh} = \frac{G'' I_0(E_0)\sigma_{coh}(E_0)\{1 - \exp[-\rho T\mu(E_0)(\csc\Psi_1 + \csc\Psi_2)]\}}{\mu(E_0)(\csc\Psi_1 + \csc\Psi_2)} \tag{2.12}$$

where G'' is a constant for a given measurement geometry and detection efficiency for the primary radiation, and $\sigma_{coh}(E_0)$ is the coherent mass scattering coefficient of the analyzed sample at the energy of the primary radiation.

Compton and Rayleigh scattered radiation are used in several quantitative procedures for both laboratory and *in situ* XRF measurements (more details are given in the following sections).

2.3 General Considerations

In this section basic parameters related to the applications of portable XRF spectrometers for quantitative *in situ* measurements are characterized.

2.3.1 Critical Penetration Depth

One of the major features of X-ray fluorescence analysis is the fact that the measured intensity of the characteristic X-rays of the analyte element originates from a layer of the sample, the thickness of which is equal to the critical penetration depth (often called information depth). This results from the attenuation of the primary radiation and characteristic X-rays in the analyzed sample. The critical penetration depth d_{crit} is defined as the sample thickness from which 99% of a fluorescence signal originates:[5]

$$d_{crit} = m_{thick}/\rho = 4.61/(\rho\mu_{tot}) \tag{2.13}$$

where ρ is the density of the sample material; $\mu_{tot} = \mu(E_0)\csc\Psi_1 + \mu(E_i)\csc\Psi_2$.

In a similar way one can derive that[5] 90% of the fluorescence signal originates from a layer of thickness $d_{90\%} = 2.303/\rho\mu_{tot}$; 80% of the fluorescence signal originates from a layer of thickness $d_{80\%} = 1.609/\rho\mu_{tot}$; and 50% of the fluorescence signal originates from a layer of thickness $d_{50\%} = 0.693/\rho\mu_{tot}$.

Based on these figures one can conclude that a sample layer of thickness of 50% of the critical penetration depth contributes 90% of the measured fluorescence signal, 35% of d_{crit} contributes 80% of the signal and 15% of d_{crit} contributes only 50% of the fluorescence signal.[5] A practical conclusion is that, whenever XRF is used for the analysis of bulk materials, the critical penetration depths should be assessed for all the elements to be determined (approximate concentrations of the elements in the bulk materials are often sufficient). The concept of the critical penetration depth is of particular importance in the *in situ* XRF analysis of heterogeneous samples (mineralogy effects), samples of irregular surface (weathering effects) or samples with surface contamination where minimal or no sample preparation is involved. Therefore, the extrapolation of results obtained for a relatively thin near-surface layer of thickness d_{crit} to estimate composition of bulk materials should be performed extremely carefully.

2.3.2 Analytical Parameters of XRF Analysis

Precision refers to a degree of agreement between the results of replicate measurements and provides an estimate of random error. The precision of a method is usually monitored by analyzing a sample containing the target analytes at various concentration levels close to the concentration expected in the samples to be analyzed. A minimum of one control sample should be run each day. When portable XRF spectrometers are applied for *in situ* analysis of soil and sediments, it is extremely important (recommended) to determine the precision of the method for the analytes at concentrations near the site action levels.[6] Based on the replicate measurements of a control sample the relative standard deviation (RSD) is calculated and used as an assessment of the precision of the method:

$$\mathrm{RSD} = (\mathrm{SD}/\text{mean concentration}) \times 100\% \tag{2.14}$$

where SD is the standard deviation calculated from replicate measurements of the concentration of the analyte.

Accuracy refers to the difference between a sample result and the reference or true value. One of the methods to assess accuracy (or bias, which is a measure of the departure from perfect accuracy), and to carry out a performance test of a portable XRF spectrometer and associated analytical procedure, is based on the analysis of a certified reference material (CRM) that contains certified amounts of the elements in a representative matrix. Another way to assess accuracy requires the analysis of the so-called confirmatory samples by using a reference (called also confirmatory) laboratory method that is considered to be a standard (approved) analytical method for the determination of the target analytes in a given matrix. The confirmatory samples (minimum 10% of total number of the analyzed samples) must be splits of the well homogenized sample materials, which should be selected from the lower, middle and upper range of the concentrations measured by the portable XRF spectrometer and/or *in situ* analytical procedure. They should also include samples with the analytes at concentrations near the site action levels. The results obtained using a portable XRF instrument and the results of the confirmatory analysis should be evaluated with a least-squares linear regression analysis. The correlation coefficient (r^2) defines a quality of the XRF results: if the correlation coefficient is between 0.7 and 0.9 the portable XRF is suitable for providing screening level data; a correlation coefficient larger that 0.9 indicates that the portable XRF data and confirmatory data are statistically equivalent at a 99% confidence level.

Detection Limit (DL) is the lowest concentration level that can be determined as being statistically significant from an analytical blank.[7] In laboratory XRF applications, the DL is defined as the concentration or amount of the analyte that gives a measured signal (net peak intensity of the characteristic X-rays) equal to three times the standard deviation of background in the energy region of the characteristic X-ray line. This simple definition of the DL (often called interference-free DL) gives values that are not fully adequate for *in situ* measurements using portable XRF instruments. Instead, two different approaches are applied. In the first approach, the DL is calculated as 3 times the standard deviation of the results for low-concentration samples with the concentration of the analyte 5–10 times higher than the estimated DL. This so-called precision-based DL can be determined for replicate analyses of (a) suitable site specific calibration standards (SSCS) representative of the samples to be analyzed, (b) appropriate certified reference materials (CRMs) or (c) a clean sample matrix spiked with the analytes of interest to represent the low concentrations.[7] Based on these results, one can also calculate the quantification limit for the method as 10 times the standard deviation. In a second approach, the DL is determined by analysis of the low concentration outliers on data cross plots for the $\log_{10}$ transformed portable XRF results *vs.* $\log_{10}$ transformed results obtained by a reference (confirmatory) method.[7] For all analytes, these cross plots exhibit a region below the DL where the linearity of the relationship disintegrates. The value of the DL corresponds to the concentration where the linear relationship disintegrates, representing the so-called field or performance-based detection limit.

2.3.2.1 Blank Samples

Field portable XRF analysis requires both instrument blanks and method blanks. An instrument blank is used to verify that no contamination exists in the spectrometer or on the probe window. In the determination of metals in soils, Piorek[8] recommends the use of silicon dioxide powder, a Teflon block, a quartz block, clean sand or lithium carbonate for the instrument blank. The instrument blank should be analyzed daily, before and after analyses, and additionally whenever contamination is suspected. In the ideal case, no concentrations above the detection limits should be found in the instrumental blank. If any measured concentrations exceed the DLs, then the probe window and the test samples (instrument blanks) should be checked for contamination. A method blank is used to monitor for laboratory-induced contaminants (introduced, for example, during sample preparation) or interferences. In the analysis of soil, the method blank can be "clean" silica sand or lithium carbonate that undergoes the same sample preparation procedure as the analyzed samples. The method blank should be analyzed at least daily and the frequency of analysis depends on the data quality objectives of the project. To be acceptable, a method blank must not contain any analyte at a concentration above the DL. In cases where the analyte's concentration exceeds the DL for that particular analyte, the cause of the problem must be identified and all samples reanalyzed.[8] For other *in situ* XRF measurements, the composition of the instrument and method blanks should be adjusted to the composition of the analyzed samples.

2.3.3 Total Uncertainty of *In Situ* XRF Measurements

In general, the total uncertainty of XRF measurements should include the contributions from the individual uncertainties introduced at all stages of the measurement process. Therefore, the total measurement variance σ^2_{tot} can be given as the following sum:[9]

$$\sigma^2_{\text{tot}} = \sigma^2_{\text{sample representation}} + \sigma^2_{\text{sample collection}} + \sigma^2_{\text{sample handling}} + \sigma^2_{\text{sample preparation}} + \sigma^2_{\text{analysis}} \tag{2.15}$$

As can be seen from Equation (2.15) the uncertainty due to the analytical stage itself, σ_{analysis}, is only one of the sources of uncertainty, and may be a minor, or even negligible, contributor to σ_{tot}. In most *in situ* XRF applications, a major contribution to the total measurement uncertainty comes from sample representation (see next section for details) as well as sample collection, handling and preparation. Therefore, even an extremely accurate analysis stage supported by a sophisticated quantitation procedure cannot improve the overall measurement uncertainty substantially. An assessment of the individual contributions of the different sources of uncertainty is advisable to better define the requirements (limitations) of the analytical method and, consequently, to select a proper quantification procedure.

2.4 Factors Influencing Accuracy

In *in situ* XRF analysis, one has to consider a number of the influencing factors that are not present in laboratory XRF measurements, where the analyzed samples presented to an XRF spectrometer are usually ideally homogeneous and flat. The existence of the influencing effects and the extent of their influence on the XRF results depends on the type of *in situ* measurements performed (*in situ* unprepared, *in situ* prepared, intrusive unprepared or intrusive prepared).[6,8]

2.4.1 Physical Matrix Effects

These effects are related to variations in the morphology or physical character of the sample, including such factors as uniformity, homogeneity, moisture content, particle size and surface conditions.

2.4.1.1 Particle Size Effects

To illustrate the importance of the particle size and heterogeneity effects let us consider the determination of chromium in soil.[10] Assuming an irradiated area of the sample is 1 cm^2, a critical penetration depth in silica matrix of 0.2 mm and detection limit of 200 mg kg^{-1}, one can calculate that the amount of chromium in the analyzed layer is 8.4 μg. This small amount of chromium may or may not be evenly distributed in the analyzed layer. In an extreme case one can assume that chromium is present as a single grain of Cr metal 135 μm in diameter. Should this single particle be located deeper than 0.2 mm from the surface no characteristic X-rays would be detected. A similar zero result would be observed if the grain of Cr were located at the surface but outside the 1 cm^2 area irradiated by the primary beam. All those results that might be classified as outliers are only because Cr may be a single grain. Even in the case where the analyte is present as very fine particles, its characteristic X-ray intensity (and concentration) will depend on how the fine particles are distributed within the grains of matrix. Another manifestation of the particle size effect is the dependence of the characteristic X-ray intensity on the shape and size of the particles.[3,11,12] If a sample contains fine particles, the XRF gives a higher concentration of the analyte than for a sample containing larger grains (despite the concentrations of the analyte in both samples being the same). To reduce such errors, which are the largest for low atomic number elements such as K, Ca, Ti, V, Cr, and Mn, granular samples should be ground and sieved to a uniform and small particle size fraction. This may not be possible using *in situ* PXRF protocols.

Another concern in *in situ* XRF analysis is heterogeneity of the analyzed materials. If possible, a preliminary (often simple) procedure to homogenize the material should be applied. This effort is of particular importance during validation of XRF results through confirmatory analysis by using an established laboratory analytical technique.

2.4.1.2 Surface Irregularity Effects

Quantitative results obtained in the analysis of geological and archaeological samples by using portable XRF spectrometers are affected by surface irregularity effects. The discrepancy arises for samples that have irregular shaped surfaces, because the portable instrument is calibrated by using flat calibration samples, and the surface of the analyzed sample is not positioned exactly in the reference plane of the analyzer unit. Such surface irregularity effects were studied experimentally by Potts *et al.*[13] These authors demonstrated that, as a result of the inverse square law, the detected fluorescence X-ray intensities, in particular for low-Z elements, are systematically lower than those observed from flat samples. The experiments were performed for a set of rock reference materials at different analyzer-to-sample surface distances. Even small discrepancies between the reference plane and analyzed surface had a significant effect on the measured X-rays. The fall in the intensity of the fluorescent lines ranged from 72 to 87% of its value at 0 mm for an air gap of 2 mm, for analytes such as K, Ca, Ti, Fe, Sr, Zr and Ba; equivalent data for an air gap of 4 mm were 49–64%. Ignoring the surface irregularity effects indicates that considerable errors in quantitative XRF analysis would be observed. A theoretical model to study the surface irregularity effects has been proposed by Liangquan *et al.*[14] These authors considered the following four types of geometrical structure for the rock surface: convex, concave, plane and undulating. Surface irregularity effects for the characteristic X-rays were described by a so-called unevenness factor that is a function of the following three parameters: the source–sample distance, the surface peak–valley amplitude, and the frequency number of the convex and concave surfaces within the effective detection area of the XRF instrument.

2.4.1.3 Mineralogy Effects

In conventional laboratory XRF, appropriate sampling and sample preparation procedures are applied to make the sample taken for analysis homogeneous and fully representative of the bulk material. By contrast, *in situ* XRF measurements are often performed directly on the analyzed material, *e.g.*, rock, with no sample preparation. In such a case, the analytical results will depend on the mineral assemblage present in the excited volume (information volume), which is defined as the critical penetration depth multiplied by the irradiated area of the sample; the associated effects are called mineralogy effects. While discussing the mineralogy effects, one has to consider two additional parameters. The first is the grain size compared to the volume from which the fluorescence signal originates – the larger the grain size and the smaller the excitation volume, the greater the observed discrepancies in the results.[5] Another parameter that has to be considered in the context of the mineralogy effects is the excitation–detection efficiency. This parameter is connected with the measurement geometry of the portable XRF spectrometer, and results from the fact that the sample is excited in a non-uniform way, and the characteristic X-rays are not detected uniformly. The excitation–detection efficiency can be determined experimentally[5,15] or assessed theoretically by using the Monte Carlo calculations.[16] The parameters considered above, *i.e.*, the mineralogy effects and the grain size that characterize the analyzed materials, and the excitation–detection efficiency that characterizes the portable XRF spectrometer, affect the degree to which an individual XRF determination is representative of the bulk composition of the sample. Potts *et al.*[5] proposed a method to quantify the effects of sample mineralogy and grain size as well as the distribution and position of the minerals in the excited volume on the precision of the individual determinations for a wide range of representative rock samples. The authors have also provided guidelines concerning the number of portable-XRF determinations that must be averaged from an individual rock to obtain more representative results of the bulk material with an acceptable precision (for more details see subsequent sections).

2.4.1.4 Moisture Effects

Moisture content may affect the accuracy of analysis of soil and sediment samples. According to Piorek,[8] the moisture content of soil $<20\%$ does not influence the accuracy of the portable-XRF measurements and can safely be neglected. However, moisture content may be a major source of uncertainty when analyzing samples of surface soil or sediment saturated with water; in such a case drying of samples before the analysis is required. When the analytical results are calculated on a dry sample basis, moisture content has to be determined, *e.g.*, by using a moisture gauge. XRF logging applied for the determination of mineral concentrations in wet boreholes is also affected by the presence of the drilling fluid that creates an absorption and scattering layer between a detector window of the XRF probe and the wall of the borehole. Liangquan *et al.*[17] considered these effects and developed an experimental procedure to correct for the presence of water in wet boreholes during XRF logging. The correction is based on the measurements of the intensities of two scattered peaks of the primary radiation, originating from the wall covered with a layer of water. The scattered primary radiation intensities at two different energies are used to determine the thickness of the water layer between the wall of the borehole and the detector window, which in turn is used to correct the characteristic X-rays of the analytes for the absorption in the water layer. The correction method was successfully applied for the determination of Zn and Pb concentrations in wet boreholes by XRF logging with up to a 6 mm water layer.[17]

2.4.2 Chemical Matrix Effects

Chemical effects result from the dependence of the intensity of the characteristic X-rays of the analyte on the overall composition of the analyzed sample [see Equations (2.1), (2.3) and (2.4)]. These effects (also called matrix effects) manifest themselves as either absorption or enhancement of the intensity of analyte X-ray fluorescence lines. The so-called interfering elements that cause the matrix effects are minor or major elements present in the analyzed sample (usually of thick or intermediate thickness). One example of absorption matrix effects is the determination of zinc (Zn) in samples containing relatively high concentrations of iron (Fe). In such a case the characteristic X-rays of Zn are strongly absorbed by Fe, thereby reducing the detected Zn characteristic X-ray intensity. As a result, analyzed samples containing the same concentrations of Zn and different concentrations of Fe provide different measured Zn characteristic X-ray intensities. The same elements present in the analyzed samples can be used to explain enhancement matrix effects. These effects are observed for Fe, which is additionally excited by the characteristic X-rays of Zn, called in this case the enhancement element; the additional excitation is relatively effective because of the proximity of the Zn characteristic X-rays (8.6 and 9.6 keV) to the K absorption edge of Fe (7.1 keV). As a result of the matrix enhancement effects, the analyzed samples containing the same concentrations of Fe and different concentrations of Zn, provide different measured Fe characteristic X-ray intensities. In both cases, *i.e.*, absorption and enhancement effects, a suitable matrix correction method has to be used to produce correct analytical results (see next section for details).

2.4.3 Spectral Interferences

These effects result from the following two sources: (a) insufficient energy resolution of the detector or (b) overlap of the characteristic X-ray lines of two or more elements in cases where their energies are almost identical. The degree to which a detector can resolve two different peaks depends on the detector energy resolution (for details see Chapter 1 in this book). In general, if the energy difference between the two peaks in a spectrum is smaller than the resolution of the detector (defined as full-width at half-maximum, FWHM), then the two peaks cannot be fully resolved by the detector. The most common spectrum overlaps are K_α–K_β interferences that involve the K_β X-ray line of element with atomic number Z with the K_α X-ray line of element with atomic number $(Z + 1)$. Other spectral interferences can arise from the overlap of other characteristic X-ray lines such as K–L, K–M, and L–M as well as from escape peaks, pileup and sum peaks.[3] To deal with the spectral interferences in the energy-dispersive X-ray fluorescence analysis, several spectrum evaluation procedures have been developed; their principles and practical use are discussed, for example, in ref. 3. However, in each case it is advisable first to identify all possible spectral interferences and then apply a spectrum evaluation software package. As a result of spectrum evaluation, the energies and net peak areas of the characteristic X-rays of the elements present in the analyzed sample can be extracted from the experimental data. In the next step(s), various correction procedures for the interference effects are used to properly convert the net peak intensities into the elemental concentrations (see subsequent sections for details).

2.5 Correction Procedures for Physical Matrix Effects

2.5.1 Correction for Surface Irregularity Effects

A most common method for correction of the surface irregularity effects is based on the Compton and Rayleigh scattered primary radiation. Potts *et al.*[13] showed experimentally that the intensities of the characteristic X-rays of analytes present in silicate reference materials fall off with the increased air gap (analyzer-to-sample surface distance) in a similar way as the sum of Rayleigh and Compton scatter peaks coming out from the primary radiation of the radioisotope sources. The scattered primary radiation can be used as normalizing (correcting) factors to compensate for surface irregularity effects in the following way:[13]

$$I_{\text{corrected}} = I_{\text{measured}} \times B_{\text{reference}}/B_{\text{measured}} \tag{2.16}$$

where I_{measured} is the measured intensity of the characteristic X-rays from a sample of irregular surface which gives the intensity of the scattered radiation B_{measured}, and $B_{\text{reference}}$ is the intensity of the scattered radiation from an equivalent reference flat sample of similar composition to the analyzed sample. Since, in general, the intensity of the scattered radiation depends also on the composition of the analyzed samples the effectiveness of the correction varies for different sources of the primary radiation. The authors[13] showed that a smallest variation of the scatter peak intensity with the sample composition was observed for a ^{55}Fe excitation source, thereby favoring the use of this source for the correction of the surface irregularity effects. In contrast, the scattered radiation intensity for a ^{109}Cd excitation source showed a substantial variation with composition of the analyzed materials, indicating that the effective correction can be performed when the reference flat sample matches closely the composition of the analyzed sample; this effect obviously generates some problems and limitations in applying the correction procedure. Another limitation, which defines the applicability range of the whole procedure, comes from a comparison of the variations of the scattered radiation intensities with the air gaps with similar variations for the characteristic X-rays of the analytes. For analysis of silicate rocks the behavior of the scattered and characteristic X-rays is similar for the air gaps (surface irregularity) up to about 3 mm only. At larger air gaps the correction for the surface irregularity effects becomes less effective. Its major physical limitations come from the following three factors:[13] (a) increased attenuation of the characteristic X-ray intensities in air, in particular for low-*Z* elements (*e.g.*, K, Ca, Ti); (b) increased intensity of the scattered radiation with the increase in the scattering angles for larger distance between the analyzer and sample; and (c) increased contribution from (spurious) scattering in air with the increase of the sample–analyzer distance, in particular for low–energy primary radiation.

A similar method for the correction of the surface irregularity effects was developed by Liangquan *et al.*[14] The authors proposed a theoretical model for the description of the so-called unevenness factors for the characteristic and scattered radiation as a function of three parameters describing the surface geometrical structure of typical field rock samples. Based on the unevenness factors the ratios of the intensity of the characteristic X-rays of Fe, Zn, and Pb to the intensity of the scattered radiation of U L_β from a ^{238}Pu primary radiation were calculated and then confirmed by experiments. The correction procedure for the surface irregularity effects appeared to be particularly effective when (a) the energies of the characteristic and scattered radiation are close to each other, (b) the ratios of the characteristic and scattered radiation intensities are averaged over the specific area of interest and (c) convex and concave regions are absent from the area under investigation.[14]

A simple method for the analysis of the irregularly shaped homogeneous samples was proposed by Wang[18] and adapted for wavelength-dispersive XRF (WDXRF) analysis by Bos *et al.*[19] The method is based on a special calibration procedure in which the relative instrumental calibration factors $R_{ik} = I_i/I_k$ are experimentally determined based on a set of the calibration samples, where I_i and I_k are the intensities of the characteristic X-rays of the analyte and of the chosen reference element, respectively. The correction method was developed and applied for WDXRF analysis as well as for quantitative EDXRF analysis of irregularly shaped homogeneous samples such as ancient metal objects of archaeological interest.[20] Quantitative XRF analysis of irregularly shaped samples can also be performed by using a mask technique[16] based on the use of a properly optimized (through Monte Carlo simulation) centered annular X-ray opaque mask, placed over the window of the X-ray detector. The experiments showed that the mask technique provides constant efficiency values (calibration factors) for Ca, Ti, V, and Zr within $\pm 5\%$ for the variations of the sample–analyzer distance from 3 to 12 mm. However, a significant disadvantage of using a mask is that a considerable reduction in peak absolute efficiency is found.

2.5.2 Correction for Mineralogy Effects

As already mentioned, analytical results for rocks and archaeological lithic artifacts are affected by sample mineralogy (size, distribution and position of minerals) combined with the size of the excited volume and spatial variations in excitation–detection efficiency. As shown by Potts *et al.*[5] these effects can realistically be quantified only by experiments performed for representative samples. To assess the effects of sample mineralogy on the precision of field portable XRF measurements the authors selected the following five rock materials with different texture from fine- to coarse-grained: dolerite (fine-grained, grain size <1 mm), quartz andesite (fine-grained, grain size in the range 1–3 mm), microgranite (fine-grained, grain size in the range 3–4 mm), medium-grained granite (containing 3 mm crystals of whitish feldspar, 1–3 mm crystals of quartz, and 1–3 mm grains of mafic minerals) and coarse-grained granite (containing various minerals in the range 3–35 mm). To avoid contribution from the surface irregularity effects, the XRF measurements were carried out for slabs or blocks with a flat surface by using a portable XRF analyzer with three excitation sources (^{55}Fe, ^{109}Cd, and ^{241}Am) and a mercury iodide X-ray detector. Based on the results obtained for randomly selected non-overlapping ten points, the concentrations for all the elements of interest (K, Ca, Ti, Mn, Fe, Rb, Sr, Y, Zr, Nb, Pb, Ba, La, Ce), the average concentrations C_{mean} and the standard deviations s and s_{mean} were calculated. Since $s_{mean} = s/n^{0.5}$ (n is the number of measurements), the relative standard deviation R for the average concentration is given by:

$$R = 100(\%)(s/n^{0.5})/C_{mean} \quad (2.17)$$

From Equation (2.17) one can calculate the number of individual measurements n required for a given sampling precision, defined by the relative standard deviation R for the mean concentration:[5]

$$n = [(100 \times s)/(R \times C_{mean})]^2 \quad (2.18)$$

As expected, the lowest sampling precision was demonstrated by the dolerite. In this case an average of eight determinations was required to achieve a 2% relative standard deviation of the mean concentration. To get the same relative sampling precision (2%) for the coarse-grained granite an impractically large number of determinations (512) would be necessary. The results (specific for a given portable XRF instrument) illustrate a critical contribution of the mineralogy effects to sampling precision in *in situ* XRF measurements of rocks, archaeological stone artifacts, *etc.* For any new application of a portable XRF analyzer, similar measurements on representative samples are recommended, to evaluate several replicate measurements that must be averaged to achieve the required relative standard deviations of the mean.

2.6 Correction Procedures for Chemical Matrix Effects

In laboratory XRF analysis, chemical matrix effects are often the major contributors to the total uncertainty of the analytical results. Therefore, to reduce the total uncertainty and to improve the accuracy of the XRF measurements, an effective correction for the chemical matrix effects becomes an important part of the whole analytical procedure. As mentioned before in *in situ* XRF measurements, there are many sources of uncertainty that come from the nature of the analyzed materials (heterogeneity effects) combined with minimum or lack of sample preparation. Nevertheless, even for such materials, a correction for the chemical matrix effects has to be applied. The type of correction for the chemical matrix effects depends on the analyzed materials, contributions of other uncertainty sources and on the purpose of the analysis (which defines the acceptable total uncertainty). The most common and often used correction procedures for chemical matrix effects in *in situ* XRF measurements are characterized below (some of the procedures are also presented in other chapters of this book). One has to mention that most of the procedures are not specific for *in situ* measurements and can also be applied for laboratory XRF analysis.

2.6.1 Analysis of Thin Samples

Determination of metal contamination in air is based on collection of air particulates on membrane filters by using an air sampler, followed by *in situ* XRF measurements. If the total mass per unit area of the material deposited on a filter is smaller than m_{thin} [Equation (2.5)] then calibration of a portable XRF spectrometer is carried out by the measurements of thin single element standard samples or certified thin-film multi-element standards. The empirical calibration factors [K_i in Equation (2.6)] are used to convert the intensity of the characteristic X-rays into the total mass of the analyte element on the filter, and these data can then be converted into the concentration of the element in air.[21] The contents of the analyte elements in the thin film standard samples can be determined, *e.g.*, by using a gravimetric method. Another method for the preparation of standard calibration filters for on-line monitoring of aerosols was developed by Harmel *et al.*[22] The thin film standard filters were prepared by using an aerosol generator that produced particles with aerodynamic diameters similar to those of real aerosols; the filters were then analyzed by methods such as AAS, ICP-OES and ICP-MS. Calibration of an energy-dispersive XRF spectrometer can also be performed by using samples of intermediate thickness as well as thick calibration samples.[23]

2.6.2 Analysis of Intermediate Thickness Samples

In cases where the total mass per unit area of an analyzed sample fulfills the relation given by Equation (2.9), one has to consider chemical matrix effects. The mass per unit area of the analyte m_i is given by:[3]

$$m_i = \frac{I_i}{K_i} A_{\text{cor}} \tag{2.19}$$

where A_{cor} is the absorption correction factor:

$$A_{cor} = \frac{m[\mu(E_0)\csc\Psi_1 + \mu(E_i)\csc\Psi_2]}{1 - \exp\{-m[\mu(E_0)\csc\Psi_1 + \mu(E_i)\csc\Psi_2]\}} \tag{2.20}$$

The absorption correction factor A_{cor} represents the combined attenuation of the primary and fluorescent radiation in the whole sample, and can be determined individually for each sample by emission–transmission (E-T) experiments.[3] These involve the measurement of X-ray intensities from a thick multi-element target located adjacent to the back of the analyzed sample with and without the analyzed sample in position. If I_i, I_i^{T}, and I_i^0 are the net intensities from the analyzed sample alone, from the sample plus target, and from the target alone, respectively, then:

$$\exp\{-m[\mu(E_0)\csc\Psi_1 + \mu(E_i)\csc\Psi_2]\} = \frac{I_i^{\mathrm{T}} - I_i}{I_i^0} = N \tag{2.21}$$

After a simple transformation, the mass per unit area of the analyte m_i can be calculated from:

$$m_i = \frac{I_i}{K_i}\left(\frac{-\ln N}{1 - N}\right) \tag{2.22}$$

The emission–transmission method can be applied only for the homogeneous samples where the total mass per unit area m is smaller than the critical mass m_{crit} given by:

$$m_{crit} = \frac{-\ln N_{crit}}{\mu_{tot}} \tag{2.23}$$

where N_{crit} is the critical transmission factor defined by Equation (2.21) (in practice $N_{crit} = 0.1$ or 0.05).[3]

In the E-T method, one additional measurement carried out with a multi-element target combined with an interpolation and/or extrapolation procedure (applied for the relationship of lnln N^{-1} *vs.* ln E) enables one to calculate the absorption correction factors for all the elements to be determined.[24] However, such a simple procedure gives accurate results only for trace element analysis. In cases where the analyzed materials contain minor and/or major elements, the accuracy of the results obtained by using the E-T method might deteriorate as a result of the discontinuities in the relationship between mass attenuation coefficient and energy at the absorption edge energies of the minor/major elements present in the sample (which is reflected in the discontinuities in the relationship between lnln N^{-1} and ln E). To correct for the presence of the minor and/or major elements in the samples, a simple iterative procedure was introduced into the E-T method to take into account the discontinuities in the absorption properties of the analyzed materials.[25]

The E-T method can also be used in the XRF analysis of air filters in monitoring of air pollution. Air particulates collected by using various air samplers form homogeneous deposits on filters and represent ideal targets for XRF analysis. Loaded air filters are usually analyzed as thin samples for which no absorption correction is required. When the criterion for thin samples [Equation (2.5)] is not fulfilled then the sample becomes of intermediate thickness, and the E-T method can be applied for the absorption matrix correction. To correct for the absorption of the primary and characteristic X-rays in the filter itself, a simple modification to the E-T measurements, and the calculation of the absorption correction factors is required.[26] In addition to measurements of the loaded filter with ($I_i^{T,l}$) and without ($I_i^{0,l}$) the multi-element target, measurements on the unloaded filter with ($I_i^{T,ul}$) and without ($I_i^{0,ul}$) the multi-element target are carried out. Next, the modified transmission factor N_{mod} is calculated:[26]

$$N_{mod} = \frac{I_i^{T,l} - I_i^{0,l}}{I_i^{T,ul} - I_i^{0,ul}} \tag{2.24}$$

The overall uncertainty of the results obtained by using the emission–transmission method depends on three major factors: uncertainty of the characteristic X-ray intensity I_i (mostly due to counting statistics and spectrum fitting procedure), uncertainty of the absorption correction factor A_{corr}, and uncertainty of the calibration factor K_i. A simple procedure to assess the contribution of the different sources of uncertainty as well as to calculate the overall uncertainty of the E-T method has been described.[27]

As already mentioned, the E-T method gives the best results only for samples characterized by a homogeneous distribution of the analyzed material over the sample area[28] and that do not show the particle size effects.[3]

2.6.3 Analysis of Thick Samples

Thick samples [see Equation (2.7)] comprise the majority of samples analyzed by the XRF method. For this type of sample, various quantitative procedures are now available, including relatively simple methods based on scattered primary radiation, fundamental parameters methods, and empirical and theoretical influence coefficients methods. All these correction methods are often applied in laboratory XRF analysis.

2.6.3.1 Methods Based on Scattered Primary Radiation

The intensity of Compton scattered primary radiation I_{Com} [Equation (2.11)] depends on the absorption properties of the analyzed material, based on the mass attenuation coefficient $\mu(E_0)$ as well as on the measurement geometry, taking into account the incident and take-off angles (Ψ_1 and Ψ_2, respectively). Variation in the intensity of the Compton scattered radiation follows, approximately, the variation in the intensity of the characteristic X-rays of the analyte I_i with the changes in the absorption properties of the analyzed materials. Therefore, one can expect that the ratio of I_i to I_{Com} is less sensitive to the variation in composition of the analyzed materials, and can be used for the correction of the matrix absorption effects.[29] In the fluorescent-to-Compton correction method, the calibration (I_i/I_{Com} ratio *vs.* concentration of the analyte) is performed for synthetic calibration samples or site-specific calibration samples (SSCSs) collected in the field and analyzed by using a reference analytical method such as AAS, ICP-OES or ICP-MS. The fluorescent-to-Compton ratio method works well if the matrix variations are limited and no significant absorption edges occur between the Compton line and the fluorescent line of the analyte. In the XRF analysis of samples for which the second requirement is not met, because some absorption edges of (minor or major) matrix elements lie between the Compton line and the characteristic X-ray line of the analyte, additional corrections (called jump corrections) have to be applied.[30] Since the background intensity in an XRF spectrum is related to the Compton scattered primary radiation, the absorption matrix correction can also be undertaken using the peak-to-background ratio (where peak is net intensity of the characteristic X-ray line). The effectiveness of correction methods based on the fluorescent-to-Compton or peak-to-background ratio methods can be improved by using the intensity of Compton scattered radiation or background raised by an exponent S.[31] In general, S varies with analyte, energy of background (or energy of scatter peaks) and the matrix composition. In the ideal case, when the energy of the characteristic X-rays of the analyte is higher than the absorption edges of the major matrix elements, the values of S should be in the range from 1.0 (Compton scattering alone) to 2.0 (Rayleigh scattering alone). For any two samples with different matrices a and b, and the same concentration of the analyte, the optimum value of S that gives the most effective correction for the matrix absorption effects is calculated from Equation (2.25):[31]

$$(I_{i,b}/I_{i,a})/(I_{bkg,b}/I_{bkg,a})^S = 1 \tag{2.25}$$

where $I_{i,a}$ and $I_{i,b}$ are the intensities of the characteristic X-rays of the analyte for samples with matrix a and b, respectively; $I_{bkg,a}$ and $I_{bkg,b}$ are scattered intensities for samples a and b, respectively. The optimum value of S can also be calculated theoretically.[31] One has to mention that the scattered primary radiation used for the correction of the matrix absorption effects also reduces the influence of other interfering effects such as surface irregularity and moisture effects (see previous sections).

2.6.3.2 *Fundamental Parameters Methods*

The fundamental parameters method is based on a fully theoretical approach that enables the calculation of the theoretical net X-ray intensities emitted by each element from a sample of known composition.[3,32,33] The equations usually consider excitation by the primary photons and secondary fluorescence (enhancement). Applications of a fundamental parameters method for analyzing unknown samples include two steps: calibration and analysis. During the calibration process, which accounts for the instrument-related factors, pure elements can be used as calibration samples. The measured (net) intensities for the unknown sample are divided by the measured (net) intensities for the corresponding pure elements measured under identical excitation and detection conditions. The relative intensities are also calculated by using the fundamental parameters programs. For each characteristic X-ray line, the measured relative intensities are plotted against the calculated relative intensities. The slope of the obtained straight line is the proportionality factor between the predicted (calculated) and measured relative intensities. The analysis step includes: (a) first estimation of the sample composition (concentrations of the elements are normalized to 100%), (b) calculation of the theoretical relative intensities and their conversion into the measured relative intensities using the proportionality factors obtained from calibration, (c) next estimation of the composition of the sample, based on the differences between the measured and calculated relative intensities. Steps (b) and (c) are repeated until convergence for the relative intensities or concentrations of all the elements (compounds) is obtained. Major weaknesses of the fundamental parameters method are related to evaluation of the first estimate of the composition (often only a very rough evaluation is made), and the constraint of normalizing the concentrations to 100%. The second weakness limits the application of the fundamental parameters methods to samples for which the characteristic X-rays of all the elements present are measured (*e.g.*, alloys, steels) or for which the low-*Z* matrix is known or can reasonably be defined. In many cases, however, the analyzed samples include very light elements, such as C, N, O, *etc.* for which the characteristic X-rays are not recorded by the portable XRF analyzers and their concentrations are unknown (the light elements create the so-called "dark" matrix of the analyzed sample). Looking at the expressions for the intensity of Compton and Rayleigh scattered radiation [Equations (2.11) and (2.12), respectively], one can notice that these two quantities depend on the overall composition of the analyzed samples, including dark matrix elements. Based on this observation the so-called backscatter fundamental parameters (BFP) methods were developed.[34–37] These methods utilize coherently and incoherently scattered peaks of primary radiation to characterize the dark matrix of the analyzed sample, usually in such a way that two additional elements representing the dark matrix are determined. The BFP methods require experimental determination of three constant calibration factors for fluorescent, coherent and incoherent radiation. The BFP algorithms include correction for the absorption and enhancement effects, and can be used for samples of any thickness, including those of thick and of intermediate thickness. Detailed description of all steps in the BFP methods is outside the scope of this book; details can be found in refs 34–37. Selected examples of practical applications of portable XRF analyzers combined with quantification based on the fundamental parameters methods are described in refs 38–40.

2.6.3.3 Influence Coefficient Methods

Another group of mathematical methods for the chemical matrix effects correction is based on the influence coefficients that quantify the total matrix effects of an interfering element j on the analyte i in a given sample.[3,41–43] Since the correction increases with the concentration C_j of the element j in the sample, the correction term is given by:[41]

$$(\cdots + x_{ij}\, C_j + x_{ik}\, C_k + \cdots) \tag{2.26}$$

where x_{ij} and x_{ik} are any type of influence coefficients; subscripts ij describe the effect of the matrix element j on analyte i. There are two different types of influence coefficient: empirical and theoretical.[3,41,43] Each category can be determined from binary and multi-element calibration samples. Brief characteristics of each category of influence coefficients are given below.

Empirical Binary Influence Coefficients (b_{ij}). The influence coefficients b_{ij} can be determined from the Lachance–Traill algorithm:[44]

$$C_i = R_i(1 + \Sigma_j\, b_{ij}\, C_j) \tag{2.27}$$

where R_i is the ratio of the intensity of the characteristic X-rays for analyte in the analyzed sample to the intensity of the pure analyte i, and C_j is the concentration of the matrix element j. For binary samples Equation (2.27) becomes:

$$C_i = R_i(1 + b_{ij}\, C_j) \tag{2.28}$$

which can be rewritten as:

$$[(C_i/R_i) - 1] = b_{ij} C_j \tag{2.29}$$

Based on measurements carried out for a set of binary calibration samples containing the analyte i and the matrix element j in different proportions, the influence coefficient b_{ij} is determined as a slope of a straight line $[(C_i/R_i) - 1]$ *vs.* C_j. The empirical binary influence coefficients in the Lachance–Traill algorithm are constants, *i.e.*, they are independent of concentrations as well as of the presence and nature of other matrix elements; this limits the concentration range over which the matrix effects can effectively be eliminated.

Empirical Multi-element Influence Coefficients (e_{ij}). One of the algorithms based on the influence coefficients e_{ij} is that proposed by Lachance and Traill[44] [see Equation (2.27) with e_{ij} instead of b_{ij}]. After a simple transformation one can obtain the following expression:

$$I_i = I_{i,\ \text{pure analyte}}\, C_i - \Sigma_j\, ei_j(I_i\, C_j) \tag{2.30}$$

where $I_{i,\ \text{pure analyte}}$ is the intensity of the characteristic X-rays of the pure analyte sample. The empirical coefficients e_{ij} are determined by linear multiple regression analysis using the measured intensities and the concentrations for multi-element calibration samples that cover the concentration ranges of the elements of interest; the best results are obtained when the compositions of the calibration samples match the compositions of the analyzed materials (both standard and analyzed samples are of the same type and prepared in the same way). In situations where n analytes are determined, it is recommended that at least $2(n + 1)$ multi-element calibration samples are used. Empirical coefficients correct not only for the chemical matrix effects but can also "include" a correction for other effects such as particle size effects, mineralogical effects, and surface roughness effects; they are also effective in reducing the errors introduced by poor sample preparation and lack of homogeneity of the analyzed materials. On the other hand, empirical coefficients require the preparation and measurements of sufficient number of calibration samples, and can only be applied effectively to the unknown samples that are similar in composition to the samples used for calibration. The values of empirical coefficients depend on the composition of the calibration samples and often have no physical meaning. Despite these limitations, empirical multi-element influence coefficients are used for the *in situ* XRF analysis of soil and sediment samples; influence coefficients e_{ij} determined by using site-specific calibration samples (SSCS) give the best results.

Theoretical Binary Influence Coefficients (a_{ij}). This concept is based on the assumption that the total matrix effects on the analyte i can be expressed as the sum of the effects of each matrix element j, calculated independently; a complex sample can thus be considered to consist of a series of binary mixtures. A series of influence coefficients is calculated from hypothetical compositions of binary mixtures that make up the analyzed samples. An example of the theoretical approach where the influence coefficients a_{ij} are used is the Claisse–Quintin algorithm,[43,45] which for multi-element samples is based on Equation (2.31):

$$C_i = R_i[1 + \Sigma_j(a_{ij} + a_{ijj}C_{\mathrm{M}})C_j + \Sigma_j\Sigma_{(k>j)}a_{ijk}C_jC_k] \tag{2.31}$$

where a_{ij}, and a_{ijj} are binary influence coefficients, a_{ijk} is the "crossed" ternary coefficient, C_{M} is the matrix concentration (sum of all elements in the sample except the analyte i).

Theoretical Multi-element Influence Coefficients (α_{ij}, ε_{ij}). This approach is based on the theoretical expressions deduced by Rousseau[43,46,47] in which the correction is based on Equation (2.32):

$$C_i = R_i = \frac{1 + \sum_j \alpha_{ij}C_j}{1 + \sum_j \varepsilon_{ij}C_j} \tag{2.32}$$

where α_{ij} and ε_{ij} are the multi-element influence coefficients correcting for absorption and enhancement effects, respectively. These coefficients depend on the total matrix composition, and are calculated theoretically for each sample.[46,47] Although the mathematical expressions that are used to calculate the influence coefficients are exact, some uncertainties in the calculated values exist as the result of the uncertainties of the fundamental parameters, including mass attenuation coefficients, fluorescence yields, transition probabilities, incident spectra, *etc.* A calibration procedure[48] can reduce the contributions of the uncertainties as well as accounting for all the instrumental parameters.

Before completing this section on the influence coefficients methods, it is worth mentioning relatively simple correction algorithms in which the chemical matrix effects are corrected using the intensity of characteristic radiation (rather than the concentration) of the interfering elements. A good example of the intensity correction is the algorithm proposed by Lucas-Tooth and Price,[3,49] based on Equation (2.33):

$$C_i = B_i + I_i\left(k_0 + \sum_{j \neq i} k_{ij} I_j\right) \tag{2.33}$$

where B_i is a background term, k_0 and k_{ij} are the correction coefficients, determined by a least-squares method applied for the experimental results for calibration samples. Since the corrections for the chemical matrix effects are undertaken using the intensities of the characteristic X-rays of the interfering elements, the method is applied in a limited range, usually for the determination of one or two analytes. However, the intensity correction method is very simple and fast because calculations of the composition of the analyzed samples do not require any iterations.

2.6.4 Other Methods Applied for the Chemical Matrix Effects Correction

Determination of lead in paint requires a correction for the absorption of the lead fluorescence signal by overlying layers of non-lead material; the correction can be based on the ratio of the intensities of Pb L_β to Pb L_α X-ray lines.[9] Since the characteristic X-ray lines of different energies (12.6 and 10.5 keV, respectively) are absorbed in a different way by the overlying layers of non-lead material, the ratio of Pb L_β to Pb L_α characterizes both the absorption properties and thickness of the layers. In practice, the attenuation indicator D defined as (Pb L_β/Pb $L_\alpha)^\kappa$ (where κ is a parameter that varies only slightly with the composition of the overlying material and with the measurement geometry) is determined, and next used for the correction of the intensity of the Pb L_β characteristic X-rays. The value of the attenuation coefficient D is in the range from 1 (analysis of immediate surface layers) to 10 (analysis of heavily shielded, deeply buried lead).[9]

Quantitative analysis of metal alloys, often of irregular surface, can be based on the ratios of the characteristic X-rays of analytes to the characteristic X-rays of an element that characterizes matrix, *e.g.*, copper K_α in case of bronzes or gold L_α in the analysis of materials containing gold[50] (the method requires normalization of the concentrations to 100%).

A portable XRF analyzer can be calibrated by using spiked background materials, *e.g.*, soil samples, with a known amount of compounds that contain the analyte; this technique has been used to determine chromium and other metals in contaminated soils.[51] Spiking of calibration samples gives satisfactory results if the following assumptions are met: (a) sample matrix at the site is identical to the calibration samples, (b) moisture content and density are uniform, and (c) absorption and enhancement effects for the analyzed materials are similar to those in the calibration samples. To test the quality of the analytical results obtained by using a portable XRF spectrometer, comparative analysis for selected unknown samples (*e.g.*, 10% of a total number of the analyzed samples) is recommended, by using a reference laboratory analytical technique.

An alternative approach for quantitative *in situ* XRF analysis is based on the partial least-squares regression (PLSR) multivariate calibration method.[52,53] PLSR is particularly suitable for a wide range of sample types with a complex matrix where both the chemical matrix effects and spectral interferences occur.[54] Moreover, the method can be integrated into an expert system in support of XRF analysis.

References

1. R. Jenkins, R.W. Gold and D. Gedcke, *Quantitative X-Ray Spectrometry*, Marcel Dekker, New York, 1981.
2. R. Tertian and F. Claisse, *Principles of Quantitative X-Ray Fluorescence Analysis*, Heyden, London, 1982.
3. R.E. Van Grieken, A.A. Markowicz, (Eds.), *Handbook of X-Ray Spectrometry – Second Edition, Revised and Expanded*, Marcel Dekker, New York–Basel, 2002.
4. R. Cesareo, *X-ray Physics: Interaction with Matter, Production, Detection*, La Rivista del Nuovo Cimento della Societa Italiana di Fisica, Bologna, 2000.
5. P. Potts, O. Williams-Thorpe and P.C. Webb, Geostandards Newsletter, *J. Geostandards Geoanal.*, 1997, **21(1)**, 29.
6. "Field Portable X-Ray Fluorescence Analyzer, HNU Systems SEFA-P", Report EPA/600/R-97/144, US Environmental Protection Agency, March 1998.
7. Nomenclature, symbols, units and their usage in spectrochemical analysis-II, *Spectrochim. Acta*, 1978, **33B**, 242.
8. S. Piorek, X-Ray Fluorescence Spectrometry – Determination of metals in soils by field-portable XRF spectrometry, in *Current Protocols in Field Analytical Chemistry*, eds. V. Lopez-Avila, D. Barcelo, W. Beckert, S.C. Goheen, K. Jinno, L.H. Keith and J.H. Rittenburg, John Wiley & Sons, Inc., New York, 1998, Chapter 3, 3B.1.1–3B.1.18.

9. S. Shefsky, J. Pesce and K. Martin, *Proceedings of Symposium on Field Analytical Methods for Hazardous Wastes and Toxic Chemicals, Las Vegas, 29-31 January 1997*, pp. 1–11.
10. S. Piorek, The current status and development trends if X-ray fluorescence techniques for in-situ applications, IAEA Consultants' Meeting on "In-situ Applications of X-ray Fluorescence Techniques", Vienna, Austria, 13–17 September 1999.
11. C.B. Hunter and J.R. Rhodes, *X-Ray Spectrom.*, 1972, **1**, 107.
12. J.R. Rhodes and C.B. Hunter, *X-Ray Spectrom.*, 1972, **1**, 113.
13. P.J. Potts, P.C. Webb and O. Williams-Thorpe, *J. Anal.Atomic Spectrom.*, 1997, **12**, 769.
14. G. Liangquan, Z. Ye, C. Yeshun and L. Wangchang, *Appl. Radiat. Isot.*, 1998, **49**, 1713.
15. A. Markowicz, N. Haselberger and U. Whai Zin Oo, *X-Ray Spectrom.*, 1993, **22**, 160.
16. J.M. Maia, J.M.F. Dos Santos and C.A.N. Conde, *Appl. Radiat. Isot.*, 1997, **48**, 1649.
17. G. Liangquan, Y. Zhang, Y. Cheng, S. Zhou, T. Xie and S. Hou, *X-Ray Spectrom.*, 1997, **26**, 303.
18. Y. Wang, *X-Ray Spectrom.*, 1996, **25**, 245.
19. M. Bos, J.A.M. Vrielink and W.E. van der Linden, *Anal. Chim. Acta*, 2000, **412**, 203.
20. M. Milazzo and C. Cicardi, *X-Ray Spectrom.*, 1997, **26**, 211.
21. M.B. Bernick and P.R. Campagna, *J. Hazardous Mater.*, 1995, **43**, 91.
22. R. Harmel, O. Haupt and W. Dannecker, *Fresenius' J. Anal. Chem.*, 2000, **366**, 178.
23. A. Markowicz, N. Haselberger and P. Mulenga, *X-Ray Spectrom.*, 1992, **21**, 271.
24. D.R. Giauque, R.B. Garrett and L.Y. Goda, *Anal. Chem.*, 1979, **51**, 511.
25. A. Markowicz and N. Haselberger, *Appl. Radiat. Isot.*, 1992, **43**, 777.
26. A. Markowicz, N. Haselberger, M. Dargie, A. Tajani, A. Tchantchane, V. Valkovic and P.R. Danesi, *J. Radioanal. Nucl. Chem.*, 1996, **206**, 269.
27. A. Markowicz, N. Haselberger, H.S. El Hassan and M.S.A. Sewando, *J. Radioanal. Nucl. Chem.*, 1992, **158**, 409.
28. A. Markowicz and A.A. Abdunnabi, *X-Ray Spectrom.*, 1991, **20**, 97.
29. A. Markowicz, *X-Ray Spectrom.*, 1984, **13**, 166.
30. J. Kikkert, *Adv. X-Ray Anal.*, 1983, **26**, 401.
31. L.G. Livingstone, *X-Ray Spectrom.*, 1982, **11**, 89.
32. J.W. Criss and L.S. Birks, *Anal. Chem.*, 1968, **40**, 1080.
33. T. Shiraiwa and N. Fujino, *Jpn. J. Appl. Phys.*, 1966, **5**, 886.
34. K.K. Nielson, *Anal. Chem.*, 1977, **49**, 641.
35. Fei He and P.J. Van Espen, *Anal. Chem.*, 1991, **63**, 2237.
36. D. Wegrzynek, B. Holynska and T. Pilarski, *X-Ray Spectrom.*, 1993, **22**, 80.
37. D. Wegrzynek, A. Markowicz and E. Chinea-Cano, *X-Ray Spectrom.*, 2003, **32**, 119.
38. S. Piorek, *Nucl. Instrum. Methods Phys. Res.*, 1994, **A353**, 528.

39. M.B. Bernick, D.J. Kalnicky, G. Prince and R. Singhvi, *J. Hazardous Mater.*, 1995, **43**, 101.
40. G. Vittiglio, K. Janssens, B. Vekemans, F. Adams and A. Oost, *Spectrochim. Acta Part B*, 1999, **54**, 1697.
41. R.M. Rousseau, *Adv. X-Ray Anal.*, 1994, **37**, 639.
42. M. Bos and J.A.M. Vrielink, *Anal. Chim. Acta*, 1998, **373**, 291.
43. R.M. Rousseau, *The Rigaku J.*, 2001, **18**, 8.
44. G.R. Lachance and R.J. Traill, *Can. J. Spectrosc.*, 1966, **11**, 43.
45. F. Claisse and M. Quintin, *Can. J. Spectrosc.*, 1967, **12**, 129.
46. R. Rousseau, *X-Ray Spectrom.*, 1984, **13**, 115.
47. R.M. Rousseau and J.A. Boivin, *The Rigaku J.*, 1998, **15**, 13.
48. R.M. Rousseau, J.P. Willis and A.R. Duncan, *X-Ray Spectrom.*, 1996, **25**, 179.
49. H.J. Lucas-Tooth and B.J. Price, *Metallurgia*, 1961, **64**, 149.
50. R. Cesareo, G.E. Gigante, P. Canegallo, A. Castellano, J.S. Iwanczyk and A. Dabrowski, *Nucl. Instrum. Methods Phys. Res.*, 1996, **A 380**, 440.
51. R.W. Puls, D.A. Clark, C. Carlson and J. Vardy, *Ground Water Monitoring Remediation*, 1994, **14**, 111.
52. M.J. Adams and J.R. Allen, *Analyst*, 1998, **123**, 537.
53. P. Lemberge and P.J. Van Espen, *X-Ray Spectrom.*, 1999, **28**, 77.
54. I. Szaloki, S.B. Toeroek, Chul-Un Ro, J. Injuk and R.E. Van Grieken, *Anal. Chem.*, 2000, **72**, 211R.

CHAPTER 3

Contaminated Land: Cost-effective Investigation within Sampling Constraints

MICHAEL RAMSEY

Department of Biology and Environmental Science, University of Sussex, Falmer, Brighton, BN1 9QG, UK

3.1 Introduction

In situ analysis with PXRF has enormous advantages for the investigation of contaminated land in terms of the reliability of the assessment made and speed with which it can be achieved. There are, however, several significant limitations to the application of the technique that have to be considered. To explain these strengths and weaknesses, it is first necessary to explain the objectives of site investigations, and the analytical capabilities that are required for effective assessment.

Industrialization in most countries across the world has left a legacy of land that has been affected by a wide range of contaminants. The legal definition of whether this land is "contaminated" varies from country to country, but such a definition relies, initially at least, on the estimation of the concentration of the contaminant present in the land, usually in the soil. If the concentration present (on a dry basis) exceeds some threshold value then a hazard is said to exist. In some countries the definition of contaminated land is based on the assessment of risk, rather than on the hazard alone. For example, in the UK, an exposure route needs to be demonstrated between the contaminant and some receptor (*e.g.* human), and a value of exposure must then be estimated to assess whether it exceeds a different threshold value for the exposure (*e.g.*, a tolerable daily intake). If this threshold is exceeded, a site-specific risk assessment will be

Portable X-ray Fluorescence Spectrometry: Capabilities for *In Situ* Analysis
Edited by Philip J Potts and Margaret West

required to decide whether a "risk of significant harm" has been identified, and the land can be classified as contaminated. Soil Guideline Values (SGVs) have been derived as threshold values for hazards in soil using this approach.[1]

The role of PXRF is largely in the initial quantification of the hazard. Most contaminated land investigations utilize analytical measurements that are made in the laboratory, based on samples that have been brought back from the field, and then dried, ground, sub-sampled and prepared in several ways. One of the main strengths of PXRF is that it can measure contamination *in situ*, without any of these time-consuming preparations. The operating principles of PXRF have been described in other chapters in this book. This chapter concentrates on the *in situ* application of PXRF to contaminated land, considering its strengths and its weaknesses, particularly in terms of the uncertainty that exists in the measurements.

Whilst the examples given in this chapter describe investigations on contaminated land using a portable system fitted with sealed radioisotopes as the means of excitation, it is recognized that systems are now available that benefit from miniature X-ray tube excitation. Additionally, the integration of a global positioning system (GPS) within the instrument to facilitate the spatial mapping of contaminants, and further miniaturization,[2] has demonstrated the effective application of PXRF in the assessment and reliable remediation of contaminated land.

3.2 Typical *In Situ* Applications of PXRF

A typical *in situ* application of PXRF to a contaminated site requires a detailed map of the site and a desktop survey of the site history to establish the nature and approximate location of any possible sources of contamination. An initial sampling scheme can then be designed for a preliminary reconnaissance of the site. Prior to the commencement of the survey, the PXRF instrument will need to be calibrated and the bias of the instrument estimated using reference materials (discussed below). At each location where a measurement is required, any surface vegetation needs to be removed using a spade to take off the top few centimetres of soil and vegetation. A few practitioners also recommend raking to homogenize the soil (ref), and even leaving the soil to dry in hot climates (ref), but most investigators analyse with original heterogeneity and soil moisture. The spade cut usually creates a reasonable level surface upon which the PXRF instrument can be placed (Figure 3.1), but further compression or levelling has been suggested by some workers. The exact positioning of the instrument within the prepared area is somewhat arbitrary, and is a potential source of uncertainty. Some practitioners take several measurements within this area, reporting the average.[3] Others take duplicate readings, with a fixed distance between them, to estimate the uncertainty in the measurement. This uncertainty arises due mainly to the heterogeneity that exists in the soil within the location at the scale of the particular surveying method.[4] The level surface provides an area for the measurement of the contaminants, but it also gives a

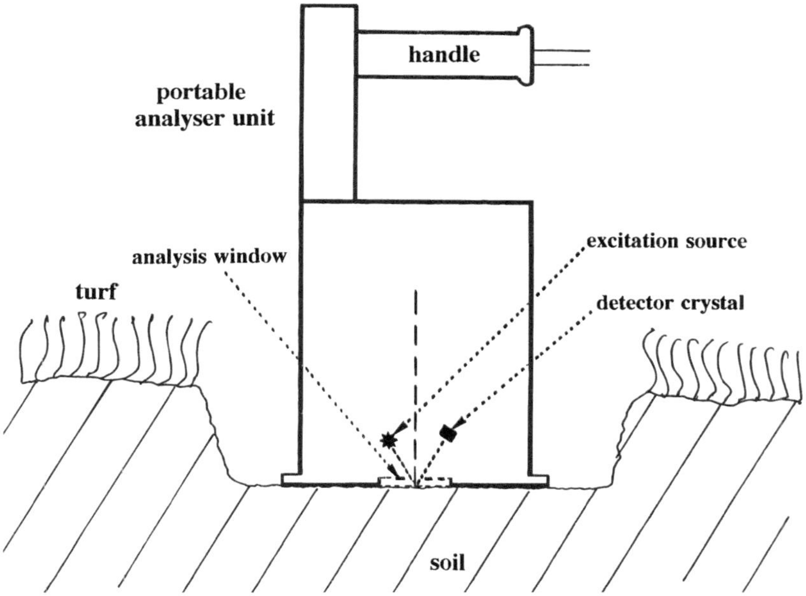

Figure 3.1 Schematic representation of PXRF instrumentation for *in situ* analysis of soils. (From Argyraki *et al.*[4] Reproduced by permission of The Royal Society of Chemistry.)

stable position for the instrument so that it won't fall over and cause a potential safety hazard. The counting times selected for each source are usually selected to be a few minutes in total. For the sources used to excite the contaminant elements of interest (*e.g.* ^{109}Cd), there is a compromise to be made between the achievement of the required detection limits and the total number of measurements needed during the visit to the site, as discussed below. The counting time required for the measurement of the major elements and backscatter peaks, if required for matrix effect correction, can often be kept to a minimal duration (*e.g.* 20 s for ^{55}Fe and 5 s for ^{241}Am).

The measurement of concentration for a wide range of analyte elements are usually reported simultaneously at the end of the counting period, often with an estimate of the uncertainty based solely on the counting statistics.

3.3 Advantages and Disadvantages of *In Situ* PXRF

Table 3.1 lists the main advantages and disadvantages of the *in situ* use of PXRF for contaminated land investigation. One of the obvious advantages of this method is to give virtually *immediate* measurements after a few minutes at the location. All of the measurements from the initial sampling design can be obtained within a few hours for a typical site. This, in turn, allows the investigator of the site to "see" the contamination in real time as the investigation

Table 3.1 (a) Advantages and (b) disadvantages of *in situ* PXRF for contaminated land investigation.

(a) Advantages

1. Immediate estimates of metal contamination of soil *in situ*.
2. Rapid analysis time for suite of several metals virtually simultaneously (*e.g.* 2 min). Measurement time may need extending if concentrations are low.
3. Iterative sampling designs possible in which a follow-up survey can be implemented to investigate apparent "hot spots", during *one* site visit.
4. Detection limits low enough to measure significant contamination of several metals of interest (*e.g.* Ba, Co, Cr, Cu, Mo, Ni, Pb, Sn and Zn).
5. Can investigate *in situ* heterogeneity of the metal across the site, with special resolutions down to the centimetre scale.
6. Good analytical precision (*e.g.* ~1%) but measurement precision determined by heterogeneity of sample site (*i.e.* sampling precision).
7. Acceptable levels of overall measurement uncertainty for some elements.

(b) Disadvantages

1. Small depth of analysis (0.04–8 mm). But excavation at the site allows *in situ* analysis of depth profiles.
2. Slightly longer time required on site compared with simple sample taking (potentially twice as long) – but no preparation time.
3. Substantial capital cost of equipment.
4. Safety regulations – time and expense required for compliance with use of instruments incorporating radioactive source (varies between countries).
5. Very small sample size ($<1\,cm^3$) gives potentially poor representivity. The consequence is a large measurement precision.
6. Potentially large estimates of bias in measured concentration will arise unless corrections can be made for soil moisture, pore space volume and surface roughness of the soil surface analysed.
7. Detection limits not low enough to quantify some elements at typical background concentrations in soils (*e.g.* Cd at $0.1–0.5\,\mu g\,g^{-1}$).

progresses, and to "zoom in" on areas of high concentration. Areas that exceed the threshold concentration value of the hazard, or "hot spots", can thus be delineated rapidly. Furthermore, the determination of all analytes is virtually *simultaneous*. This means that in the same period required to survey the spatial distribution of one element, measurements are gained for a range of other contaminants that can also be assessed, if the detection limits of that particular element by PXRF are appropriate (see later discussion).

The field sampling design employed for lab-based analysis is usually entirely predetermined before the site visit. It often uses an even coverage of samples across each sub-area of the site that is considered to have an equal probability of containing contamination, based on the review of the site history. In this traditional approach, a second site visit is required to investigate in detail areas around locations that give high concentrations in the samples from the first visit, to delineate hot spots. Using PXRF this *iterative* approach to sampling can be employed within one site visit (Figure 3.2), and it can facilitate two, three or even four phases of iteration.

This advantage of "seeing" contamination whilst at the site applies equally to all techniques of *in situ* chemical analysis, but also to some extent to any

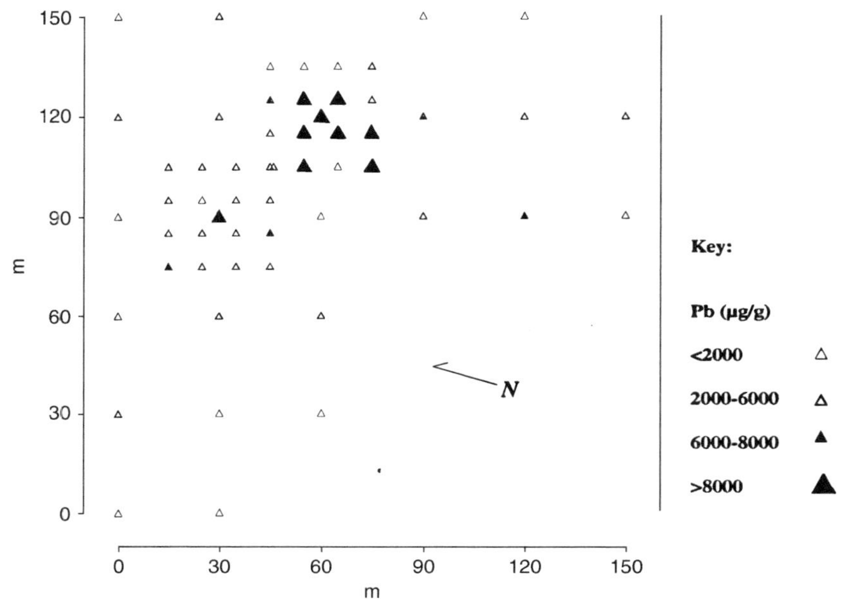

Figure 3.2 Spatial mapping of Pb concentration determined by PXRF, showing the iterative sampling design with initial 30-m grid followed by the 10-m grid centred on candidate "hot spots" (Pb > 8000 $\mu g\, g^{-1}$). The eastern "hot spot" is confirmed by a cluster of high values whereas the western "hot spot" was shown not to be part of a "hot spot" covering any extended area. (From Argyraki *et al.*[4] Reproduced by permission of The Royal Society of Chemistry.)

techniques of "on site" analysis. In "on site" analysis the sample is measured at the site, often in a field lab, but only after it has been taken from its exact location, and often prepared in some way. An extra advantage of *in situ* analysis using PXRF is that the "true spatial distribution of the contamination at a small scale can be investigated". Most workers have attempted to minimize small-scale heterogeneity as a source of unwanted uncertainty. One of the unique capabilities of PXRF is that it can be used to investigate this small-scale variation, as an analogous field equivalent to the laboratory microprobe.[19]

One of the obvious disadvantages of the use of PXRF *in situ* is that the nominal depth of analysis is very small. The critical *depth* of penetration by the X-rays has been estimated to range from 0.035 mm for Ca, 0.4 mm for Pb to 8 mm for Ba (see for example Ref. 5). In the initial sampling protocol described above, this limitation is very serious. A low measured concentration that apparently shows the site to contain no hazard could miss the fact that there is serious contamination present at depth. Alternatively, a site with contamination from recent aerial deposition could report very high concentrations of a contaminant, but this could be limited to the top few mm, and the bulk of the soil could be uncontaminated. One way around this problem is to use the

PXRF on core samples from drilling at locations across the site.[6] In this way the profile of the contamination with depth can be estimated by placing the PXRF directly onto slices of the core.

A less serious disadvantage is that an investigation will take significantly *longer*, perhaps twice as long, when measurements are being made on site, than when the samples are only being collected and removed for subsequent lab analysis. This extra time should be more than offset by the fact that no time will be required for the chemical analysis in the laboratory for the PXRF survey.

The purchase *cost* of the instrument, although less than for the larger lab-based techniques, is often an additional capital investment for an organization (£20–40k, 2007 prices). There are also significant running costs, including the periodic replacement of the radioactive sources every few years for machines using that type of excitation device. Notably, X-ray tubes have a finite life.

The last of the evident disadvantages of the current design of PXRF instruments, especially those using a radioactive source, is the need to comply with *safety* legislation. All users of PXRF instrumentation will require appropriate training to comply with ionizing radiations regulations. In addition, licence restrictions on the management and use of PXRF containing radioactive sources place extra constraints on the application of these instruments. Although these regulations are justified in terms of improved protection of the operators, their assistants and potentially the general public, they have probably had a negative effect on the widespread application of PXRF in some areas of the world.

The less obvious disadvantages of *in situ* PXRF relate to the uncertainty of the measurements, and this requires a discussion of the appropriate terminology.

3.4 Uncertainty in PXRF Measurements

All the discussion so far has assumed that the analytical technique will provide a "true" estimate of the concentration of the contaminant. This assumption is, however, never wholly justified. All measurements of concentration, by any analytical technique, are only estimates and have an associated value of uncertainty. Using an informal definition, this is the range of concentration within which the true value is considered to lie, with a specified confidence interval. The uncertainty in the *measurements* is caused by the precision and the bias of the *method*. The precision can be used to quantify the random component of the uncertainty, and the bias to quantify the systematic component.

The *precision* of any analytical method can be quantified either as the standard deviation of several replicated measurements (s_{meas}) or as the relative precision ($200s_{meas}/\bar{x}$, at approximately the 95% confidence interval), where $\bar{x}$ is the mean value of those replicates. The relative precision value becomes larger as the analyte concentration decreases towards zero. When the relative precision reaches 67%, the concentration value complies with one definition of the *detection limit* (Figure 3.3). This concentration value is also equal to three

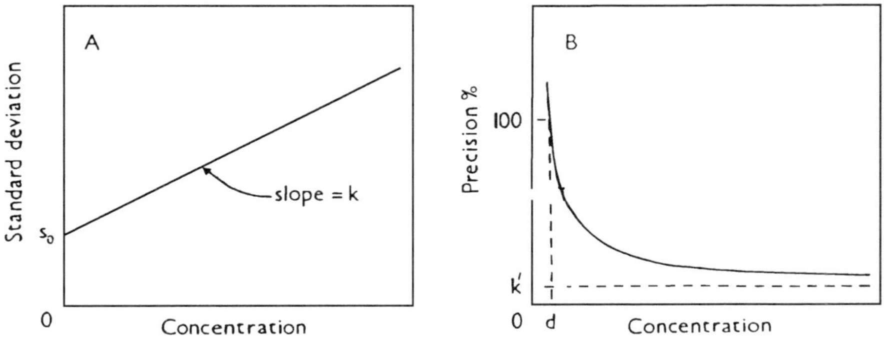

Figure 3.3 Schematic diagram showing how the (A) precision, as standard deviation and (B) relative precision of an analytical method ($200s/\bar{x}$) increases as the concentration of the analyte decreases. The concentration at which the relative precision is 67% is defined as the detection limit (*d*, where $\bar{x} = 3s$).

standard deviations of the analytical noise at zero concentration ($3s_{\text{meas}}$). Traditionally, concentration values below the detection limit are not reported, and are often not even recorded.

The *bias* of any analytical method is often expressed as the difference between a measured value of concentration (c_{meas}) and the "true" value. Alternatively, this can be expressed as the relative bias, where the bias is calculated as a percentage of the "true" value. The "true" value is usually represented by the certified value (c_{cert}) of a certified reference material (CRM) that has a similar matrix composition to the samples. Ideally, the bias should be characterized over the whole range of analyte concentration, using several CRMs with concentration values varying from the background concentration to the highest value likely to be encountered in samples. Estimates of bias made in this way are traceable to the officially recognized certifying body, comply with the definition of bias given by ISO, and are therefore legally defensible.

There is, however, an important extra factor in the estimation of bias for an *in situ* PXRF method. The CRMs are dry, very fine powders (< 50 μm) that are usually prepared for XRF analysis as compressed pellets with negligible pore space between the sample grains and with a very flat surface for measurement. In contrast, *in situ* measurements are made on soil that has *field moisture* (typically 5–50%), a coarse open texture with a large proportion of pore space and an uneven surface for measurement. The bias estimated by the first method is, therefore, not likely to be fully representative of the bias for the *in situ* measurements. Arguably, most of this extra "bias" in not really bias, as water is a real component of the sample when measured *in situ*. However, threshold values, which have mainly been developed for investigations based on laboratory measurements, are usually quoted for soils that have been dried to some extent (*e.g.* air dried).

The question then remains as how to measure this overall "bias" in a PXRF method when operated in the *in situ* mode. One option is to measure the concentration of the contaminant in the soil by a second method that is directly traceable to CRMs. The "bias" of the PXRF method can then be estimated by comparison against this second reference method. Several examples of this approach have been reported,[20,21] using lab analysis of dried powders by ICP-AES or AAS after acid digestion as the "reference method". The values of bias obtained by a regression of PXRF against the reference method vary widely between different studies, and range from around –60%[4] to "not statistically different from zero".[3] As well as the causes for bias already discussed (*e.g.* moisture, roughness, pore space) another potential cause is in sampling bias. In one study, the PXRF was estimated to represent the top 1 mm of soil, whereas the ICP-AES analysis was based on a dried soil sample from the top 150 mm taken from the same location. Clearly, these two samples are not measuring the same test portion. In this particular case, however, no systematic change in analyte concentration with depth was detected, when the PXRF was used to monitor how the Pb concentration varied every 6 mm down a 62 mm core.[4] Incidentally, this also demonstrates the capability of PXRF to monitor small-scale geochemical variations with depth.

A second option for the estimation of one component of this "bias" is to measure the water content of the soil. This approach has been applied using *ex situ* measurements of soil moisture,[4] but it would be more convenient to make such measurements *in situ*, using a field portable moisture probe of some kind. The contaminant concentration measurements could thereby be calculated back to a dry basis on site. However, this would not allow for any "bias" caused by surface roughness or pore spaces in the soil.

A third option for the estimation of the overall bias is to use a reference sampling target (RST). An RST is the equivalent for sampling to CRM for chemical analysis. For contaminated land, an RST consists of an area of land with an accepted value for the concentration for one or more elements, with a specified uncertainty value. This value of concentration can either be the mean concentration for a site, or a series of values for particular locations within the site. An RST can be an existing contaminated site, which has been characterized by an inter-organizational sampling trial, such as a sampling proficiency test.[7] Alternatively, an RST can be constructed synthetically to have a known concentration of specified elements in a particular spatial distribution. Such a synthetic RST has recently been constructed and used for the estimation of bias and uncertainty.[18] In general, however, RSTs are still at the "proof of concept" stage and not yet widely available.

The *range* of contaminants that can usefully be measured by PXRF is limited, and can be established by consideration of their detection limits. None of the organic chemicals potentially found in contaminated land can be detected by XRF and require the use of other analytical techniques, such as various forms of chromatography. PXRF can detect a wide range of inorganic contaminants that are considered to have the potential to cause "significant harm" by different regulators around the world (*e.g.* As, Ba, Cd, Co, Cr, Cu, Hg, Mo, Ni, Pb, Se, Sn, Zn). Many of these elements, however, when using current PXRF equipment, have detection limits that are higher than the highest threshold concentration values used over several counties to define the hazard (Table 3.2).

Table 3.2 Comparison between some typical detection limit concentration values, worldwide median background concentrations in soil, and some regulatory threshold values for the classification of contaminated land from different counties. Comparison between these values, shown in Figure 3.4, allows judgements to be made on whether PXRF is feasible for the estimation of concentration (all in $\mu g\,g^{-1}$).

Element	*PXRF typical detection limit*	*Canada: residential*[a]	*Canada: industrial*[a]	*SGV-UK:*[b] *residential, with (or without) plants*[e]	*SGV-UK:*[b] *commercial/industrial*	*Dutch:*[c] *target*	*Dutch:*[c] *intervention*	*International high threshold*	*Soil: world median*[d]
As	60	12	12	20	500	29.0	55	500[b]	8
Ba	9	500	2000			160.0	625	2000[a]	300
Ca	70	*	*			*	*	*	200000
Cd	250	10	22	1, 2, 8[f] (30)	1400	0.8	12	1400[b]	1
Co	100	50	300			9.0	240	300[a]	10
Cr	1080	0.4(CrVI)	1.4(CrVI)	130(200)	5000	100.0	380	5000[b]	43
Cu	50	63	91			36.0	190	190[c]	15
Fe	110	*	*			*	*	*	21000
Hg	80	6.6	50	8 [15[e]]	480	0.3	10	480[b]	0
K	150	*	*			*	*	*	11000
La	8	*	*			*	*	*	33
Mn	200	*	*			*	*	*	320
Mo	4	5	40			3.0	200	200[c]	3
Ni	116	50	50	50(75)	5000	35.0	210	5000[b]	17
Pb	39	140	600	450	750	85.0	530	750[b]	17
Sb	30	*	*			*	*	*	2

Table 3.2 (*Continued*).

Element	*PXRF typical detection limit*	*Canada: residential*[a]	*Canada: industrial*[a]	*SGV-UK:*[b] *residential, with (or without) plants*[e]	*SGV-UK:*[b] *commercial/industrial*	*Dutch:*[c] *target*	*Dutch:*[c] *intervention*	*International high threshold*	*Soil: world median*[d]
Se	45	1	3.9	35(260)	8000	*	*	8000[b]	0
Sn	40	50	300			*	*	300[a]	10
Sr	4	*	*			*	*	*	67
U	10	*	*			*	*	*	1
V	50	130	130			*	*	*	57
Zn	63	200	360			140.0	720	360[a]	36

[a]Canadian soil quality guidelines for the protection of the environment and human health (1996) http://www.ec.gc.ca/ceqg-rcqe/English/Pdf/soil_protocol.pdf. Updated (2002) http://www.ccme.ca/assets/pdf/e1_061.pdf.
[b]Soil Guideline Values from DEFRA/EA (refs. 1, 8 *etc.*).
[c]Netherlands Ministry of Housing, Spatial Planning and Environment.[17]
[d]Rose Hawkes and Webb.[9]
[e]All values also apply to land use as allotments, except a different value for mercury ($15\,mg\,kg^{-1}$).
[f]These Cd values depend on the pH of the soil (1, 2 and $8\,\mu g\,g^{-1}$, for pH 6, 7 and 8, respectively).

For example, a typical detection limit for As by PXRF is 60 μg g^{-1} in soil, whereas one UK soil guideline value for further investigation for residential land use is 20 μg g^{-1}. PRXF cannot be used, therefore, to classify land for As against this SGV, but it could be used to classify soils against the SGV of 500 μg g^{-1} set for commercial or industrial use.[8]

Generally, the particular elements that can be determined by PXRF at concentrations below their respective threshold values can be judged by a comparison between the threshold and the detection limit values (Figure 3.4b). The elements considered (As, Ba, Cd, Co, Cr, Cu, Hg, Mo, Ni, Pb, Se, Sn and Zn) are those that are potentially quantifiable at the highest threshold concentration levels set across the world (Table 3.2). All 13 of these elements are detectable at these very high concentration values, because they lie below the

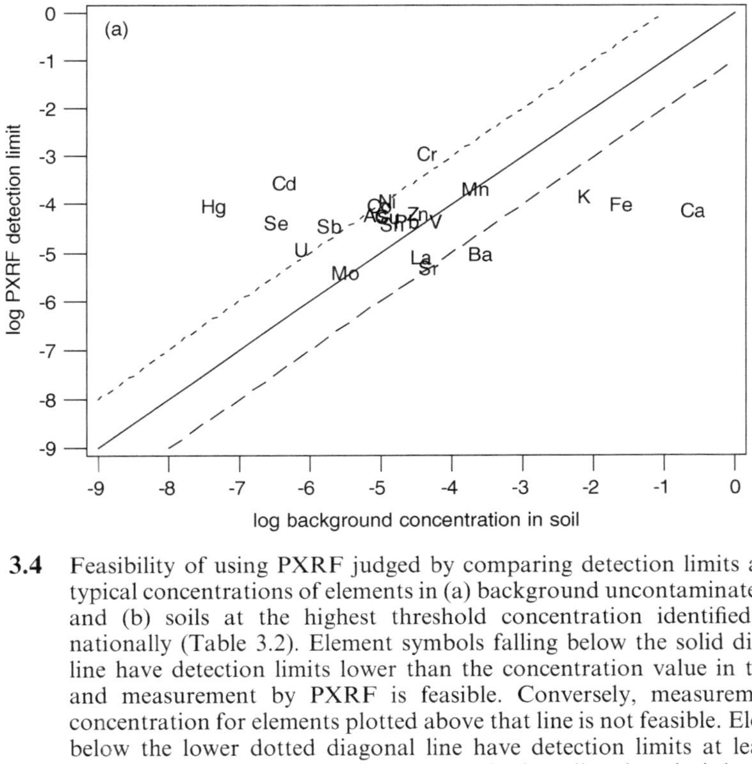

Figure 3.4 Feasibility of using PXRF judged by comparing detection limits against typical concentrations of elements in (a) background uncontaminated soils and (b) soils at the highest threshold concentration identified internationally (Table 3.2). Element symbols falling below the solid diagonal line have detection limits lower than the concentration value in the soil and measurement by PXRF is feasible. Conversely, measurement of concentration for elements plotted above that line is not feasible. Elements below the lower dotted diagonal line have detection limits at least ten times lower than the concentration value in the soil and analysis by PXRF will have good analytical precision. Elements above the solid line, but below the upper diagonal line, will have high levels of analytical uncertainty, but analyses may be useful for some purposes. Using these criteria, only one regulated element (Ba) is detectable in background soils, whereas 13 regulated elements (As, Ba, Cd, Co, Cr, Cu, Hg, Mo, Ni, Pb, Se, Sn and Zn) are detectable at their highest threshold values. Detection limit values are those for a Spectrace 9000 with a 200 s count time. Units of concentration are the log (to base 10) of the ratio of the analyte mass to soil mass (*e.g.* −6 is equivalent to 1 μg g^{-1} = 1 ppm).

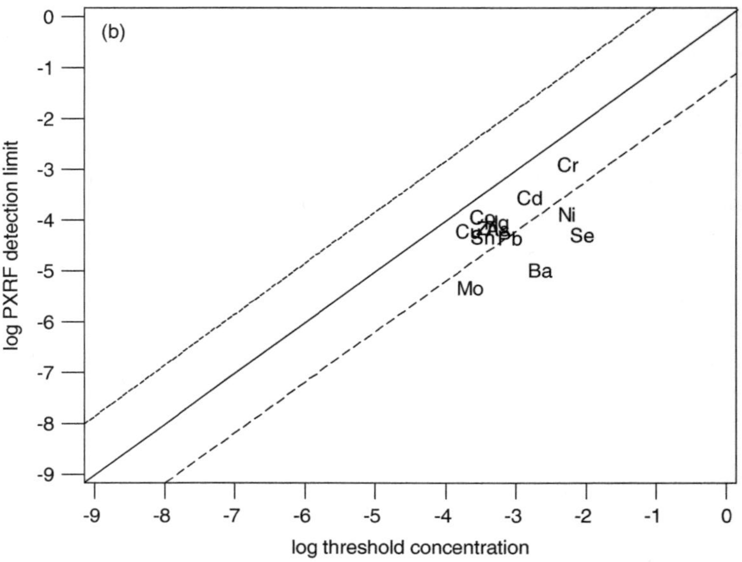

Figure 3.4 Continued.

diagonal line of equality on Figure 3.4b. Only five of these elements (Ba, Mo, Ni, Pb and Se) lie below the lower diagonal line, indicating that they will have good instrumental precision, because the threshold concentration is at least ten times higher than the detection limit.

In addition to this capability, it would also be useful to be able to measure element concentrations down to the levels at which they occur naturally. This would enable contamination to be detected above these "background levels", even when the concentration is not high enough to cause the land to be classified as "contaminated". These values of background concentrations in soil vary greatly, often due to the different bedrock present beneath the soil, but median values (Table 3.2) have been published.[9] PXRF is not currently capable of quantifying most toxic elements of interest in soils at these background levels, but it should in principle be able to quantify Ba, Ca, Fe, K, La, Mn, Sr and V (Figure 3.4a). To give a better instrumental precision, it is preferable that the background concentration should be larger than the detection limit by a factor of ten, but this is the case only for Ba, Ca, Fe, K and Sr, only one of which have ever been given a threshold value (Ba).

3.5 Factors Controlling the Precision, and hence Detection Limits

Although it is necessary that the detection limit lies below the threshold value, it is better ideally that it should be at least a tenth of the threshold value. This gives a relative analytical precision of around 10%, which is similar to that set in the

MCERTS[10] criteria for lab-based measurements. Using this criterion, the only elements that are measured acceptably by PXRF are Pb, Cr, Cu, Mo, Pb and Zn at the highest threshold concentration levels set across the world (Table 3.2 and Figure 3.4b). It is always possible, however, that the detection limit for any element can be reduced by increasing the *counting time* in the PXRF. An increase in count time by a factor of x, however, only reduces the detection limit by a factor of $\sqrt{x}$. This is because the random error on the XRF counting statistics are related to the square root of the number of counts.[11] Moreover, the longer the counting time for each measurement, the fewer the number of measurements that can be made during one site visit. The compromise that is often used, to give the optimal amount of information with acceptable levels of uncertainty, is to use a total count time of typically around 200 s, but this can be extended if required.

In the laboratory, every effort is made to ensure that the analytical test portion is homogeneous. In this case the precision of the measurement method (s_{meas}), ignoring uncertainty from field sampling, is essentially the same as that for the analytical procedure (s_{anal}), giving:

$$s_{\mathrm{meas}} = s_{\mathrm{anal}}$$

The situation for *in situ* analysis is different in one other important respect. The measurement precision is not only limited by the analytical procedure. For *in situ* PXRF in general, and for contaminated land applications in particular, the analytical test sample is *heterogeneous*. In this case the precision of the method, when viewed as a whole, should include a contribution from the precision of the field sampling (s_{samp}), giving:

$$s_{\mathrm{meas}} = \sqrt{(s_{\mathrm{samp}}^2 + s_{\mathrm{anal}}^2)} \tag{3.1}$$

From a practical view point, this definition is also intuitively satisfactory. In the field it is very evident that the process of sampling, by the act of placing the PXRF instrument at a particular place within the sampling location, is the necessary first part of the measurement process.

The consequences of this definition are far reaching. The sample volume analysed by PXRF is typically very small (*e.g.* 0.2 cm^3 for Pb), due to the fixed area of analysis (*e.g.* 5 cm^2) and a limited critical depth of penetration by the X-rays (*e.g.* 0.04 cm for Pb). This means that heterogeneity will often be high, especially for media like contaminated soil, and s_{samp} will therefore also be high. The random component that limits the measurement uncertainty will therefore become s_{samp} and not s_{anal}. The effective precision of PXRF becomes limited, therefore, by the heterogeneity of the sample, rather than by the noise of the analytical signal.[4] Interestingly, this is also true of measurements by many high precision *ex situ* laboratory-based techniques, where the total uncertainty of the measurement is limited by the uncertainty of the sampling, rather than that of the analysis.

A second consequence of Equation (3.1) is in terms of deciding on the level of uncertainty in a measurement that is acceptable. This question can be used to judge the fitness-for-purpose (FFP) of an analytical method. FFP has been defined as "the property of data produced by a measurement process that enables a user of the data to make technically correct decisions for a stated purpose".[12] In the laboratory, the FFP criterion often applied is that the relative precision of a method, at 95% confidence, should be below 10% (*e.g.* MCERTS[10]). In terms of analytical quality control (AQC), this means that batches where the precision is larger than 10% are rejected and re-analysed. A second FFP criterion that has been proposed relates the precision required for any particular site to be investigated. It suggests that the variance of the measurement (s^2_{meas}) should not contribute more than 20% to the total variance ($s^2_{\text{total}} = s^2_{\text{meas}} + s^2_{\text{geochem}}$), where s^2_{geochem} describes the true geochemical distribution of an element between sampling points, which would be seen in the absence of any measurement errors. Using this criterion it is possible to show that measurement methods with values of relative precision as large as 80% can be fit for purpose.[13] This is despite the fact that such measurements would be considered to be at concentrations below the detection limit, based on the formal definitions given above.

Using Equation (3.1) to define measurement precision and the second FFP criterion, it is theoretically possible to make use of measurements made with PXRF at concentration levels that are below the traditional detection limit values. The experimental design required to estimate the various random components of the system (s_{anal}, s_{samp}, s_{geochem}, s_{total}) is based on the taking of duplicated samples, followed by robust analysis of variance.[13]

3.6 Less Obvious Advantages and Disadvantages of *In Situ* PXRF

Given an appreciation of the uncertainty of measurements by PXRF, it is possible to comment on the less obvious advantages and disadvantages of the technique.

In terms of advantages, it is possible to have acceptable levels of overall measurement uncertainty for some elements, even though the detection limits for the element by PXRF are nominally not low enough to equal the threshold value of the hazard. Conversely, because the instrumental precision of PXRF is good (*e.g.* 1%), at concentrations well above the detection limit, instrumental precision is not usually the limiting factor controlling the uncertainty of the measurements (*i.e.* the sampling precision is the limiting factor). The count times can be reduced, therefore, in some cases, without significantly affecting the overall uncertainty of the measurement.

Regarding the disadvantages, the small size of the sample analysed *in situ* does cause high levels of measurement uncertainty (*e.g.* 50%), but the method can still be shown to be fit for some purposes. However, this may not be a particular disadvantage of *in situ* measurement. When sampling is included in the definition of the measurement process, even *ex situ* measurement of contaminated sites can have similarly high levels of uncertainty. This idea was tested in a comparison between *in situ* PXRF and *ex situ* ICP-AES for the investigation of the same site.[4] Both systems had almost identical measurement variance (15% and 14% of total variance respectively), because they are both limited by the uncertainty from the field sampling. For the spatial delineation of hot spots, where the variation in the concentration of the contaminant across the site may vary by several orders of magnitude ($s_{geochem}$ is large) then the method (s_{meas}) will contribute <20% to the total variance, and can be shown to be FFP. Nevertheless, in terms of the existing US EPA Statistical Criteria for Classifying Data Quality, PXRF would generally rate Level 1, Qualitative Screening as the RSD is greater than 20%.[14]

As one of the advantages, however, there has recently been a revision of the approach to field-based analysis by EPA, and a growing appreciation that the quality of decisions on contaminated land can improve, even though the quality of the field-based analysis is not as good as the lab-based analysis.[15] The overall cost of investigation and remediation can be reduced substantially by use of PXRF to improve decisions on whether a soil is actually contaminated before, during and after the remediation process, not just for the site investigation. In this context, when financial considerations are used in the assessment of fitness-for-purpose, it has been shown that *in situ* PXRF, despite its higher uncertainty, can be more FFP than *ex situ* lab-based analysis.[16]

Among the disadvantages, values for the overall bias between the concentration values measured in the field on moist soil by PXRF and those measured on dried lab samples are often large (*e.g.* −60%). If the value of this bias is known (and reasonably constant) it can be used to either correct the measurements (*e.g.* for field moisture), or to improve the estimate of the measurement uncertainty.

3.7 Future Developments

A lowering of detection limits for a wide range of elements is the most pressing requirement in the near future. This would increase the number of elements that can be quantified, with acceptable levels of uncertainty, at threshold concentrations on contaminated sites. If there were, for example, a general improvement by a factor of ten, then determination of Cu, Ni, As and Ba at background concentrations could become practical (Figure 3.4a). Further improvements could make the determination of Cd, Co, Se, Sb and even Hg possible.

Overcoming the safety problems associated with the use of radioactive sources would also be a real boost to the wider use of PXRF. As a range of small-scale, low power consumption X-ray tubes have become commercially available, this objective is now being achieved. A safety hazard associated with X-rays that are being generated *in situ* remains, but that hazard would not be present at times when the instrument is switched off, facilitating transit between sites, and even between countries.

The apparent bias that is caused by soil water and surface roughness during *in situ* measurement of contaminant concentration often causes serious under-estimation in comparison with regulatory threshold values (which are based on dried soils). If an integrated probe were available with the PXRF to measure soil moisture, then most of this apparent bias could be removed by recalculation to a dry basis. Routine application of the Rayleigh scatter peaks could also be applied to reduce bias due to surface roughness of the sample.[11]

The next generation of instrumentation, with improved performance, is likely to become a routine tool for the assessment of metal contaminated sites. The improvements need to be mainly aimed at lower detection limits for a wider range of analytes, and fewer safety constraints. There also needs to be the acceptance by the operators of PXRF instruments and the regulators that measurement uncertainties, if routinely measured and reported, need not be a limitation to its effective use in the assessment and reliable remediation of contaminated land.

References

1. DEFRA/EA (2002a). CLR 7 Assessment of Risks to Human Health from Land Contamination: An Overview of the Development of Soil Guideline Values and Related Research, Department for Environment, Food and Rural Affairs and the Environment Agency.
2. S. Piorek, Field-portable X-ray fluorescence spectrometry: past, present and future, *Field Anal. Chem. Technol.*, 1997, **1**(6), 317–329.
3. Johnson et al (1995) Effective XRF field screening of lead in soil.
4. A. Argyraki, M.H. Ramsey and P.J. Potts, Evaluation of portable XRF for in-situ measurements of lead on contaminated land, *Analyst*, 1997, **122**, 743–749.
5. P.J. Potts, (2002) Sampling and Analysis of Contaminated Land. Course Notes, Imperial College (unpublished).
6. R.W. Puls, D.A. Clark, C. Carson and J. Vardy, Characterization of Chromium-Contaminated soils using field-portable x-ray fluorescence, *Groundwater Monitoring*, 1994, **14**(3), 111–115.
7. A. Argyraki, M.H. Ramsey and M. Thompson, Proficiency testing in Sampling: Pilot study on Contaminated Land, *Analyst*, 1995, **120**, 2799–2804.
8. DEFRA/EA (2002) *SGV 1 Soil Guideline Values for Arsenic Contamination*, Department for Environment, Food and Rural Affairs and the Environment Agency.
9. Rose, Hawkes and Webb, *Geochemistry in Mineral Exploration*, 2nd edn, Academic Press, London, 1979.
10. MCERTS, www.mcerts.net/ (UK Environment Agency site).
11. P.J. Potts, *A Handbook of Silicate Rock Analysis*, Blackie, Glasgow, 1987, p. 252.
12. M. Thompson and M.H. Ramsey, Quality concepts and practices applied to sampling – an exploratory study, *Analyst*, 1995, **120**, 261–270.

13. M.H. Ramsey, Sampling as a source of measurement uncertainty: techniques for quantification and comparison with analytical sources, *J. Anal. Atomic Spectrom.*, 1998, **13**, 97–104.
14. EPA http://www.epa.gov/etv/pdfs/vrvs/01_vs_tnspectrace_ninek.pdf (accessed 24.09.04).
15. D.M. Crumbling, C. Groenjes, B. Lesnik, K. Lynch, J. Shockley, J. van Ee, R. Howe, L. Keith and J. McKenna, Managing uncertainty in environmental decisions, *Environ. Sci. Technol.*, 2001, **35**(9), 405A–409A.
16. P.D. Taylor, M.H. Ramsey and P.J. Potts, Balancing measurement uncertainty against financial benefits: a comparison of *in situ* and *ex situ* analysis of contaminated land, *Environ. Sci. and Technol.*, 2004, **38**, 6824–6831.
17. Netherlands Ministry of Housing, Spatial Planning and Environment (2000) *Circular on target values and intervention values for soil remediation: DBO/1999226863*, Directorate-General For Environmental Protection, Department of Soil Protection (IPC 625), PO Box 30945, 2500 GX The Hague. 9 February 2000. Published in the Netherlands Government Gazette No. 39 on 4 February 2000 http://www2.vrom.nl/Docs/internationaal/annexS_I2000.pdf.
18. M.H. Ramsey, S. Squire and M.J. Gardner, Synthetic reference sampling target for the estimation of measurement uncertainty, *Analyst*, 1999, **124**(11), 1701.
19. P.D. Taylor, M.H. Ramsey and P.J. Potts, Spatial contaminant heterogeneity: quantification with scale of measurement at contrasting sites, *J. Environ. Monitoring*, 2005, **7**, 1364–1370.
20. R.P. Swift, Evaluation of a field-portable X-ray fluorescence spectrometry method for use in remedial activities, *Spectroscopy*, 1995, **10**, 31.
21. C.A. Kuharic and W.H. Cole, FPXRF, EDXRF and ICP comparison of Pb contaminated soils from Leadville, Colorado, *Advances in X-Ray Analysis*, 1995, **38**, 725.

CHAPTER 4

Coatings, Paint and Thin Film Deposits

STANISLAW PIOREK

Thermo Niton Analyzers, LLC, 900 Middlesex Tpk., Bldg. 8, Billerica, MA 01821, USA

4.1 Introduction

Coatings play an increasingly important role in many areas of industry. There is an increased interest in various methods of non-destructive measurement of coating thickness. Recent developments in such technologies as composite materials, optical filters, superconductors, consumer and industrial electronics, as well as semiconductor device production, have propelled the research and design of both coatings and methods for their applications. This, in turn, has increased the need for non-destructive, rapid and accurate methods of measurement of coating thickness and composition.

4.2 What is a Coating?

A coating may be defined as a layer of one material applied onto the surface of another material in such a way that both adhere permanently to each other. Usually, the layer of coating is thinner, often very much thinner, than the substrate. Initially, coatings were mainly applied for preventive and decorative reasons (such as paints, the most commonly encountered coatings). The role of the coating was to preserve the substrate from elements and extend the service life of coated object. In contrast, a thin layer of oil or water applied between two surfaces in contact reduces the frictional forces between them and allows for ease of motion of one surface against the other. This lubricant

Portable X-ray Fluorescence Spectrometry: Capabilities for *In Situ* Analysis
Edited by Philip J Potts and Margaret West

layer can also be considered as a coating. In examples mentioned above (paint, lubricant) the coating was "sacrificed" to protect a more expensive substrate. Sometimes, however, the situation may be reversed. For example, some less expensive jewellery is often plated with a very thin layer of a more precious material such as gold, to create the appearance of an object made of solid gold.

Contemporary coatings are many, and they serve many diverse purposes. Examples include ferric oxide on magnetic tape for audio and video recording, selenium on Mylar® for electrostatic copying, composite metallic layers on polymer films, thin metallic films on catalytic electrodes, silicone coating on paper, copper film on printed circuit boards, thin antiglare coatings on glass or thin hard protective coating on optical lenses – to name just a few. These very specialized coatings are often very thin when compared to a typical layer of paint (micrometres *versus* millimetres) and are, therefore, referred to as "thin films".

4.2.1 Brief Overview of Major Non-destructive Methods of Coating Thickness Measurement

The diversity of combinations of coatings, films and substrates is served by various methods specifically developed for the measurement of coating thickness. Depending on the coating, its thickness and the substrate combination, one may use magnetic induction, microresistance, the eddy current method, interferometry, Beta backscatter, X-ray or gamma backscatter or X-ray fluorescence.

From an analytical stand point, it is practical to separate coatings into single- and multilayer. Multilayer coatings can be measured by essentially one method, X-ray fluorescence (XRF), while several techniques, including XRF, can tackle the much simpler problem of a single layer coating. Although many of these coatings, especially the very thin and optically transparent, are amenable to measurement *via* optical methods such as infrared interferometry or ellipsometry, a great number of them are also easily quantified using the XRF technique – specifically those that include a combination of metal on non-metallic substrate or *vice versa*.

4.2.1.1 Magnetic Methods

These methods are used to measure non-destructively the thickness of a non-magnetic coating on magnetic (ferrous) substrates. Two principles of operation are used:

- magnetic pull-off
- magnetic or electromagnetic induction.

Magnetic Pull-off. In a magnetic pull-off gauge, the force of attraction between the permanent magnet of the gauge and magnetic substrate is a measure of thickness of the coating separating the two. Such gauges are rugged and inexpensive. They are sensitive to surface roughness, substrate thickness and composition of the substrate alloy. Their typical measurement error (tolerance) is about ±10% relative.

Magnetic or Electromagnetic Induction. The gauges based on this principle feature a probe that generates a magnetic field, and a sensor of magnetic flux density. When the probe approaches the magnetic substrate, the detector measures the magnetic flux density at the probe, which is related to the distance of the probe from the substrate. That distance is determined by the coating thickness. If the source of the magnetic field of the probe is a permanent magnet, then the Hall-effect element or magneto-resistor is used to measure flux density. If the source of magnetic field of the probe is an electromagnetic coil, a second electromagnetic coil is used as the sensing element. Magnetic induction based instruments can deliver typical accuracy of ±1% relative.[1]

4.2.1.2 Eddy Current Method

This method is used to measure the thickness of nonconductive coatings on nonferrous metal substrates. A fine wire coil supplied with a high frequency alternating current (typically above 1 MHz) is used to generate an alternating magnetic field at the tip of the probe. When the probe is brought close to the conductive surface of a substrate, it will induce eddy currents in it which in turn modify the electromagnetic field of the probe. The magnitude of eddy currents and subsequent change of magnetic field is a measure of the distance of the probe from the substrate (that is of the coating thickness). Typical accuracy of eddy current gauges is about ±1% relative.[2]

4.2.1.3 Ultrasonic Method

Typically, this method has been used to measure the thickness of coatings on non-metallic substrates such as wood, plastic, ceramics, *etc*. However, contemporary applications of the ultrasonic method can also allow metallic substrates and up to three layers of coatings to be measured.[3] This method is based on the measurement of time delay of the ultrasonic pulse reflected from the boundary between the coating and the substrate. It is also referred to as a pulse–echo technique. Its typical accuracy is of the order of 3% relative,[4] with the thickness measurement ranging up to several mm, depending on the coating–substrate combination.

4.2.1.4 Microresistance

Microresistance is used to measure the thickness of a conductive coating on a nonconductive substrate, such as the thickness of copper in through-holes on printed circuit boards. This technique requires precise measurement of the resistance of the copper cylinder that forms the plated through-hole on the board.

Once this resistance is known, the average thickness of copper plating is calculated using the dimensions of the hole and Ohm's Law.

4.2.1.5 Ellipsometry

In ellipsometry, we utilize changes in the polarization state of light when it is reflected from a sample. When a thin film changes its thickness, then its reflection properties for the light will also change, and the measure of these changes in reflection properties allows us to deduce the actual changes of the film thickness. Ellipsometry is applicable to thin films, ranging in thickness from essentially zero to several thousand ångströms.

4.2.1.6 Methods Based on Ionizing Radiation

Essentially there are three non-destructive techniques for the measurement of coating thickness that utilize interaction of ionizing radiation with matter:

- beta particle backscatter
- gamma- or X-ray backscatter
- X-ray fluorescence.

Of these three we will discuss in detail the last one as it is readily available with practically all portable, hand-held X-ray fluorescence analyzers. In some instances, the X-ray scattering approach may also be employed with these instruments.

The *beta backscatter* method is based on phenomenon of the scatter of electrons (beta particles) off a surface.[5] Usually, an isotope of ^{14}C is used as source of beta particles. For a given excitation energy of the source, the intensity of β-particles backscattered off a thick target is proportional to the atomic number, Z, of the scatterer. Therefore, for effective measurement of coating thickness applied over a substrate, the atomic number of the coating must differ from that of the substrate, at least by five units (or about 20% relative). Consequently, this method would not be suitable to measure a coating of nickel ($Z=28$) over a copper ($Z=29$) substrate. However, it would perform very well when measuring a tin coating on steel. If the ^{14}C isotope (maximum beta energy of 155 keV, half-life 5760 years) is used as excitation the source, then coatings up to $3\,mg\,cm^{-2}$ can be measured. Note that coating thickness is expressed in mass per unit area rather than in absolute, linear thickness. This is typical of methods based on ionizing radiation, as the intensity of scattered or fluorescent radiation is primarily dependent on the so-called "surface density" or mass per unit area of the sample. By dividing measured mass per unit area by specific density of the coating, a linear coating thickness can be obtained.

The *gamma-* or *X-ray scattering* methods also use the intensity of the scattered radiation as a measure of the thickness of the scattering layer.

However, the intensity of scattered gamma or X-rays is also a strong function of composition of layer, or, to be exact, of the average atomic number of the scattering material. This opens up the possibility to not only measure the thickness of a layer of material but also its average atomic number (composition). Gamma or X-rays impinging on a sample can be scattered with loss of energy (so-called incoherent or Compton scattering) or without energy loss (so-called coherent scattering). It is Compton scattering that is very dependent on an average atomic number of the material. One of the more interesting examples of the application of this method is in the identification of the corrosion of steel reinforcement under the aluminium skin of aircraft. When a steel element is corroded, a void is created in place of a solid metal. The void, which is filled with air and with oxides of iron, changes the intensity of scattered radiation. Typically, isotope sources are preferred in this method as they emit monoenergetic X or gamma rays that produce well-defined Compton scattered peaks.

The *X-ray fluorescence* method utilizes the phenomenon of generating characteristic X-rays in material irradiated with a sufficiently energetic beam of gamma or X-rays from an external source. The characteristic X-rays may be induced in both the coating and the substrate. Thus, the coating thickness can be determined either by measuring the intensity of X-rays excited in the coating or by measuring the intensity of X-rays excited in the substrate as they are attenuated by a coating of increasing thickness. The exciting radiation may be generated by an X-ray tube or by certain isotopes. X-Ray fluorescence is applicable to coatings that are too thin for the β-backscatter method.

4.3 XRF Method for Coating Thickness

4.3.1 Theory

For clarity, we limit our discussion of the XRF method for coatings to instances of a single layer of coating over a single layer of substrate. Modern coating systems may have many different layers of different composition and thickness. Such complex, multilayer systems can be successfully analyzed by dedicated XRF spectrometers using fundamental parameters-based software, which can handle up to seven different layers of varying chemical composition. However, even those instruments require that the order of layers is known *a priori*. This is so because the number of variables exceeds the number of independent equations that can be created for such a multilayer system to solve it. Many excellent contributions on this interesting subject can be found in volumes of *Advances in X-ray Analysis*.[6–14]

Figure 4.1 illustrates typical measurement geometries utilized in testing the coating thickness by XRF. Geometries (a) and (b) are the most popular, especially with isotope based instruments. Geometry (c) is the one that is most often found in X-ray tube based instruments. One of its advantages is that it de-emphasizes the radiation scattered from the sample (mainly substrate). Its other

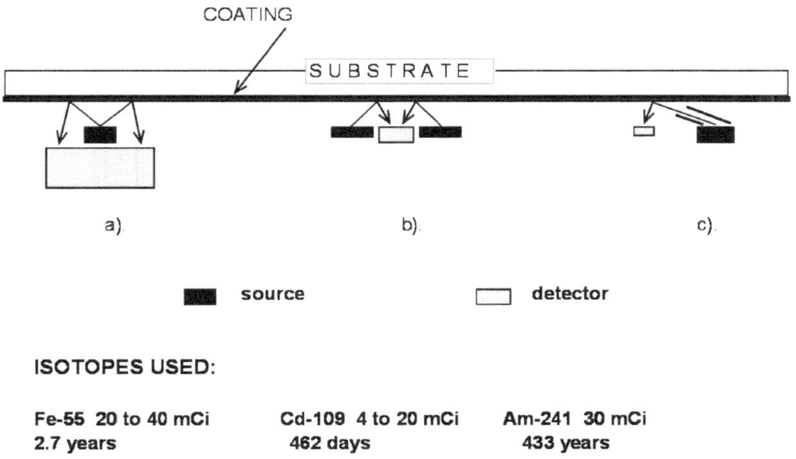

Figure 4.1 Typical measurement geometries used in XRF measurements for coating thickness.

advantage is that it allows for the measurement of thinner coatings; radiation travelling at an angle through the surface layer must go through thicker medium than when travelling perpendicular to the layer.

As previously noted, coating thickness can be determined by measuring either the intensity of X-rays excited in the element in the coating or the intensity of X-rays excited in the element of the substrate. In the first case, an increase in X-ray intensity corresponds to the increase of coating thickness; the thicker the coating, the more intense its X-rays are. In the second case, an increase in coating thickness is accompanied by a decrease in the intensity of X-rays from the substrate as they are attenuated by the layer of coating. A determination as to which approach is better depends on the specifics of the application.

Whichever approach is used, the most important requirement assuring feasibility of the measurement must be met – the element measured can only be present either in the coating or in the substrate, but never in both. Often, when measuring coating thickness on thin plastic films, it is advantageous to place the coated film on or against a flat sheet of metal backing and measure the intensity of element in the backing through the film. This method also allows the measurement of the thickness of the uncoated plastic film or paper.

Additionally, we require that the chemical composition of either the coating or the substrate remains constant.

4.3.1.1 X-rays Excited in Coating

Figure 4.2 shows schematically a single coating on a substrate and exciting and emitted radiation.

Assuming a coating of one element over a substrate of another element, the intensity I_{fc} measured at the energy of characteristic X-rays excited in the

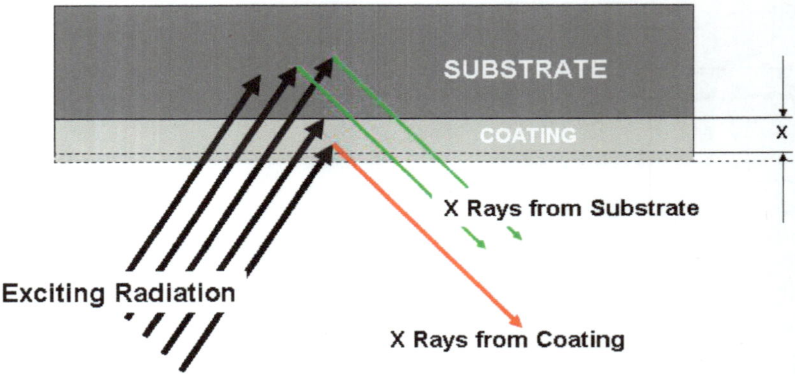

Figure 4.2 Generating X-rays in coating and substrate.

coating increases according to Equation (4.1):

$$I_{fc} = I_{So} + (I_{\infty} - I_{So})\left[1 - e^{-(\mu_e^c + \mu_c^c)m}\right] \quad (4.1)$$

where I_{∞} is the saturation intensity (*i.e.*, for an infinitely thick coating layer) of the fluorescent radiation emitted by the element in the coating; I_{So} is the intensity measured from the substrate when the coating thickness is equal to zero; m is the mass per unit area of the coating ($m = dx$), in g cm^{-2}, where d is specific density of coating (g cm^{-3}), and x is its thickness (cm); $\mu^c{}_e$ is the mass absorption coefficient of the coating at the energy of exciting radiation (cm^2 g^{-1}); and $\mu^c{}_c$ is the mass absorption coefficient at the energy of the fluorescent radiation from the coating (cm^2 g^{-1}).

To be exact, the mass absorption coefficients in the exponent should be divided by sine functions of incident and exit angles, respectively. As it is written, Equation (4.1) assumes that both exciting as well as characteristic radiation are perpendicular to the coating surface (unlike in Figure 4.2 above). This minor simplification makes the formula less cumbersome without changing its functional character.

Note that we used in the exponent the mass absorption coefficient without explicitly showing the linear thickness, x. This is because X-ray absorption primarily depends on mass per unit area of the absorbent. Additionally, very often we do not know exactly the density of the absorbent, as it very much depends on the method of production. For example, a hot dipped tin coating will have a different specific density than that produced by electrolytic deposition or sputtering. These are the reasons why "mass per unit area" is frequently used in coating industries.

4.3.1.2 X-rays Excited in Substrate (Backing)

In this case, we measure the intensity of radiation at the energy of characteristic X-rays from the element in the substrate. The appropriate formula is as follows:

$$I_{fS} = I_{\infty} + (I_{So} - I_{\infty})e^{-(\mu_e^c + \mu_{fS}^c)m} \quad (4.2)$$

where this time I_∞ is the saturation intensity of the X-rays emitted by an element from the substrate when the coating thickness is infinite, and $\mu^{c}{}_{fS}$ is the mass absorption coefficient of the coating for fluorescent X-rays emitted from the substrate ($cm^2\,g^{-1}$); all other symbols have the same meaning as in Equation (4.1).

If we rearrange terms in the equations, they assume the following forms. Equation (4.1) is now:

$$\frac{I_{fc} - I_{So}}{(I_\infty - I_{So})} = \left[1 - e^{-(\mu_e^c + \mu_c^c)m}\right] \tag{4.3}$$

and Equation (4.2) now has a form:

$$\frac{I_{fS} - I_\infty}{(I_{So} - I_\infty)} = e^{-(\mu_e^c + \mu_{fS}^c)m} \tag{4.4}$$

Note that denominator in the left-hand side of each equation represents the maximum change of the fluorescent intensity from either coating [Equation (4.3)] or substrate [Equation (4.4)]. The numerators show the measured fluorescent intensities reduced by their analytical backgrounds. In other words, the left-hand side of each equation shows the normalized fluorescent intensities, while the right-hand sides are pure exponential functions. This transformation allows for convenient graphical representation of each equation (Figure 4.3).

4.3.1.3 Range of Measurable Coating Thickness

Arbitrarily, the range of measurable coating thickness is determined as the range of values for which the ratio of normalized intensity as in Figure 4.3 falls between 0.05 and 0.95. This is because thinner coatings usually produce an X-ray signal that is too weak to be measured reliably, while past the 0.95 ratio the X-ray signal becomes insensitive to thickness variations. Therefore, the simple criteria for maximum and minimum measurable thicknesses, m_{max} and m_{min}, respectively, can be written as:

$$m_{max} = \frac{\ln(20)}{(\mu_e^c + \mu_c^c)} = \frac{3}{(\mu_e^c + \mu_c^c)}\ (g\,cm^{-2}) \tag{4.5}$$

$$m_{min} = \frac{\ln(1.05)}{(\mu_e^c + \mu_c^c)} = \frac{0.05}{(\mu_e^c + \mu_c^c)}\ (g\,cm^{-2}) \tag{4.6}$$

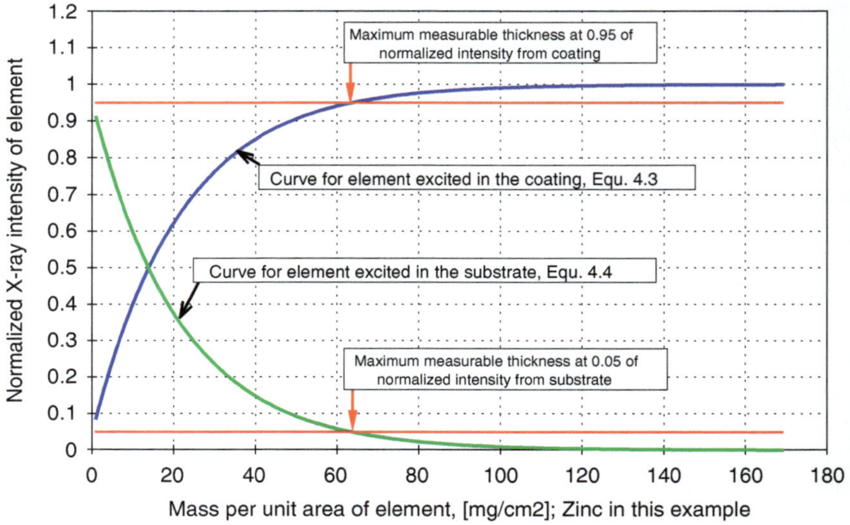

Figure 4.3 Typical shape of calibration curves for determination of coating thickness by X-ray fluorescence; the curves are plotted for the element zinc.

Incidentally, the same criteria can be used for the gamma- or X-ray backscatter techniques.

With most metallic coatings, thicknesses from a few 10^{-5} mm to a few 10^{-3} mm can be measured, while for less dense non-metallic coatings the measurable thickness is larger by at least a factor of 10.

If the specific density of a simple, one-element coating is known, then the maximum measurable thickness can be approximated from the curve in Figure 4.4. For example, we can determine the maximum thickness of a pure zinc coating measured *via* the intensity of zinc K_α line in a coating. The zinc atomic number is 30, hence from the plot in Figure 4.4, we read the maximum mass/unit area for zinc coating, m, to be about 65 mg cm^{-2}; therefore, for specific density of Zn, $d_{Zn} = 7.14$ g cm^{-3}, the linear thickness of the zinc coating, x, will be, $x = m/d_{Zn}$, or $0.065/7.14 \cong 0.01$ cm.

4.4 Selection of Optimum Analytical Conditions

Optimization of analytical conditions has a direct influence on the quality of results. For example, the type and required range of coating thickness to be measured will dictate the selection of excitation energy (type of radioisotope or high voltage of the X-ray tube). Similarly, the choice of using secondary X-rays either from coating or substrate will have an impact on sensitivity and precision of the measurement. We will discuss these issues by referring to examples illustrated in the figures below.

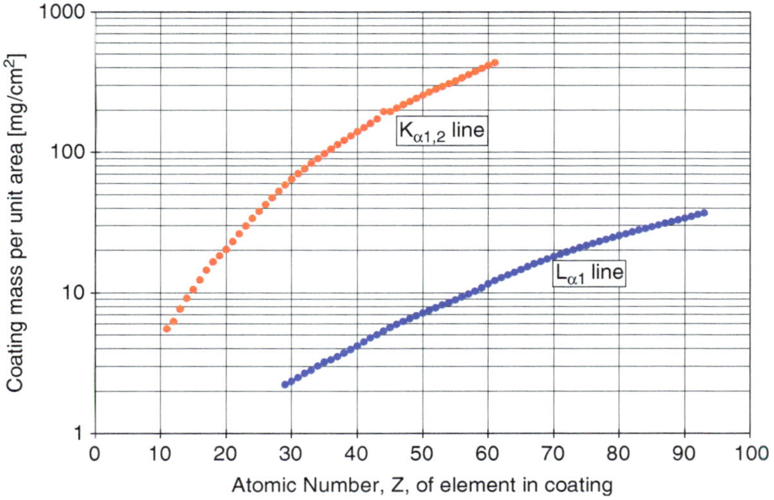

Figure 4.4 Maximum coating thickness measurable by Kα or Lα1 X-rays excited in coating as a function of atomic number, *Z*, of an element.

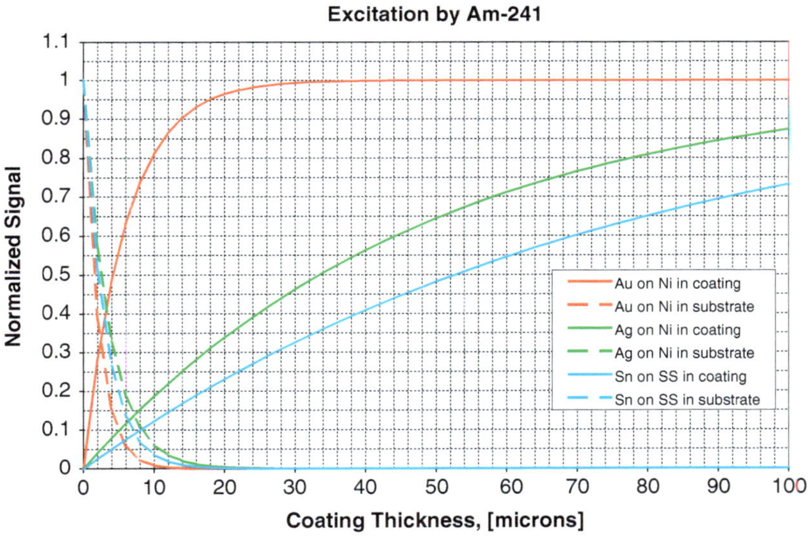

Figure 4.5 "Calibration" curves for gold or silver on nickel and for tin on stainless steel.

Figure 4.5 shows examples of the measurement of gold or silver on a nickel substrate, and tin on stainless steel, all using excitation by 59.95 keV gamma radiation from an ^{241}Am radioisotope source (30 mCi). Nickel plating with gold or silver is common in electronic applications while steel-makers often plate the alloy with tin to prevent corrosion.

Table 4.1 Properties of radioisotope sources used in portable XRF analyzers.

Radio-isotope	*Half-life (years)*	*X- or γ-ray energy (keV)*	*Photons per disintegration*
^{55}Fe	2.7	Mn K X-rays (5.9, 6.5)	0.28
^{109}Cd	1.3 (464 days)	Ag K X-rays (22.1, 25); γ-rays, at 88.03	1.02; 0.04
^{241}Am	432.7	γ-rays, at 59.54; Np L X-rays (11.9–22.2)	0.36; 0.43

4.4.1 Coating Range and Excitation Source

It is easy to see that to measure a wide range of tin coating thicknesses one needs to use tin Kα X-rays excited in the coating. Then, to excite these X-rays, we need to use a source of sufficiently high energy, such as ^{241}Am or an X-ray tube working at a high voltage (HV) of at least 35 kV. However, should we only be concerned with very thin coatings of tin, we might consider a measurement *via* iron Kα X-rays excited in the steel substrate, which would also allow us to use a less energetic source, such as ^{109}Cd or an X-ray tube working at a lower HV.

Another valuable observation that can be made by studying Figure 4.5 is that when we analyze a coating of one metal whose X-rays have energy similar to the element in the substrate, there is little difference between using excitation in the coating or in the substrate. This is shown by the example of gold over nickel. Both curves allow pretty much the same range of thickness. The situation changes if we consider the case of silver over nickel. Here, Kα X-rays of silver at 22.1 and 25 keV are much more energetic than the 8.47 keV of nickel Kα X-rays. Consequently, measurement with silver lines offers an order of magnitude larger (wider) range of measurable thickness than nickel does with its Kα X-rays.

Table 4.1 lists isotope sources used in portable XRF analyzers that can also be employed in coating thickness measurements.

4.4.2 Sensitivity and Precision of Measurement

4.4.2.1 Sensitivity (S)

The sensitivity of an analytical method is formally defined as a ratio of change of the measured signal to the change of the concentration of element that caused that signal change; the larger this ratio, the more sensitive the method. Mathematically, sensitivity is expressed as the derivative of the instrument response curve over the corresponding change of concentration of analyte. For example, in our case the response curve is the intensity of fluorescent radiation of the element in a coating, I_{fc}, as function of mass per unit area, m, of the element in that coating:

$$I_{fc} = f(m) \tag{4.7}$$

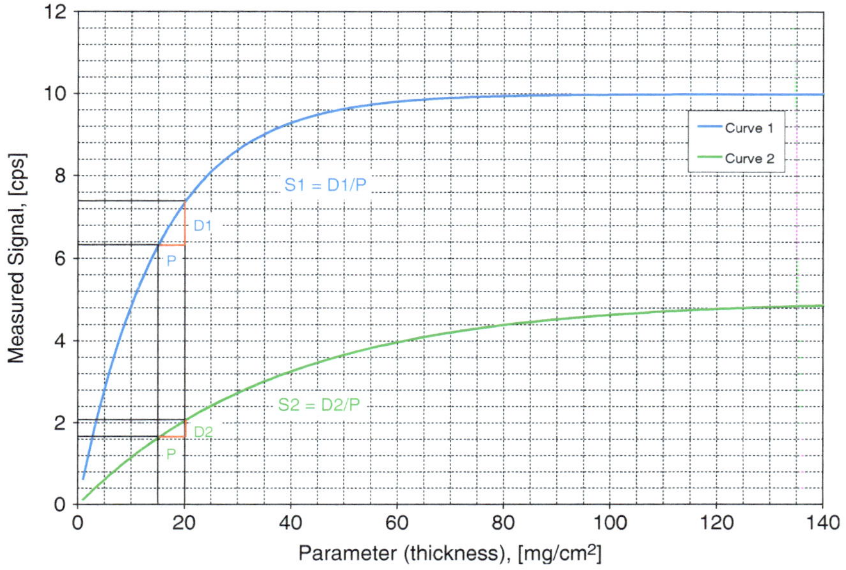

Figure 4.6 Sensitivity of calibration (analytical) curve; cps = counts sec^{-1}.

Then the sensitivity, S_{fc} is:

$$S_{\text{fc}} = \frac{\partial I_{\text{fc}}(m)}{\partial m} \cong \frac{\Delta I_{\text{fc}}(m)}{\Delta m}. \tag{4.8}$$

In other words, sensitivity is equal to the local slope of the calibration curve. This is illustrated in Figure 4.6. If we calculate local sensitivity at 15 mg cm^{-2} for curve 1 we obtain $S_1 = 0.21$ cps mg^{-1} cm^2. For curve 2, $S_2 = 0.08$ cps mg^{-1} cm^2, almost three times smaller.

When we look at Figure 4.5 again we can easily see that the gold on nickel curve offers higher sensitivity than silver on nickel. How can one benefit from that? For example, there are coatings that are not of single elements, such as zinc phosphate on steel. The stoichiometry of the coating is constant and, therefore, one may use either phosphorus or zinc X-rays for analysis. This type of coating has thicknesses that are well within the capability of phosphorus and zinc X-rays. However, phosphorus offers better sensitivity than zinc, and therefore the coating thickness is usually measured *via* phosphorus X-rays.

4.4.2.2 Precision

The precision of a measurement is affected by several factors, one of them being the proper selection of the operating section of the calibration curve. This is shown in Figure 4.7.

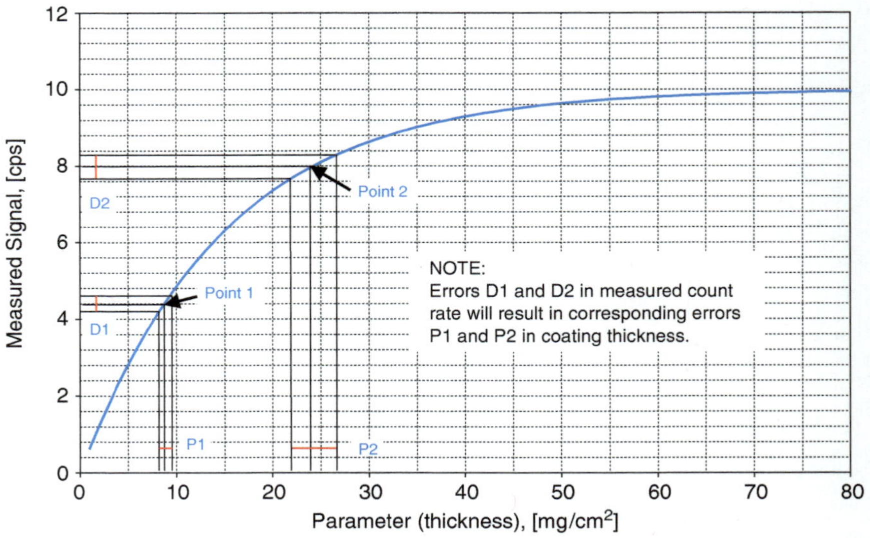

Figure 4.7 Dependence of measurement error on calibration curve; cps = counts sec^{-1}.

Let us consider two points on the curve shown in Figure 4.7. Assuming that the measured coating yields in a given measurement time a signal of 4.4 cps, with an error of ±0.2 cps, this will project itself as per the curve shown at $8.5 \pm 1.0\,mg\,cm^{-2}$ of coating thickness (Point 1 in Figure 4.7). That is about 11.7% of relative error. If, however, the intensity measured from the coating is now 8.0 cps its error will increase due to counting statistics to about ±0.3 cps, which in turn will convert into $24 \pm 2.5\,mg\,cm^{-2}$. This corresponds to about 10.5% of relative error. It can be seen that while the relative error was reduced, its absolute value increased 2.5 times and not 1.5 times, only because of the different curvature (or sensitivity) of the calibration curve at Point 2 than at Point 1. Should this not be acceptable, one simple remedy to reduce the absolute error, P2, is to extend the measurement time as this would make the initial error, D2, in signal count rate smaller.

4.5 Typical Examples

Table 4.2 lists some typical coating thickness applications that can be addressed using a portable X-ray Fluorescence Analyzer with radioisotope (and also with X-ray tube) excitation.

The most common type of coating measurement application is that of a metallic coating on metallic substrate. Examples would be the case of zinc-plated steel, hot-dipped zinc, tin-plated steel, cadmium-plated steel, and the like. The coating may be measured either by using X-rays induced in the

Table 4.2 Selected examples of coating thickness measurements by X-ray emission spectrometry (XRF).

Description	*Range of thickness (μm)*[a]	*X-rays used*	*Radioisotope used*	*Typical relative error (% rel.)*
Electroplated Sn on steel	0.02–1	Fe Kα from substrate	10 mCi ^{109}Cd	2
Hot-dipped Sn on steel	0.5–15	Sn Kα	10 mCi ^{241}Am	2
Electroplated Zn on steel	0.5–13	Fe Kα from substrate	10 mCi ^{109}Cd	2
Hot-dipped Zn on steel	0.5–70	Fe Kα in substrate	10 mCi ^{109}Cd	2
Cd on steel	3–8	Cd Kα	10 mCi ^{241}Am	1–2
Pt electroplated on titanium	0.05–1.3	Ti Kα in substrate	40 mCi ^{55}Fe	2
Ti evaporated on Kovar	0.15–0.25	Ti Kα	20 mCi ^{55}Fe	5
TiO_2-based paint on polyester backing	20–100	Ti Kα	20 mCi ^{55}Fe	5
Ag in photographic emulsion	0.10–40 mg cm^{-2}	Ag Kα	30 mCi ^{241}Am	3
Ink on paper	0.50–4	Fe Kα	10 mCi ^{109}Cd	5
Ti on glass	1000–5000 Å	Ti Kα	20 mCi ^{55}Fe	<1
LaOBr on polyester film	20–60 mg cm^{-2}	Br Kα	10 mCi ^{109}Cd	1
$YTaO_4$ on polyester film	30–100 mg cm^{-2}	Sn Kα[b]	10 mCi ^{241}Am	<1
Ru on titanium	0.05–1.2 gm ft^{-2}	Ru Kα	15 mCi ^{109}Cd	<5

[a]unless stated otherwise.
[b]Sn plate was used behind the polyester to measure Sn Kα attenuation by the coating.

coating or in the substrate. For example, a hot-dipped tin coating on steel is better measured *via* Sn K_α X-rays induced in the coating since the hot-dipped process usually produces much thicker layers than electroplating. In contrast, zinc coating on steel produced by electroplating may be measured either by Zn K_α X-rays in the coating or by Fe K_α X-rays induced in the substrate and absorbed by the overlaying zinc.

An excellent example of the measurement of paint thickness on galvanized steel is discussed in ref. 10. The authors use Fe K_α X-rays for paint thickness measurement and Zn K_α X-rays to compensate for large thickness variations of zinc plate over steel.

Many coatings are applied over very thin films of polyester. These coatings can be measured effectively and accurately if a proper backing material is used under the measured film. The role of the backing material is twofold: First, it provides repeatable measurement conditions from sample to sample. Second, it blocks the stray radiation from spreading, which otherwise would pass unobstructed through the thin plastic film. An example illustrating such a case is the measurement of titanium coating on Mylar using an ^{55}Fe source with copper plate used as backing. The Mn K-series X-rays from the source very effectively excite Ti Kα X-rays but none of the copper X-rays. Alternatively, the presence of a 2 mm thick plate of copper behind the plastic completely stops source radiation and provides repeatable measurement conditions.

Other coating applications where XRF is used include conductive coatings on plastic, metallized plastic foils for food packaging and solar power cells using thin coatings of a metallic alloys on a polymer substrate.

By now, it should be obvious that, in addition to the coating material, the thickness of the plastic film, as such, can be measured using the intensity of X-rays induced in the backing material placed under the film. For example, a thickness of the polyester film (also some types of paper) can be measured online using the intensity of the iron Kα X-rays induced in the steel roller supporting the stretched film on the production line.

The ASTM Method B 568[15] is an example of the application of XRF to the measurement of nickel coating thickness, and evidence of industry acceptance of the method.

Figure 4.8 shows an experimentally obtained graph of a calibration curve for nickel sulfamate coating on steel. We can easily observe in the inset how the nickel peak intensity increases with coating thickness while at the same time the iron peak intensity decreases.

Notably, the excellent precision of XRF measurements is frequently a source of controversy because the XRF can – as the result of its excellent

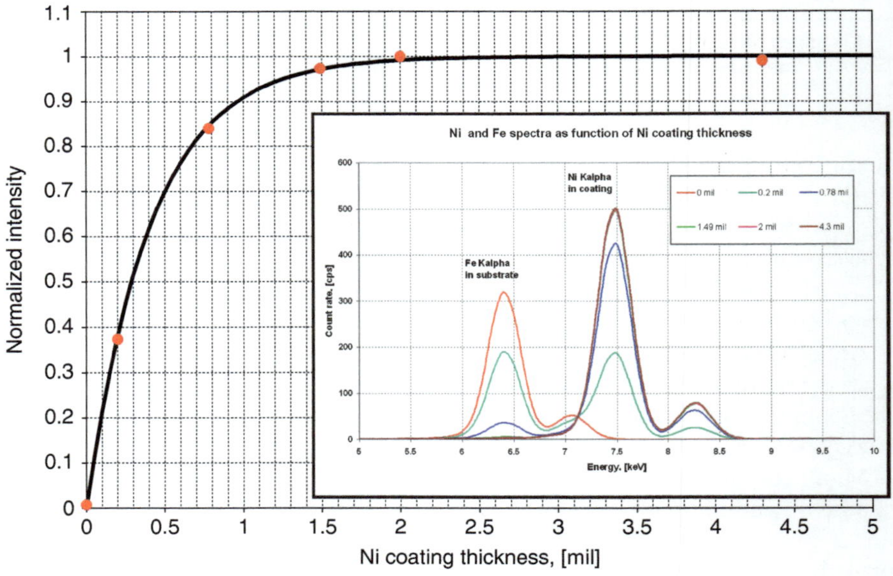

Figure 4.8 Calibration curve for nickel sulfamate on steel.

precision – pinpoint and map inhomogeneity of the coating, much to the surprise of the coating manufacturer. One has to be very aware of this issue. The situation is often further complicated when the particular measurement is constrained by the exacting requirements (traditions) of the applicable ASTM Method.

4.6 Special Cases of Coating Measurements

There are two important applications of XRF analysis that stand on their own and are routinely performed with portable analyzers, and which – in principle – resemble the measurement of coating thickness. These are the measurement of lead in applied paint and the analysis of air particulates collected on filter paper.

4.6.1 Lead in Applied Paint

It has been long known that chronic exposure to even low levels of lead may result in serious problems in young children, including impairment of the central nervous system, behavioural disorders and mental retardation. The presence of this highly toxic element in the environment is strictly regulated, and lead-containing paint has been permanently banned in the USA since 1976. According to the US Federal Regulations[16] any lead-based paint (LBP) hazards equal to or greater than 1.0 mg of lead per square centimetre must be abated, as well as any applied to surface paint that contains more than 0.5% lead by dry weight. In addition, there may not be any surface lead contamination in residential buildings that would yield more than 100 μg of Pb per square foot (or 0.108 $\mu g\,cm^{-2}$) in a dust wipe test. Recently, the danger posed by LBP has become the focus of attention in many European countries. France, for example, introduced guidelines for acceptable levels of lead in paint and followed these with enforceable regulations.[17] Other European countries are expected to follow France's example in the near future. The preferred method for testing paint for lead is field-portable X-ray fluorescence (PXRF).

4.6.1.1 Specifics of the XRF Method as Applicable to Analysis of Lead in Paint

Analysis of lead in applied paint amounts to the determination of mass per unit area of lead in $mg\,cm^{-2}$, which in practice ranges from 0 to about 10 $mg\,cm^{-2}$. From the graph in Figure 4.4 it follows that for atomic number 82 the maximum measurable mass of lead per unit area is about 800 $mg\,cm^{-2}$ for the $K\alpha$ line and about 25 $mg\,cm^{-2}$ for the $L\alpha$ line of lead. We can see now that the layer of paint containing lead may be treated as a coating for the purposes of XRF analysis.

Painted surfaces are likely to be covered over their life time with more than one layer of paint, which means that a lead-containing layer of paint may be covered with other layers that may or may not contain lead. This creates difficulties and forces us to use the K-series lines of lead rather than L-series, as the former are much less absorbed by overcoats of paints. This is the main reason why traditional lead-in-paint analyzers used radioisotope ^{57}Co, which with its 120 keV radiation can excite K-series X-rays of lead (in the 70–88 keV range). Unfortunately, this desirable analytical feature can easily result in mistakes of false high lead readings caused by lead tubing hidden behind an otherwise lead-free drywall. Additionally, the primary radiation from the ^{57}Co source scattered from the analyzed object creates an unusually high analytical background under K-lines of lead whose magnitude is very much dependent on the type of substrate to which the paint is applied.[18] These errors could be minimized by using L-series of lead X-rays (at 10.5 and 12.6 keV), which afford much better precision and sensitivity for lead measurements than the K-series, if it were not for the fact that they are less penetrating and can cause the analyst to miss lead buried under heavy overcoats of nonleaded paint. On the other hand, the use of L-series radiation of lead is not feasible with ^{57}Co due to its poor efficiency of excitation of the L-series X-rays of this element.

Recently, these problems have been satisfactorily resolved by judicious use of the ^{109}Cd isotope as the excitation source. This isotope emits the bulk of its photons in the 22–25 keV energy range, which is perfect for excitation of the lead L-series X-rays. In addition, it also emits about 4% of its output as 88.03 keV photons, which are just right for the excitation of K-series of lead X-rays (lead K_{abs} edge is 88.01 keV). Because the highest energy emitted by this isotope is only 88.03 keV, it is also much safer in use than ^{57}Co. Another important advantage of ^{109}Cd over ^{57}Co is its longer half-life (463 days compared to 270 days), which translates into a lower maintenance cost.

The use of ^{109}Cd isotope allows the opportunity for the simultaneous excitation of both K- and L-series of lead X-rays. Lead produces two major L-series lines, L_α at 10.5 keV and L_β at 12.6 keV. The more energetic L_β line is absorbed by a layer of paint less than the weaker L_α line. Therefore, it is reasonable to expect that the ratio of the intensities of the two lines will vary with the thickness of the overlaying layer of paint or, to put it differently, with the depth at which lead layer is buried. This idea has been employed in one of the lead paint analyzers.[19,20] The so-called "Depth Index", *D*, is expressed as:

$$D = \left(\frac{I_\beta}{I_\alpha}\right)^\kappa \tag{4.9}$$

where I_β *and* I_α are measured intensities of 12.6 and 10.5 keV energies of lead, respectively, and exponent κ is a constant that varies only slightly with the composition of the overlying material. The depth index, *D*, is multiplied by the measured intensity I_β and a calibration constant to quantify lead concentration in mg cm^{-2}. Real lead paint samples yield a depth index from 1 (surface lead) to

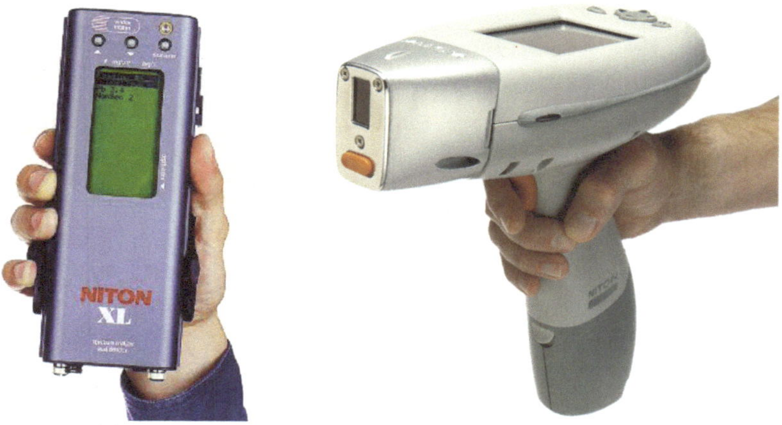

(a) (b)

Figure 4.9 Niton Models XL-300 (a) and XLp-300 (b) Series Lead in Paint Analyzers.

above 10 (deeply buried lead). Precise quantification of lead by this method is possible for a depth index less than about 8. At higher values, only the semi-quantitative determination of lead is viable.[19,20]

All contemporary XRF analyzers for lead in paint provide direct readout of the lead mass per unit area.[21,22] Figure 4.9 shows the embodiments of a handheld X-ray analyzers for lead in paint. These Niton Model XL-300 Series Lead Analyzers utilize ^{109}Cd isotope as excitation source. The information derived from the combined intensities of K and L series X-rays allows the instrument to not only measure lead in paint at any practical depth but also to determine the depth at which the lead paint may be buried. This is accomplished by using the "depth index" concept described above. Using an up to 1.48 GBq (40 mCi) active ^{109}Cd source the measurement time can be as short as 1 s per sample. The instrument precision for measuring lead in paint is 0.05 mg cm^{-2} under optimal conditions, but never worse than 0.1 mg cm^{-2}.

4.6.2 Air Particulates on Filter

The first attempts to use Energy Dispersive XRF for the analysis of air particulates collected on filter media date to the early 1970s.[24] Presently, this method is used routinely throughout the world for monitoring pollution of ambient air. The success of XRF in this application comes from the unique combination of sample type and features of the XRF method. Samples of air particulates are obtained by pumping ambient air through an appropriate filter. Thus, instead of being dispersed in air, air particulates are collected and simultaneously pre-concentrated on the filter medium in the form of a thin layer/film. As we will see, such a layer is an ideal sample configuration for XRF analysis, and may be considered as a special case of the thickness measurement of the coating on a thin, low-scattering substrate.

4.6.2.1 Criterion of Thin Film Sample and its Analytical Benefits

Let us look at the general equation expressing X-ray intensity I_{fi} of the analyte i as function of its concentration, w_i, in sample and mass thickness, m, of the latter:

$$I_{fi} = GI_0\varepsilon_{K,L}\omega_{K,L}\frac{\tau_{i,E}\cdot w_i}{\mu_1+\mu_2}\left(1-\frac{1}{j_{K,L}}\right)[1-e^{-(\mu_1+\mu_2)m}] \tag{4.10}$$

where I_{fi} is the intensity of characteristic X-rays of series K or L of element i excited by an external source of energy E, in cps; I_o is the intensity of exciting radiation of energy, E, in cps; G is the geometry of measurement coefficient; $\varepsilon_{K,L}$ is the detector efficiency for X-rays of K or L series of element i; $\omega_{K,L}$ is the fluorescence efficiency for K or L series of characteristic radiation of element i; $j_{K,L}$ is a ratio (jump factor) of photoelectric absorption coefficients just above to that of just below the absorption edge for element i, for K or L series; $\tau_{i,E}$ is the photoelectric absorption coefficient for element i at energy E; μ_1 is the total mass absorption coefficient of the sample for exciting radiation of energy E ($cm^2\,g^{-1}$); μ_2 is the total mass absorption coefficient of the sample for analyte K or L series X-rays ($cm^2\,g^{-1}$); m is the mass per unit area of the sample ($g\,cm^{-2}$), w_i is weight fraction (concentration) of analyte in sample.

For simplicity of notation the above equation assumes that the incident and emitted radiation is perpendicular to the surface of the sample and that the sample thickness is negligible compared to the distance from the detector to sample and from the source to sample.

This equation acquires a much simpler form for sample mass thickness, m, either very close to zero (thin sample) or very large ("infinitely thick" sample). Our immediate interest is to investigate the case of a thin sample. If we look again at Figures 4.3 or 4.8 we notice that, at the beginning, the calibration curve follows almost a straight line, starting from zero and only later following the curvature of exponential function. From the analyst viewpoint, a linear calibration curve is much more desirable since it is easier to produce and to use. If we expand the exponential in Equation (4.10) into a series around $(\mu_1+\mu_2)\cdot m=0$ and keep only the first-order terms, then Equation (4.10) reduces to:

$$I_{fi} = GI_0\varepsilon_{K,L}\omega_{K,L}\tau_{i,E}w_i m \tag{4.11}$$

Note that now the X-ray intensity of the analyte is a function of its weight fraction alone, and that the terms containing mass absorption coefficients cancelled out. Physically this means that in a "thin sample" the matrix effects (absorption and enhancement) typical of the bulk sample analysis are negligible.

We may now ask the following question:

"What would be a maximum sample thickness that would not introduce to our measurement a relative error greater than δ should we use a simplified, linear calibration curve equation rather than the exact, exponential one?"

Mathematically, our question/condition can be written for each analyte as follows:

$$\frac{I_{\text{approx}} - I_{\text{exact}}}{I_{\text{exact}}} \leq \delta \tag{4.12}$$

where I_{exact} is the X-ray intensity of analyte in sample expressed *via* exact Equation (4.10), and I_{approx} is the X-ray intensity of analyte in the sample expressed *via* a simplified, linear relationship, as in Equation (4.11), and δ is the acceptable value of relative error (divided by 100%).

To solve this condition we again use a series expansion of the exponential function in the expression for I_{exact}, keeping the second-order terms, and after simple algebra obtain the result we call a "criterion of thin sample":

$$m(\mu_1 + \mu_2) \leq \frac{2\delta}{1+\delta} \tag{4.13}$$

This relationship is plotted in Figure 4.10 up to δ of 0.20 (or 20% relative).

As can be easily seen, if by using a straight line calibration we accept the relative error of 5% ($\delta = 0.05$), then the maximum sample thickness should fulfil the condition:

$$m(\mu_1 + \mu_2) \leq \frac{0.1}{1.05} \cong 0.1 \tag{4.14}$$

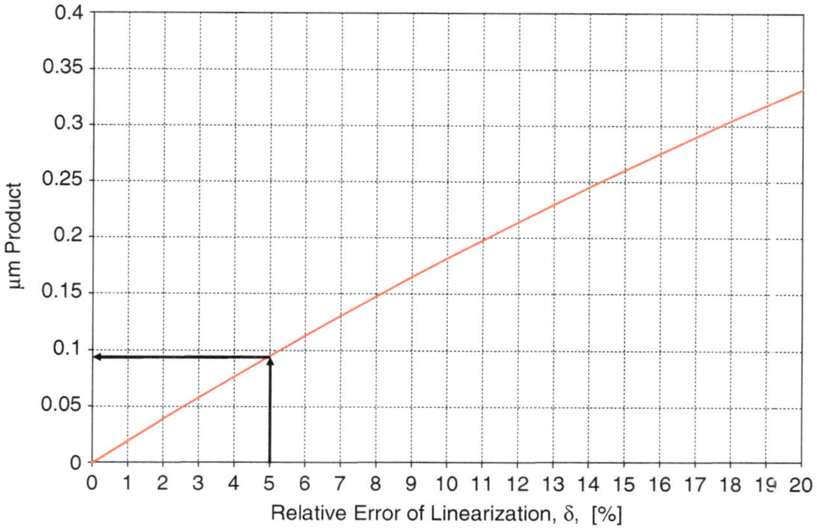

Figure 4.10 Graphical interpretation of "thin sample" criterion.

or finally:

$$m(\mu_1 + \mu_2) \leq 0.1 \tag{4.15}$$

This last equation is our practical "criterion of thin sample". It was first reported in this numerical form in ref. 23, and since then many analysts have adopted it in their practice.[24,25] Obviously, this condition must be fulfilled for each analyte in the sample. Air particulates collected on filters usually fulfil the criterion[15] for elements with $Z > 20$ (calcium). Typically, this means that the surface mass of the sample varies between 1 and 3 mg cm^{-2}. For example, for the analysis of lead in air particulates on cellulose filter paper the maximum mass per unit area (or mass thickness) of a sample varies between 1 and 2 mg cm^{-2}.

The specific value of mass thickness depends on the element determined, and the energy of excitation source and composition of the sample matrix. For example, for 5% m/m of FeS_2 in silica matrix (SiO_2), using 18 keV excitation, the critical surface mass of the sample for iron Kα X-rays is 1.6 mg cm^{-2}. This, in the author's experience, is below the maximum level found in even a very polluted, industrial area air, which yields deposits that rarely exceed 1 mg cm^{-2}.

4.6.2.2 *Signal-to-Background Ratio in Thin Samples*

Another benefit of using a thin sample approach is improved "signal-to-background" ratio or the ratio of the intensity of characteristic X-rays to the intensity of scattered radiation. It is, after all, the radiation incoherently scattered off the sample that gives the rise to the spectral background in XRF analysis. Let R_{thin} be the ratio for a thin sample of the fluorescent intensity of an analyte to the intensity of incoherently (Compton) scattered radiation, and let R_∞ be that for an "infinitely" thick sample. Using Equations (4.10) and (4.11) we obtain:

$$R_{\text{thin}} = \frac{\tau_{i,E}}{\mu_3} \tag{4.16}$$

and:

$$R_\infty = \frac{\tau_{i,E}(\mu_1 + \mu_3)}{\mu_3(\mu_1 + \mu_2)} \tag{4.17}$$

where μ_3 is mass absorption coefficient of sample for incoherently scattered radiation (cm^2 g^{-1}), and the meaning of other parameters is as previously defined.

If we now take the ratio R of R_{thin} to R_∞ and notice that μ_1 is numerically very close to μ_3 we obtain:

$$R = \frac{R_{\text{thin}}}{R_\infty} \cong \frac{1}{2}\left(1 + \frac{\mu_2}{\mu_1}\right) \tag{4.18}$$

For example, if we excite the iron Kα line with 25 keV X-rays of silver from a silver anode X-ray tube or ^{109}Cd isotope we will have $\mu_2 = 45.2$ cm^2 g^{-1} and

$\mu_1 = 2.2\,cm^2\,g^{-1}$, which will yield a ratio R of about 10. Thus, by analyzing iron in a thin sample rather than a bulk one, we can realize an almost ten-fold reduction of background, which in turn will improve the detection limit of the method.

4.6.2.3 *Typical Performance*

As mentioned above, energy-dispersive XRF has been used for the last 25 years by various governmental and international institutions for routine monitoring of ambient air pollution. This type of monitoring is normally performed using the laboratory EDXRF systems. However, it has been widely understood that work place air pollution must also be monitored to protect workers in such occupations as welding and painting.[25] This has been addressed in the USA by NIOSH (National Institute for Occupational Safety and Health) and OSHA (Occupational Safety and Health Administration), who promulgate using portable XRF for routine monitoring of work place air.[26–29] Small, 37 mm diameter membrane filters mounted in sampling cartridges are carried by workers for the period of shift, during which a small pump pulls air through a membrane filter at $2\,L\,min^{-1}$, for 8 hours. Table 4.3 shows performance of Field Portable XRF in the analysis of air particulate deposits from personal samplers. As can be seen, the limits of detection achievable by portable XRF analyzers are below the permissible values. Figure 4.11 shows the spectrum of simulated air particulates collected on a membrane filter.

Table 4.3 Performance of Field Portable X-ray Analyzers in monitoring air particulates in work place air.

Element	*Det. limit*[a] ($\mu g\,cm^{-2}$)	*Det. limit*[b] ($\mu g\,cm^{-2}$)	*Det. limit*[c] ($\mu g\,cm^{-2}$)	*PEL*[d] in air ($\mu g\,m^{-3}$)[e]
Cr	0.43	3.5	1.0	1000
Mn	0.45	2.0	1.0	50
Fe	0.45	3.0	17	10^4 (as Fe_2O_3)
Ni	0.29	1.5	1.0	1000
Cu	0.29	2.5	0.8	100–1000
Zn	0.27	2.0	0.8	$1–5 \times 10^3$
As	0.16	1.0	0.9	10
Se	0.16	0.5	0.3	200
Pb	0.24	0.6	1.3	50
Cd	0.07	10.0	N/A	100–3000
Sn	0.05	N/A	N/A	2000

[a]For $10\,mm^2$ by 5 mm thick Si(Li) detector, 10 mCi ^{109}Cd.
[b]For $5\,mm^2$ by 0.5 mm thick Si PIN diode detector, 10 mCi ^{109}Cd, 60 s measurement per sample.
[c]For $5\,mm^2$ by 0.5 mm thick Si PIN diode detector, X-ray tube at 35 kV, 10 µA, 60 s measurement per sample.
[d]Permissible Exposure Limit in work place air, PEL (as per 29 CFR 1910.1000, 1 Jan. 1977).
[e]To convert $\mu g\,cm^{-2}$ into $\mu g\,m^{-3}$ divide each entry by 0.15. This is based on a flow of $1\,m^3$ of air through a filter area of $7.55\,cm^2$.

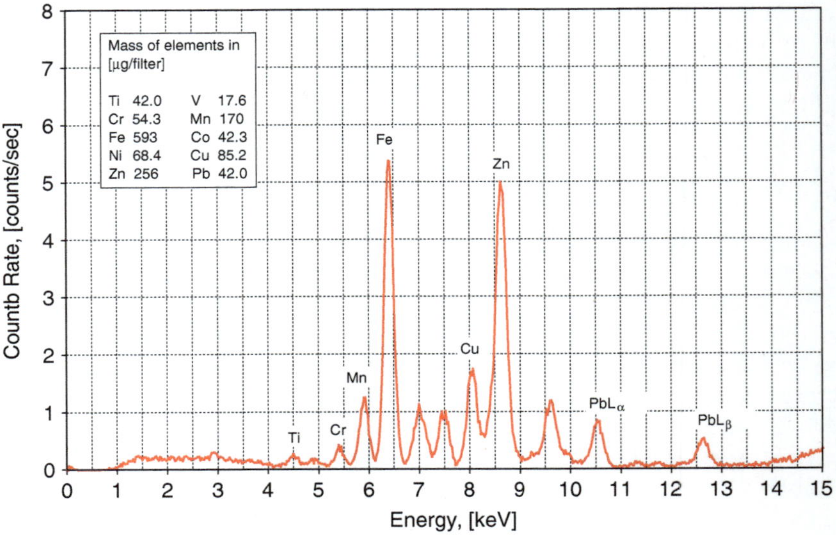

Figure 4.11 Spectrum of simulated air particulate collected on a cellulose ester membrane, excited with an Ag-anode miniature X-ray tube (200 s acquisition time, 40 kV, 20 μA).

Incidentally, the same technique of analyzing thin layer samples can be used for measurement metallic ions in water, by cycling the water through an ion-exchange membrane. Piorek and Pasmore[30] have reported the detection limits for transition metals in water at less than 0.1 mg L^{-1}, if 250 mL of water was cycled seven times through a 37 mm diameter ion-exchange membrane, followed by 200 s analysis with a 0.37 GBq (10 mCi) ^{109}Cd source.

Another interesting application of thin film type measurement is in archaeometry.

Figure 4.12 shows a picture of an icon from the 17th century from the Bieszczady Region of Poland. Such icons had a background of the image made of silver (or less often gold) leaf. If we look carefully at points marked 952 and 954 we notice a thin horizontal line between them. Such lines are noticeable in other parts of the icon as well. The presence of these line may indicate the overlap of two layers of thin-leaf. It was very easy to confirm that this was the case by using a portable XRF analyzer that was calibrated for silver foil thickness measurements with thin film standards.[31] Measurements performed at the two points marked yielded thicknesses of 0.32 and 0.17 μm.[32] Since the ratio of these two numbers is very close to two, it is quite reasonable to assume that we actually identified the overlap of two silver leafs. A further conclusion that may be derived is that the artist was applying pieces of silver leaf starting from the bottom of the picture. Interestingly, the silver leaves used by the artists have an average thickness of 0.15 μm.

Figure 4.12 Measurement of silver leaf thickness on an icon.

4.7 Summary and Conclusions

The measurement of thickness of metallic and non-metallic coatings is an ideal application for XRF. Typical metals used for coatings, among others, are titanium, aluminium, zinc, zirconium, chromates and phosphates, *e.g.* for steel pre-treatment. TiN, for example, works well for machining iron-based materials. CrC, with its high-temperature oxidation resistance, is used in die casting while tungsten carbide/carbon (WC/C) is designed to coat and protect highly-loaded precision components, gears and gear drives, and engine components. Ruthenium and rhodium plated on Ti or Ni mesh are used for electrodes in chlorine production by electrolysis. All these coatings are easily measured with XRF.

Many types of coatings can be treated as "thin film", which makes calibration and analysis very simple. Air particulates collected on filter media are a prime example of the thin film method. In fact, the thin film method offers so many advantages over a bulk sample approach that many analysts prefer to convert a sample from its bulk form into a thin film and then analyze it. This is especially true of trace analyses.

The range of coating applications that portable XRF can address is surprisingly large – from industrial through environmental to art and archaeometry, and most recently even in the public health arena. There is also a widespread need for biocide coatings on objects of daily use in public places such as hospitals, schools, military facilities, dormitories, *etc.* The most effective biocides are those that contain ions of silver and iodine. They are applied in the form of paints or as components of other type of coatings at levels that are easily measurable with XRF. The effectiveness of such anti-microbial coatings depends on the amount of silver or iodine, and therefore there is a need to control its concentration (mass per unit area) in the coating during application process as well after its prolonged use. Given the multitude and sheer number of possible situations and objects that would require testing it is only natural to see them addressed by *in situ* Portable XRF.

References

1. Standard Test Methods for Magnetic Induction Methods: ASTM D 1186, ISO 2138, ISO 2808.
2. Standard Test Methods for Eddy Current Methods: ASTM B 244, ASTM D 1400, ISO 2360.
3. J. Buchler and T. Pagel, High precision coating measurement with ultrasound, paper presented at the 15th World Conference on Nondestructive Testing, Roma 2000, http://www.ndt.net/article/wcndt00/papers/idn158/idn158.htm.
4. Standard Test Method for Ultrasound Methods: ASTM D 6132.
5. Standard Test Method for Measurement of Coating Thickness by the Beta Backscatter Method: ASTM B 567 – 98 (Reapproved 2003).
6. M. Kaufmann, M. Mantler and F. Weber, Analysis of multi-layer thin films by XRF, *Adv. X-ray Anal.*, 1993, **37**, 205–212.
7. J. Dunne, Continuous determination of zinc coating weights on steel by X-ray fluorescence, *Adv. X-ray Anal.*, 1962, **6**, 345–351.
8. W.A.N. Severance, On-line X-ray measurements and control of silver emulsion coating weight, *Adv. X-ray Anal.*, 1974, **18**, 259–264.
9. Ohno Katsumi, Standardless thickness measurement of steel coating by X-ray spectrometry, *Adv. X-ray Anal.*, 1977, **21**, 89–92.
10. Z. Shanfield and M. Bertin, An X-ray fluorescence coating gauge for paint thickness, *Adv. X-ray Anal.*, 1977, **21**, 93–104.
11. R. Shen and A. Sandborg, Measurement of composition and thickness for single layer coating with energy dispersive XRF analysis, *Adv. X-ray Anal.*, 1982, **26**, 431–436.
12. R.L. White and C.T. Huang, Detection of ultra-thin carbon coating thickness measurements, *Adv. X-ray Anal.*, 1988, **32**, 331–339.
13. H. Kato, K. Imai and H. Tanabe, In-process coating layer analysis of galvannealed steel sheets with monochromatic incident X-rays, *Adv. X-ray Anal.*, 1991, **35B**, 1211–1218.
14. N. Matsuura, S. Kurozumi, T. Fukuzaki and T. Arai, On-line paint coating weight gauge using compton scattered X-rays, *Adv. X-ray Anal.*, 1992, **36**, 111–120.
15. Standard Test Method for Measurement of Coating Thickness by X-ray Spectrometry: ASTM B 568–98 (Reapproved 2004).
16. Lead; identification of dangerous levels of lead, Final Rule, *Code of Federal Regulations Title 40*, Part 745. 2001. pp. 1205–1240.

17. French Government Publication: J.O. Numero 133 du 11 Juin, 1999, p. 8544; "Décret no 99-483 du 9 juin 1999 relatif aux mesures d'urgence contre le saturnisme prévues aux articles L. 32-1 à L. 32-4 du code de la santé publique et modifiant le code de la santé publique (deuxième partie: Décrets en Conseil d'Etat)" NOR: MESP9921621D.
18. S. Piorek, J.R. Pasmore, B.D. Lass, J. Koskinen and H. Sipila, New developments in lead-paint film analysis with field portable X-ray fluorescence analyzer", in *Lead Poisoning, Exposure, Abatement, Regulation*, ed. J.J. Breen and C.R. Stroup, CRC Press, 1995, pp. 127–134.
19. S. Shefsky, J. Pesce and K. Martin, A new high-speed method for confirming the quality of L X-ray lead paint measurements in the field, http://www.niton.com/shef03.html (last accessed 23 February 2006).
20. L. Grodzins, *US Patents* 5,274,688 (1993); 5,390,229 and 5,461, 654 (1995).
21. NITON XLp 300 Series Lead analyzer, http://www.niton.com/documents/NITONXLp300Flyer-final.pdf.
22. The RMD LPA-1 XRF Lead Paint Spectrum Analyzer, http://www.rmd-lpa1.com/.
23. V.M. Makov, N.F. Losev and G.V. Pavlinski, *Zavod. Lab.*, 1968, **34**(12), 1459.
24. T. Florkowski and S. Piorek, Analysis of air particulates by radioisotope X-ray fluorescence for air pollution control, *Nukleonika*, 1974, **19**(10), 891–898.
25. J.R. Rhodes, J.A. Stout, S.S. Schindler and S. Piorek, Portable X-ray survey meters for *in-situ* trace element monitoring of air particulates, in *Toxic Materials in The Atmosphere: Sampling and Analysis*, ASTM STP 786, American Society for Testing and Materials, 1981, pp. 70–82.
26. C. Elskamp, OSA1- Lead (Pb) in *Workplace Air by Niton 700 Series Field Portable X-ray Fluorescence (XRF) Analyzer*, Applied IH Chemistry Team, Program Support Division, OSHA Salt Lake Technical Center, Salt Lake City, UT, April 2003.
27. ASTM E 1792_96a. Standard Specifications for Wipe Sampling Materials for Lead in Surface Dust 1996.
28. P. Eller, Lead in Surface Wipe Samples, Method 9100, *NIOSH Manual of Analytical Methods*, 4th edn, 1994, Cincinnati, OH: National Institute for Occupational Safety and Health.
29. J. Morley, Lead by Field Portable XRF, Method 7702, *NIOSH Manual of Analytical Methods*, 4th edn, 1998, Cincinnati, OH: National Institute for Occupational Safety and Health.
30. S. Piorek and J.R. Pasmore, in *Proceedings of the 3rd International Symposium on Field Screening Methods for Hazardous Wastes and Toxic Chemicals*, VIP-33, Air and Waste Management Association, Pittsburgh, 1993, p. 1135.

31. MicroMatter Co., 18218 – 18th Ave. N.W. Arlington, WA 98233, USA, www.micromatter.com.
32. S. Piorek, Current State-of-the-Art in Portable X-Ray Analyzers and Their Application for Non-destructive Investigation of Objects of Art and Cultural Heritage, in Proceedings of art'05 – 8th International Conf. on Non-destructive Investigations and Microanalysis for the Diagnostics and Conservation of the Cultural and Environmental Heritage, Lecce (Italy), May 15–19, 2005, paper A-102.

CHAPTER 5

Hazardous Substances in the Workplace

MARGARET WEST

West X-ray Solutions Limited, 405 Whirlowdale Road, Sheffield S11 9NF, UK

5.1 Introduction to Occupational Hygiene

Occupational hygiene is defined by the British Occupational Hygiene Society as

> the applied science concerned with the identification, measurement, appraisal of risk, and control to acceptable standards of physical, chemical and biological factors arising in or from the workplace which affect the health or well being of those at work or in the community.

This definition offers a challenge to those charged with the assessment of hazardous chemical substances in the workplace.

The first step to meet the challenge is the recognition of a potential or clearly revealed hazard through inspection of the workplace and identification of any symptoms or signs of disease amongst the workforce. The second, evaluation stage, is to assess the risk to the worker by identifying pathways for exposure and measurement of both the level and duration of the exposure. The final stage is to control the hazard and minimize the risk.

In the UK, the Control of Substances Hazardous to Health (COSHH) Regulations[1] provides the legislative framework for protecting the workforce. Associated with these regulations are guidance documents covering occupational hygiene issues[2] including workplace exposure limit values[3] and detailed measurement methods[4,5] for individual hazardous substances.

Portable X-ray Fluorescence Spectrometry: Capabilities for *In Situ* Analysis
Edited by Philip J Potts and Margaret West

It may be helpful at this stage to define two terms widely used in this chapter, namely, hazard and risk. A hazard describes a situation where there is a potential to cause harm, whereas risk is the chance of that harm occurring. For example, in a laboratory, an acid in a labelled sealed bottle in a fume cupboard represents a hazard; however, the same acid in a beaker on an open bench demonstrates that there is a greater possibility that the acid could cause harm, thereby revealing an enhanced risk.

Historically, most exposures to hazardous substances have been assessed using air-monitoring techniques, because inhalation was the main route of exposure and occupational standards for chemicals were based on air concentrations. However, it is now recognized that air sampling will not evaluate the total exposure to those chemicals capable of adversely affecting the body through other routes such as dermal exposure and ingestion. Workers subject to primary exposure may contribute to secondary exposure of others *via* cross contamination, *e.g.* wearing protective clothing such as lab coats, in refreshment areas or workers bringing home work clothes to be washed. Offices, washrooms, catering facilities and locations away from the primary production area need to be monitored to ensure that no contamination is missed.

This chapter discusses the capabilities of a portable XRF instrument to provide a means of rapid, *in situ* analysis for the quantification of chemical contamination and the generation of data for informed decisions on "approved" sampling strategies to satisfy the needs of the occupational hygienist.

5.2 Routes for Exposure

Chemicals may enter the body by various pathways related to the chemical and physical properties of each compound. Routes for both occupational and accidental exposure may be inhalation, dermal and oral.

5.2.1 Inhalation

Many substances may enter the body through respiration; and the lungs are perhaps the most vulnerable organ in the body, hence, this route is of prime importance in occupational hygiene. The respiratory system incorporates an effective filter and self-cleaning mechanism. However, these defence mechanisms can be overwhelmed by exposure to hazardous chemicals, particularly in the form of dusts, leading to deposition of substances in various parts of the respiratory system. This may result in serious occupational disease, such as silicosis or asthma. All particles drawn into the nose and mouth constitute the inhalable fraction, whereas the respirable fraction refers to those particles with an aerodynamic diameter of less than 7 μm, which can reach the gas exchange region of the lung. These particle size fractions are defined in BS EN 481.[6] The XRF analyst will need to take into account the effects of layer depth and particle size when measuring particulates collected on filter substrates from workplace air (Sections 5.4.1 and 5.4.2).

5.2.2 Dermal Exposure

The main pathways for dermal exposure are direct deposition on the skin from a source or ambient environment or contact with contaminated surfaces. In turn, dermal contamination may contribute to the total body burden through the oral route (Section 5.2.3). Several factors affect the permeability characteristics of the skin, *e.g.* presence of hair, calluses, abrasions and also its smoothness, degree of wrinkling and thickness and, in addition, the degree of sweating. Unlike respiratory exposure, contamination on worker's skin or clothing can contribute to secondary exposure of family members and others outside the workplace as well as co-workers not considered to be at risk. Organic lead, nickel and arsenic are known to transverse the epidural cells and enter the blood steam. Other materials are termed sensitizing agents, where contact with the skin causes dermatitis, one of the most prevalent occupational diseases in the UK today.[7]

An accidental breach of the skin can carry substances or micro-organisms through to underlying tissue.

5.2.3 Oral Exposure

Ingestion of hazardous substances occurs by inadvertent hand-to-mouth contact usually associated with activities such as eating, drinking and smoking with contaminated hands. Toxic substances are carried to the gut where they may cause systemic poisoning following absorption by gastric juices. Like the lung, the gut also acts as a selective filter but substances such as lead, organic lead and mercury can cause harm within the gut itself, without absorption.

Inhaled material may also be ingested following clearance of particles from the upper parts of the respiratory tract, back up into the throat. Such particles are most likely to be larger than those deposited in the deep gas exchange regions of the lungs.

5.3 Sampling

For most applications, the sample is brought to the laboratory. This may be a cause for concern if the sampling protocol is in doubt. Is the sample representative? Has it been contaminated during transfer to the laboratory? Is it correctly labelled? For the occupational hygienist, the collection of a representative sample of workplace air can be problematic. However, taking a spectrometer, such as a PXRF, to the workplace offers many attractions as previously mentioned (Chapter 1). The workplace provides various challenges for the collection of representative samples of dust, vapours, aerosols, gases and liquids from differing substrates. Many PXRF spectrometers can be used in two different modes; firstly, direct analysis, suitable for large samples and *in situ* measurements on surfaces and, secondly, bench mode for the analysis of filters, collected powders and liquids.

5.3.1 Air

Workplace air is subject to many variables reflecting activity capable of generating hazardous substances, *e.g.* batch production or continuous operation, temperature and pressure. The distance from the emission source and physical parameters such as release rate, air current, meteorological conditions and ventilation are also influencing factors. Fluctuations in contaminant concentrations are to be expected. However, the established sampling procedure of drawing air through a filter offers two advantages, pre-concentration and a deposit with low mass per unit area (thin layer).

Guidance for the assessment of exposure to chemical agents by inhalation for comparison with regulatory limit values and measurement strategy can be found in BS EN 689.[4] Methods for the determination of hazardous substances are published in the UK by the Health and Safety Executive, with MDHS 91[9] giving the analyst guidance on XRF measurement of metals and metalloids in the workplace. In the USA, the National Institute for Occupational Safety and Health (NIOSH) and the Occupational Safety and Health Administration (OSHA) publish methods for routine monitoring of workplace air.[10–13]

5.3.1.1 Inhalable Samplers

Inhalable samplers suitable for personal sampling are described in MDHS 14/3.[6] Some samplers are designed to collect the inhalable fraction of airborne particles on a filter such that any particulate matter deposited on the internal surfaces of the device are disregarded. Others are designed such that airborne particles, which pass through the entry orifice(s) match the inhalable convention, in this case particles deposited on internal surfaces form part of the sample. The latter sampler design (Figure 5.1) incorporates an internal filter cassette, which can be removed from the sampler to enable this material to be recovered. The reader is advised to consult the manufacturer's instructions.

A typical total inhalable sampler (Figure 5.1) is the 7 Hole Head or Man Made Mineral Fibre (MMMF) head commonly used to sample with a flow setting of $2\,L\,min^{-1}$.

5.3.1.2 Respirable Samplers

Samplers designed to collect the respirable fraction of airborne particles are defined in BS EN 481[7] and comply with European Standards. The cyclone style samplers appropriate for this work are described in MDHS 14/3.[6] The sampler should be worn on the worker's lapel as close to the mouth and nose as reasonably practicable (Figure 5.2).

For both inhalable and respirable fractions, static samplers may be used to monitor the general workplace atmosphere.

Figure 5.1 IOM inhalable dust sampler.

Figure 5.2 Worker wearing a respirable sampler.

5.3.1.3 Filters

Filters of 25 mm diameter are preferred, rather than 37 mm often used for many applications in WDXRF configurations, thereby ensuring that the whole filter can be exposed to the X-ray beam. Mixed cellulose ester membrane filters with 0.8 μm mean pore diameter are considered suitable since the airborne particles are deposited as a thin surface layer. It is also necessary for the filter design to provide high filtration efficiency, typically 99.5% for particles with a 0.3 μm diameter.

Depth effect glass fibre or paper filters are less suitable for quantitative analysis due to the variable absorption of the characteristic radiation emitted from a particle related to its depth of penetration into the filter body.

A sample loading of ~500 μg is recommended with a sampling time selected appropriate to the level of dust present in the workplace air (see Section 5.4.2). Filters should be handled at the edge with flat-tipped forceps to minimize the risk of damage or contamination.

5.3.1.4 Dual Fraction Sampler

For studies where both the inhalable and respirable dust are measured, a dual fraction sampler may be used. A porous polyurethane size-selective foam insert is placed in front of the normal filter. Figure 5.3 shows a schematic diagram of the IOM dual-fraction dust sampler where the slightly extended cassette has been engineered to house either one or two foam plugs. A study by Kenny *et al.*[14] described the advantages and limitations of this low-cost option for the occupational hygienist as part of a collaborative European project to develop a range of porous foam aerosol samplers.

The foam inserts separate the inhalable dust into thoracic and/or respirable sub-fractions. The inhalable dust concentration is calculated from the weight of dust collected in the entire cassette (filter and foam) whereas the filter alone gives the respirable dust fraction. The study examined practical problems

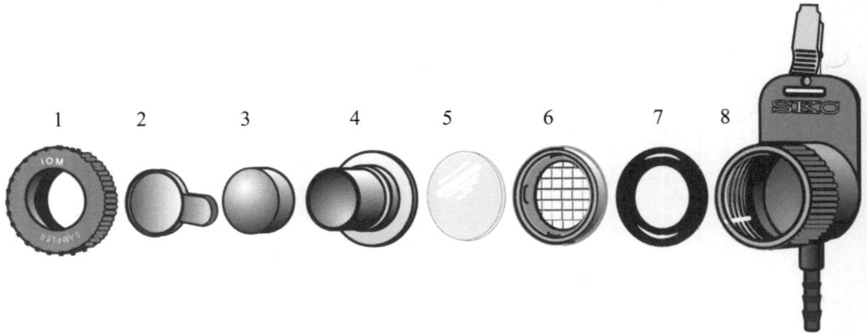

1: IOM body front part
2: Transport cover for cassette
3: Foam insert
4: Cassette front part
5: Filter
6: Cassette rear part
7: O-ring
8: IOM body rear part

Figure 5.3 Schematic diagram of a dual-fraction sampler.

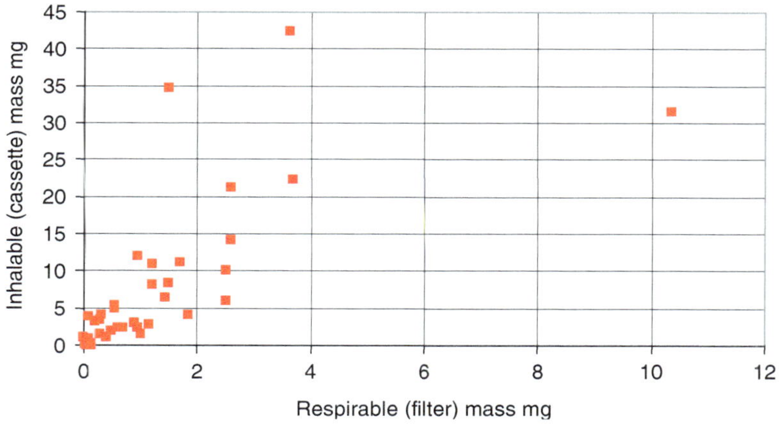

Figure 5.4 Dust weights collected in workplaces.

experienced in the workplace, limitations on particle loading and losses or movement of particles during transportation of the samples. Figure 5.4 demonstrates data collected from a range of substances from very fine to coarse materials, compact particles, ultra-fine agglomerates and fibres.

5.3.2 Dermal Contamination

Sampling of contamination on workers' skin should not cause discomfort or give rise to any side effects. Suitable sampling methods include hand washing, commercial wipes and forensic adhesive tapes. An indirect method using cotton gloves as a barrier between the contaminant and the skin is also possible, as reported in the EU 5th framework project on the evaluation and prediction of dermal adsorption of toxic chemicals.[15]

Personal protective equipment is designed to form such a barrier and offer opportunities for direct measurements on clothing.

5.3.3 Contaminated Surfaces

Primary exposures result from direct involvement with an operation and include contact with process-generated contamination. Equipment, tools, furniture and fittings used in the workplace may become contaminated because of the work activities. Surfaces frequently touched by process workers, such as control panels, switches, instrumentation, drum lids, clothing and personal items can become contaminated and offer sampling opportunities.

Cross contamination is to be expected as workers move about, and use desks and workstations where other people can also sit to complete paperwork and log information into computer systems. Doorknobs and light switches are also sites for secondary contamination. Cloakrooms where workers change into and out of work wear may have designated "clean" and "dirty" areas and are possible sites for cross contamination. Vehicle door handles, gear levers, hand brakes and steering wheels often build up contaminants and can be wipe sampled. Seating may be vacuum cleaned and tested. Workers' clothing and shoes worn in the home can also be sampled. Many of these sampling areas may be analysed directly by PXRF.

Carpets, walls and counters in the workplace, home and other buildings regularly used by the community should not be forgotten. These surfaces may accumulate substances from different environmental media, including water, soil, sediment and consumer products. Samples from secondary contaminated surfaces may comprise toxic dust and offer opportunities for re-suspension during vacuuming and dusting. The American Society for Testing and Materials has developed standardized methods[16,17] for sampling carpet-embedded dirt.

Static and mobile surfaces in the workplace offer various sampling and analytical challenges. Some surfaces such as human tissue cannot be directly analysed by XRF, but representative collection techniques offer opportunities for investigation. Gloves, hard hats and the inside of face shields can be assessed and present opportunities during employee training programmes to highlight the potential for contamination and the need to adopt safe working practices.

Particulates collected on filters, wipes and adhesive tapes and direct *in situ* analysis raise analytical issues of particle size and layer depth that will influence the interpretation of results. All sampling techniques must be assessed for blank background and controls. Certified reference materials are also needed to confirm calibration validity and routine testing in the field should comply with standard operating procedures.

5.4 Theoretical Considerations

The continued development in PXRF instrumentation offers the occupational hygienist the potential to collect analytical data from various surfaces outside the constraints of the conventional laboratory environment. The sensitivity and accuracy of measurement of a given contaminant in the workplace can be assessed for comparison with published threshold limit values. However, the conversion of measured characteristic X-ray intensities into meaningful concentration levels demands an understanding of X-ray and analytical theory so that pitfalls associated with specimen layer depth, particle size and background blanks may be avoided.

5.4.1 Specimen Layer Depth

X-Rays are absorbed as they pass through material. The emergent beam of intensity (I) is always less than the incident beam intensity (I_0) due to absorption and scattering. The intensity loss through an infinitely thin layer ($\mathrm{d}x$) is given by:

$$\mathrm{d}I = -\mu_1 I \mathrm{d}x \tag{5.1}$$

where μ_1 is the linear absorption coefficient (cm^{-1}) for an element at the energy of interest. For a finite thickness (x) Equation (5.1) is integrated and becomes:

$$I = I_0 \mathrm{e}^{-\mu_1 x} \tag{5.2}$$

The linear absorption coefficient, for the irradiated cross section expressed in cm^2, represents the fraction of the intensity absorbed per cm of sample traversed. However, in XRF analysis it is more appropriate to consider absorption per gram in the sample:

$$\mu = \mu_1 \rho^{-1} \tag{5.3}$$

where ρ is the density of the absorber ($\mathrm{g\,cm}^{-3}$). Substituting Equation (5.3) in Equation (5.2):

$$I = I_0 \mathrm{e}^{-\mu\rho x} \tag{5.4}$$

where μ is the mass absorption coefficient ($\mathrm{cm}^2\,\mathrm{g}^{-1}$) and x is the path length through the sample.

In this way we may consider the total mass absorption coefficient of a multi-element sample to be the sum of the individual mass absorption coefficients.

The relationship between emitted, fluorescent radiation of intensity (I_i) and the mass fraction of an element (i) in a specimen of thickness (x) is related to spectrometer geometry:

$$I_i = k_i c_i (1 - \mathrm{e}^{\mu\rho x})/\mu \tag{5.5}$$

where c_i is the mass fraction of element I; k_i is a constant; $\mu = \mu_1 \mathrm{cosec}\psi_1 + \mu_2 \mathrm{cosec}\psi_2$, where ψ_1 and ψ_2 are incident and take-off angles; and μ_1 and μ_2 are mass absorption coefficients for incident and fluorescent radiation, respectively.

For a homogeneous thin specimen, *e.g.* a deposit on a filter or contaminated surface, the product $\mu\rho x$ is very small, and by series expansion:

$$\mathrm{e}^{-\mu\rho x} = 1 - \mu\rho x \tag{5.6}$$

hence:

$$I_i = k_i c_i \rho x \tag{5.7}$$

This relationship is not subject to matrix effects and holds provided that $\mu\rho x < 0.1$, and is known as the thin specimen criterion as the degree of attenuation in the sample is less than 90%.

5.4.2 Particle Size

In practice, most samples collected in the workplace are not homogeneous but consist of discrete solid particles of differing size. The particle size parameter in X-ray theory is the mean physical diameter of a particle. It is not the aerodynamic diameter of a particle, which is related to particle shape and density, and, as such, may be larger or smaller than the physical diameter.

Particle size effects may be considered as the absorption of radiation within a particle (attenuation factor A):

$$I_i = Ak_i c_i \rho x \tag{5.8}$$

Using the method proposed by Criss[18] to calculate A:

$$A = 1/(1 + ba)^2 \tag{5.9}$$

where a is the particle diameter and b is a calculated coefficient related to the mass absorption of the particle.

Tables of mass absorption coefficients can be used to calculate correction factors applied to XRF measurements according to the weight, composition and size of particulates concerned. However, in practice such information is rarely available. An alternative approach is to measure analyte lines that are less affected by particle size effects and to ensure that the sample collected on a filter is within the thin specimen criterion (equivalent to approximately 500 μg loading). For heavier elements, the K spectral lines with higher energy are less affected by depth or particle size effects than the L series from the same analyte. In practice, the lower energy, longer wavelength radiation emitted by elements below sulfur in the periodic table are readily absorbed by air and, therefore, not analysed by a portable spectrometer.

5.4.3 Background Blanks

The experienced analyst will be familiar with procedures for checking background blanks, in the instrument itself and any sample collection media such as filters and wipes. Blank substrates should be treated in the same manner as specimens, with representative blanks handled and transported for measurement. Where *in situ* measurements are concerned, it may not be possible to locate a representative "clean" surface for use as a blank. In such cases, the contaminated surface should be measured, the area then wipe sampled and re-measured to ascertain the background blank (Figure 5.5). The efficiency of wiping may be tested.

It may be necessary to repeat the wipe stage until successive re-measurements confirm that all the contaminants have been removed.

Some surfaces may contain an element that is also present in the surface contamination or gives rise to interference due to line overlap effects. In such cases, the analysis of wipe samples would be the preferred route.

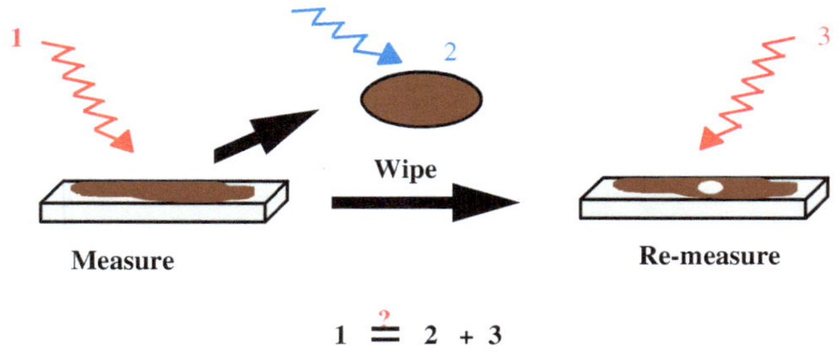

Figure 5.5 Workplace background measurement protocol.

Measurements of contaminants on thin substrates such as protective clothing, gloves and other fabrics should be made using Teflon as a backing material.

5.5 Measurements in the Workplace

To retain confidentiality, the examples given are generic and serve to give the reader an insight into the capabilities of the PXRF technique for the measurement of hazardous substances in the workplace. In general, portable spectrometers are supplied with pre-calibrated applications such as "Soils", "Lead in paint" and "Thin Film". In the analysis of dusts and particulates, where the sample is assumed to be a thin layer, the "Thin Film" application may be used. Where the dust has a large particle size and critical depth effects are concerned the "Soils" application is preferred.

5.5.1 Air Monitoring

Some 20 years ago, Rhodes *et al.*[19] described portable X-ray survey meters for *in situ* trace element monitoring of air particulates. These instruments may be described as "luggable" when compared with the lightweight systems available today. Nevertheless, the potential for *in situ* analysis of particulate deposits on filters was recognized. Detection limits for elements in the range 0.1–1.0 $\mu g\,cm^{-2}$ of deposit were reported. These values translate to 1–10 $\mu g\,m^{-3}$ for personal samplers operated at 2 $L\,min^{-1}$ over an 8-hour shift. Being 1 to 4 orders of magnitude below the (then) American Permitted Exposure Limits for workplace air contaminants, the potential for routine monitoring of breathing zone and work areas for fugitive emissions and surveys of accidental releases of toxic materials was recognized.

The technology improved and equipment, originally purchased for soil analysis and metal contamination at hazardous waste sites, was adapted for the analysis of membrane filters. Bernick and Campagna[20] used a thin film, fundamental parameter (FP) based application to compare *in situ* PXRF with

Table 5.1 Typical detection limits ($\mu g\,cm^{-2}$) for air monitoring by PXRF.

Element	^{109}*Cd excitation*	*X-ray tube excitation*
As	1.0	0.9
Cd	10	–
Cr	3.5	1.0
Cu	2.5	0.8
Mn	2.0	1.0
Ni	1.5	1.0
Pb	0.6	1.3
Se	0.5	0.3
Zn	2.0	0.8

NIOSH method 7300.[21] The NIOSH method called for chemical ashing of 37 mm diameter filters followed by AA or ICP analysis. The filters were prepared for PXRF analysis by mounting on 40 mm double open-ended sample cups between two layers of 5 μg polypropylene film. The results compared favourably with blind performance analyses of the filters after chemical ashing.

In the UK, the exploitation of PXRF for monitoring hazardous substances in the workplace was pioneered by the Health & Safety Laboratory (HSL), an agency of the Health and Safety Executive. In 1996, Dost[22] reported that the lower limits of detection of airborne inorganic contamination were found to be within the required sensitivity for the analysis of 18 elements at one-tenth of the UK occupational exposure level. Precision at higher concentration levels was generally better than 10%. West *et al.*[23] compared a portable XRF spectrometer with a conventional laboratory based, wavelength dispersive XRF spectrometer for the analysis of particulates on filters. The ability of the technique to play a key role in cost-effective sampling and analysis strategies for airborne workplace monitoring was recognized. The "Thin Film" application could be used to analyse the following elements: Ag, As, Ba, Ca, Cd, Co, Cr, Cu, Fe, Hg, K, Mn, Mo, Ni, Pb, Rb, Sb, Se, Sn, Sr, Ti, Zn and Zr. Table 5.1 shows typical detection limits for air monitoring by PXRF.

Portable spectrometers were being used by industry as well as Regulatory authorities to assess exposure to toxic substances. The time delay and cost associated with conventional sample collection and submission to the laboratory was replaced by *in situ* analysis. Zamurs *et al.*[24] demonstrated the ability of the technique to measure lead concentrations on high volume air filters during the removal of lead-based paint by abrasive blasting from bridges in New York, USA. Whilst abrasive blasting is efficient in the rapid removal of paint, large amounts of dust and lead paint particulates are produced. Shielding of the work area will reduce the release of contamination to the general environment but some release will occur, causing concern in the neighbourhood. Portable systems designed to determine lead levels in paint use a threshold of $1\,mg\,cm^{-2}$ whereas the New York Air Quality Standard for total suspended solids (TSP) for an 8-hour sampling period, at a pump flow rate of $70\,m^3$ per hour, calls for the detection of $1.4\,\mu g\,cm^{-2}$ on a filter to detect an airborne lead level of $1.0\,\mu g\,m^{-3}$.

This application required a detection limit more than 700 × less than that for the established lead-based paint testing. It was, therefore, necessary to distinguish between these two application requirements. To meet the enhanced performance requirements for the TSP study, Pb L spectra were excited using a ^{109}Cd source rather than the ^{57}Co source commonly employed to excite K spectra for lead in paint. A combination of the greatly reduced L spectra background and a ten-fold increase in signal provided a 1000-fold improvement in signal-to-background, enabling the detection of sub 1.0 $\mu g\,cm^{-2}$ levels on the TSP filters. This application of PXRF provided a timely warning of poor or compromised containment at bridge working sites.

Readers interested in the analysis of lead in paint are referred to Chapter 4, Section 4.6.1 (Special Cases of Coating Measurements, Lead in Applied Paint).

5.5.2 Contaminated Surfaces

Hard, flat surfaces such as painted or varnished wood, laminated plastic, metal and solid flooring may be directly monitored and compared with wipe samples to assess the extent of available surface contamination. *In situ* measurements provide the occupational hygienist with the ability to survey the workplace and identify any "hotspots". Movement of workers from a dust-generating area to other areas such as offices, washrooms and canteens may be monitored to demonstrate the potential for cross contamination.

Dust transferred through the air and by foot traffic to offices has been shown to build up on carpets, where the larger particulates tend to settle in the body of the pile. These particulates are then brought to the surface during vacuum cleaning and increase the surface contamination and airborne concentration, thereby posing a respiratory hazard. Background blank measurements can be taken from carpet covered by furniture.

Contamination has been found on the inside of workers gloves, particularly the heavy-duty designs that are worn for longer periods.

Surveys have shown that contamination may be transferred to the workers' home, posing a possible health risk to the community. Dusts from the workplace, carried on clothing and footwear, have been monitored in vehicles and household flooring and furniture. When such dusts contain hazardous substances, survey work has demonstrated the need for increased vigilance.

As with all analytical equipment, the data generated depend upon the calibration and operating schedule governing its use. Training is of paramount importance, particularly when the operator takes an instrument to a sample. Results are affected by surface roughness due to variations in the air gap between the sample and the analyser, as demonstrated by Gauvin and Lifshin.[25] They showed that the intensity of the measured spectra was influenced by changes in the generation and absorption of the X-rays as the beam moved across the sample. Repeat measurements with the PXRF head turned through 90° have proved effective in overcoming variations due to the geometry of the radiation source. Nevertheless, Sterling *et al.*[26] continued the task of assuring analysts of the validity of PXRF measurements in work to compare on-site PXRF measurements with laboratory AAS data, reporting no significant difference between the two sets of results.

Readers interested in a correction procedure for surface relief are referred to the work of Potts *et al.*[27] who developed a method for measuring the intensity of scatter peaks for the analysis of rock artefacts (see also Chapter 8 on The Analysis of Archaeological Lithic Provenancing).

5.6 Conclusion

In common with other applications for PXRF, the ability to generate rapid analytical data in a cost effective manner has been welcomed by industrial hygienists. Spectrometer sensitivity provides a means for the assessment of hazardous substances in the workplace with low limits of detection, in many cases an order of magnitude below the legislative limits. Measurements can be taken to survey an area to assess the extent of any hazard, enabling the industrial hygienist to satisfy the second evaluation stage of their remit.

The technique is also a useful tool in training programmes to demonstrate the effectiveness of personal protective equipment and the adoption of safe working practices.

References

1. *The Control of Substances Hazardous to Health Regulations, 2002. Approved Code of Practice and Guidance L5 (fourth edition)*, HSE Books 2002 ISBN 0 7176 2534 6.
2. *Health and Safety Executive Hygiene Series (EH) Guidance Notes*, HSE Books.
3. *Environmental Hygiene Guidance Note EH 40*, HSE Books 2005 ISBN 0 7176 29677 5.
4. British Standards Institution BS EN 689, 1996.
5. *Health and Safety Executive Methods for the Determination of Hazardous Substances*, MDHS series, HSE Books.
6. *Health and Safety Executive Methods for the Determination of Hazardous Substances*, MDHS 14/3, HSE Books, 2000.
7. British Standards Institution BS EN 481, 1993.
8. Labour Force Survey: Results 1998, European Communities, Statistical Office.
9. *Health and Safety Executive Methods for the Determination of Hazardous Substances*, MDHS 91, HSE Books, 1998.
10. Elskamp C. OSA1- Lead (Pb) in *Workplace Air by Niton 700 Series Field Portable X-ray Fluorescence (XRF) Analyzer*, Applied IH Chemistry Team, Program Support Division, OSHA Salt Lake Technical Center, Salt Lake City, UT, April 2003.

11. ASTM E 1792_96a. Standard Specifications for Wipe Sampling Materials for Lead in Surface Dust, 1996.
12. P. Eller, Lead in Surface Wipe Samples, Method 9100, *NIOSH Manual of Analytical Methods*, 4th edn., 1996.
13. J. Morley, Lead by Field Portable XRF, Method 7702, *NIOSH Manual of Analytical Methods*, 4th edn, 1998.
14. L. Kenny, K. Chung, M. Dilworth, C. Hammond, J. Wynn Jones, Z. Shreeve and J. Winton, Application of low-cost, dual-fraction dust samplers, *Anal. Occup. Hyg.*, 2001, **45**(1), 35–42.
15. Evaluation and prediction of dermal exposure of toxic chemicals EU Report QUALK4-20000-00196.
16. American Society for Testing and Materials ASTM:F608-07.
17. American Society for Testing and Materials ASTM:F1284-04.
18. J.W. Criss, *Anal. Chem.*, 1976, **48**(1), 179.
19. J.R. Rhodes, J.A. Stout, J.S. Schindler and S. Piorek, *Toxic Materials in the Atmosphere*, ASTM STP 786, American Society for Testing and Materials, 1982, 70.
20. M.B. Bernick and P.R. Campagna, *J. Hazard Mater.*, 1995, **43**, 91.
21. *NIOSH Manual of Analytical Methods*, 3rd edn., National Institute for Occupational Safety and Health, 1984, Section A, method 7300.
22. A. Dost, *Ann. Occup. Hyg.*, 1996, **40**(5), 589.
23. M. West, F.N. Meder and J. Wheeler, *Proceedings of 48th Chemists' Conference* 1996, 41.
24. J. Zamurs, J. Bass, B. Williams, R. Fritsch, D. Sackett and R. Heman, Niton website: www.niton.com.
25. R. Gauvin and E. Lifshin, *Mikrochim. Acta*, 2000, **132**, 201.
26. D.A. Sterling, R.D. Lewis, D.A. Luke and B.N. Shadel, *Environ. Res.*, 2000, **83**(2), 174.
27. P.J. Potts, P.C. Webb and O. Williams-Thorpe, *J. Anal. Atomic Spectrosc.*, 1997, **12**, 769–776.

CHAPTER 6

Alloy Identification and Analysis with a Field-Portable XRF Analyser

STANISLAW PIOREK

Thermo Niton Analyzers, LLC, 900 Middlesex Tpk. Bldg. 8, Billerica, MA 01821, USA

6.1 Introduction

6.1.1 Rationale behind Analysis of Alloys in the Field

By alloy analysis in the field we usually mean analysis and identification of the already made alloy rather than analysis performed in a foundry during alloy production. It is important to make this distinction at the outset to better understand the topic of this chapter.

A rapid, 100% Positive Materials Identification – or PMI as it is often referred to – is currently a mandatory requirement in many industries, which range from those critical to life and environment, such as nuclear and aviation, to those that are driven often by just pure economics, as in the case of the metal recycling industry. PMI is most often associated with alloys, as these are still the predominant materials of construction. From the time man discovered copper and then mastered making iron, alloys have been the mainstay of all types of construction, be it buildings, industrial plants, automobiles, consumer products, *etc.* They surround us, and both the quality as well as the safety of our lives and our environment heavily depend on them. If one were to follow the life cycle of an alloy, it would turn out that PMI accompanies an alloy at each and every stage of its life cycle (Figure 6.1). It is important to know the composition or identity of an alloy because using an alloy of inadequate characteristics (composition) may

Portable X-ray Fluorescence Spectrometry: Capabilities for *In Situ* Analysis
Edited by Philip J Potts and Margaret West

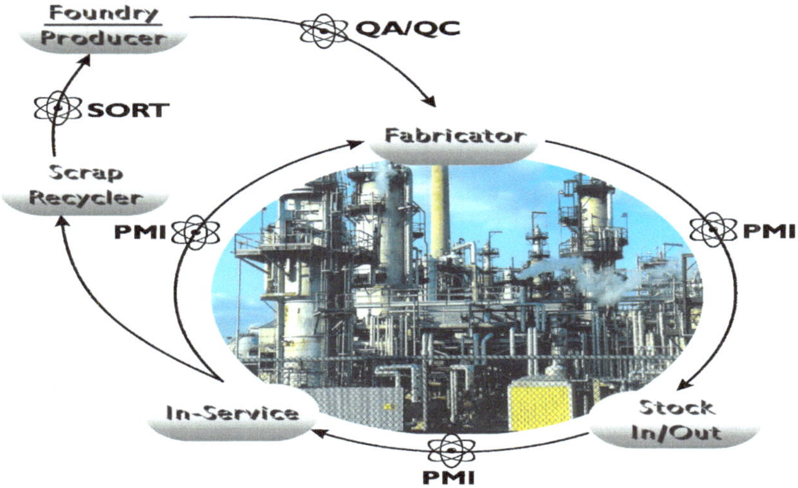

Figure 6.1 PMI-Positive Materials Identification, and alloy "life" Cycle. PMI accompanies the alloy at every stage of its "life".

result in huge financial losses or even in loss of human life. That is why a 100% PMI is required in safety sensitive industries such as the petrochemical, nuclear power, and aerospace. The obvious conclusion is that if we are to verify composition or identity of an alloy made component after it is installed, we must have the ability to perform such tests in a truly non-destructive fashion and *in situ*.

Apart from the broadly understood safety factors mentioned above there is also an important aspect of producing an alloy as economically efficiently as possible. This is, perhaps, best illustrated in Figure 6.2, which shows the distribution of concentrations of four major alloying elements, Fe, Mn, Ni, and Cr, which make up stainless steels of the 300 series. Each of the four elements in each alloy grade of the series has a specified range of concentration. Unless concentrations of all alloying elements are within their specification boundaries the alloy would not meet its composition criteria. When we examine the concentrations of these four elements in 300 series steels and compute the frequency of occurrence of a particular concentration of a given element within its specification range, we come up with a histogram similar to that shown in Figure 6.2. For example, let us say that Cr concentration range for a certain grade of steel is specified from 16 to 18% m/m. One might expect that the Cr concentration for this grade would, on average, be about 17%, or in the middle of its range. Now, an actual concentration of Cr in samples of this steel happens to be 16.38%. This means that 16.38% falls into the second tenth of the specification range ($0.38/2.00 = 0.19$, which is the second tenth of the 2% wide range). In other words, the Cr concentration is skewed to the lower boundary of its specification range. Figure 6.2 shows that this is also true for other steel grades in the group, and we make an important observation that the more expensive alloying elements, such as Cr and Ni in our example, are skewed towards the lower boundary, while less expensive

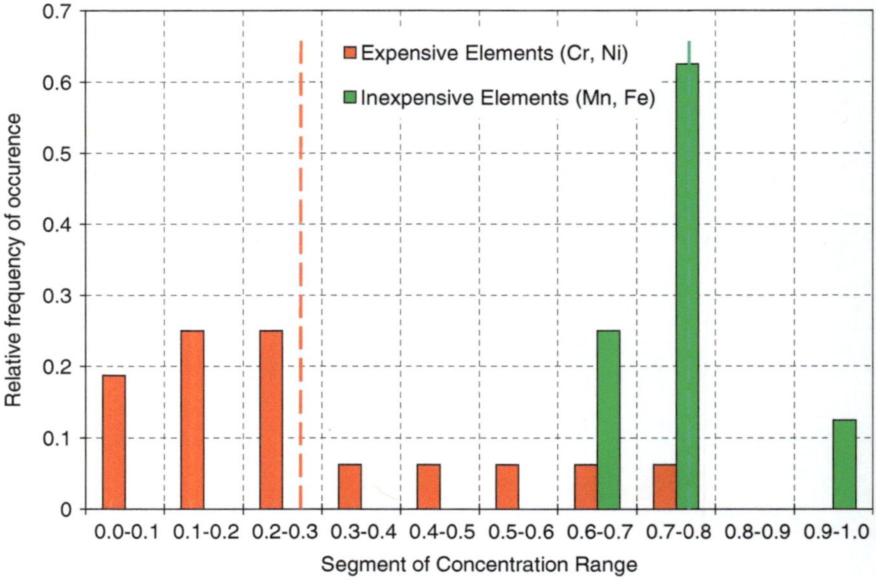

Figure 6.2 Distribution of alloying elements within their specified concentration range for 300 series stainless steels.

elements, such as Mn and Fe, are skewed towards the upper boundary of their specification range. This holds true for any alloy. Therefore, it is important to control the composition of an alloy, firstly to make sure that it is within specifications and, secondly, that it does not cost more than it has to. Alloy recyclers need to know the composition of metal scrap so they can sell it as a certain grade, while alloy smelters and producers need to be sure that alloy scrap added to the melt will not jeopardize its compliance with target specifications.

Alloy recyclers are also faced with new challenges. Just three decades ago, when energy was less expensive, alloy scrap was used only on a small scale in the production of raw alloys. However, it soon became obvious that it was less expensive, more energy efficient, and better for the natural environment to add metal scrap to the melt than to produce more of the metal from the ore. More and more alloy scrap would enter smelters and foundries. Unfortunately, for initially unknown reasons, the steel made with the admixture of steel scrap started to show an ever increasing presence of copper. As it turned out, copper entered the smelting process *via* unsorted steel scrap obtained from car wrecks, which obviously contained copper in the form of wiring and electric motors. Consequently, alloy scrap must be sorted and tested for the presence of "poisonous" elements. The fall of the "iron curtain" and the realization of a truly global economy has opened up huge and often inexpensive alloy supply from countries of the former Soviet Block. However, the alloys engineered in that previously isolated part of the world often have a quite different composition from their functional counterparts as known in the West. This fact is enough for alloy scrap sorters and dealers to need to control chemical composition of their purchases. The growing tendencies for stricter and tighter control of product quality are now officially sanctioned under the ISO 9000 rules,[1] which alone are presently reason in itself for controlling the composition of alloys at any stage of their useful life.

6.1.2 Existing Methods of Alloys Identification and XRF Analysis

Several techniques and procedures, other than classical laboratory wet chemistry analysis, have been in use for alloy sorting and alloy identification in the field. Traditional ones include:

- Colour recognition – aluminium, copper, gold-based alloys can most of the time be recognized by their colour.
- Magnetism – nickel-based alloys are typically nonmagnetic; also one can easily sort austenitic 300 stainless steels from the martensitic 400 series with a small magnet.
- Spark testing – shape, colour, length and intensity of sparks generated on metal with grinding wheel may be used for alloy sorting.
- Difference in apparent density – aluminium alloys are lighter than titanium alloys, which in turn are significantly lighter than steels.
- Chemical spot tests – the colour and rate of surface discolouration of an alloy after its exposure to acid can be used for sorting.

Any of these techniques requires an experienced operator to complete the test by making the identification decision himself, based on the results of measurements or tests and his personal experience. None of these techniques is discriminant (sensitive) and conclusive enough to provide a positive result on its own. Therefore, quite often, several of them have to be used in specific sequence to arrive at an unambiguous result. For example, one might first use visual colour recognition and weight, followed by a magnetic test and then a spark test. In addition, a conventional, full scale chemical analysis of an alloy must be followed by a search through composition tables to find the matching alloy name or grade designation. Good insight into these mostly non-instrumental techniques can be found in a published handbook.[2]

More sophisticated, instrumental methods utilize the phenomena of thermoelectricity (the Seebeck effect), optical emission spectroscopy (OES),[3–5] and X-ray fluorescence (XRF). Their advantage is that, with proper software, they can be automated so that the user does not have to be involved in the decision making process. This is specifically true for OES and XRF methods.

Over the last four decades, X-ray fluorescence spectrometry has gained the recognition of metallurgists as a major analytical tool, originally using wavelength (WDXRF) and then energy-dispersive (EDXRF) instrumentation. There probably is no metallurgical facility without at least one WDXRF spectrometer. The speed, reliability and non-destructive character of the XRF method make it suitable not only for laboratory applications, but especially for field and plant use.

These instrumental methods are more complete and sensitive enough to compositional differences between the alloys to stand on their own as analytical tools. The successful expansion of X-ray fluorescence analysis from laboratory to plant and field environments was made possible only over the last 15 or so years, by the availability of portable XRF analysers. The availability of several critical pieces of equipment were responsible for this development, namely:

- small, sealed radioisotope sources to excite the characteristic X-rays of the sample;
- small, room temperature, high energy resolution detectors;
- powerful microprocessors;
- compact, high capacity, rechargeable batteries to make the instrument independent of AC power.

In particular, developments in microprocessor technology have made it possible for portable, battery-operated X-ray analysers to perform, in real time, complex analyses of the X-ray spectra from the sample, followed by sophisticated data processing; the task previously assigned only to off-line computers. Most recently, another technological breakthrough – development of miniature, low-power X-ray tubes – has improved the performance of portable XRF analysers and has made them even easier to use. Replacing the isotope with an X-ray tube not only improves the instrument's sensitivity and analytical range but also makes it easier to transport and, overall, safer to operate because no X-rays are emitted from the X-ray tube once the power to the instrument has been turned off. The portable, microprocessor-based X-ray analysers have been a real breakthrough by combining speed and truly non-destructive character of analysis (the instrument is being brought to the sample and not otherwise) with an expert identification/sorting software, thus relieving the operator of any decision making. All that is necessary is to expose the sample for 5–10 s to the instrument's excitation source and read the final result of identification from the display or printout.

6.2 Addressing the Problem of Alloy Identification

6.2.1 Defining the Task

According to "The American Heritage Dictionary of the English Language", (4th edition, Houghton Mifflin Co., Boston, New York, 2006, "identity" entry on page 870, and "identify" entry on page 871):

> Identification of a material is defined as a process of ascertaining those characteristics of a given material by which it is definitively recognizable or known.

We will now examine the meaning of this definition in the context of alloy identification by X-ray fluorescence analysis.

Some accounts put the number of all different alloys that have been in use at some thirty plus thousand.[6] In practice, however, contemporary industry uses only about several thousand alloys, which – by any measure – is still a large number. For example, the Unified Numbering System (or UNS) data base for alloys and metals lists over 4800 alloys of all kinds, organized into 17 families.[7] Major alloy recyclers and producers publish their own lists of alloys. The ELG Company's listing contains about 1200 entries. Published in Germany and updated every year, a listing of steels, entitled "Stahlschlüssel" ("The Key to Steel"), lists well over two thousands of carbon, mild, tool, stainless and high temperature steels.[8] Nowadays, many excellent alloy databases may be found on the internet.[9]

Of all alloys in use, carbon and low alloy steels constitute about 70% of the whole alloy market. These are best analysed using optical emission spectroscopy (OES) (either laboratory or field-portable). The remainder, most of which are high-temperature alloys and steels, copper-based, nickel-based, and titanium alloys, is the domain of XRF analysis. This is still an overwhelming number of alloys to be able to differentiate.

Thorough examination of the specifications of thousands of alloys reveals that from 40–50 different elements can be involved in the alloying process, with any given alloy containing only 10–20 elements. Of this number, only about ten elements are responsible for the main characteristics of any given alloy. Thus, the number of elements to be analysed to identify an alloy is finally reduced to about 10–15. Fortunately, these elements are also those most readily measurable by X-ray spectrometry.[10] The other elements, such as carbon, oxygen, nitrogen, are present at a fraction of percent, and mainly to modify the physical properties of the alloy. Typical examples of the diversity of compositions encountered within the families of alloys are shown in Figures 6.3 and 6.4, in which a selection of the most commonly used Ni-based alloys and Stainless Steels, respectively, and their elemental percent compositions are depicted. Similar data are shown in Figure 6.5 for carbon and low alloy steels. We can see that the composition differences between the alloys of this last group are less pronounced and this is why the XRF technique is not very successful in analysing them. It is the carbon content (at several hundred mg kg^{-1} levels) that distinguishes alloys in the carbon steels group, which is the domain of OES. Figures 6.6–6.8 show the ranges of composition for these five most common alloy groups. Fourteen elements easily measurable by X-ray spectrometry are shown. It can be seen that 4–10 of these are present in any given alloy group in amounts that are useful for identification purposes. Thus, it is sufficient to monitor concentrations of about ten elements in a given alloy to be able to discriminate it from other alloys of its own group, and a seemingly impossible task becomes manageable. Now that we have defined the problem we will see how to effectively and efficiently solve the identification question using X-ray fluorescence analysis.

6.2.2 Solutions

Contemporary, portable instrumentation, whether OES or XRF, usually employs two approaches to alloy identification. The first, more traditional and straightforward, is based on performing first an accurate quantification of alloying

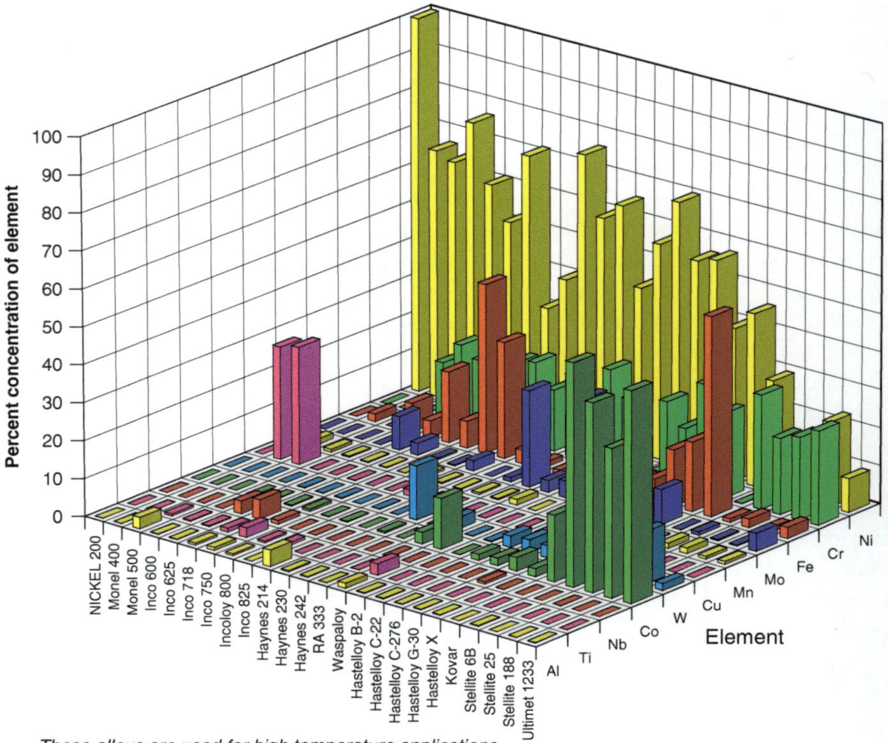

Figure 6.3 Diversity of compositions between alloys within the same family. Example of Ni-based alloys.

elements and then comparing the results with tables of specifications listing chemical compositions along with alloy designations (grade or trade names). This mode of identification is often called "identification *via* a grade table or grade ID". The ruggedness of this method greatly depends on the accuracy and precision of analysis, which requires longer rather than shorter measurement time per sample. This is especially true for XRF, the precision of which improves two-fold for each four-fold increase of measurement time. The multidimensional "vector" of elemental concentrations measured in an unknown alloy is compared with composition specification ranges of many alloys stored in the instrument as an "alloy library". The end product of this approach is chemical composition and grade identification of an alloy. Its success also depends on the calibration method used, such as empirical or *via* Fundamental Parameters (discussed below). This mode of alloy identification is favoured mainly by alloy producers who must maintain tight tolerances on alloy compositions and, therefore, need to know percent concentrations of all elements in an alloy.

The second approach is based on the concept of "alloy spectral signature" or "pattern recognition" in which spectral features of the unknown alloy are compared to obtain a match with previously stored spectral features of known,

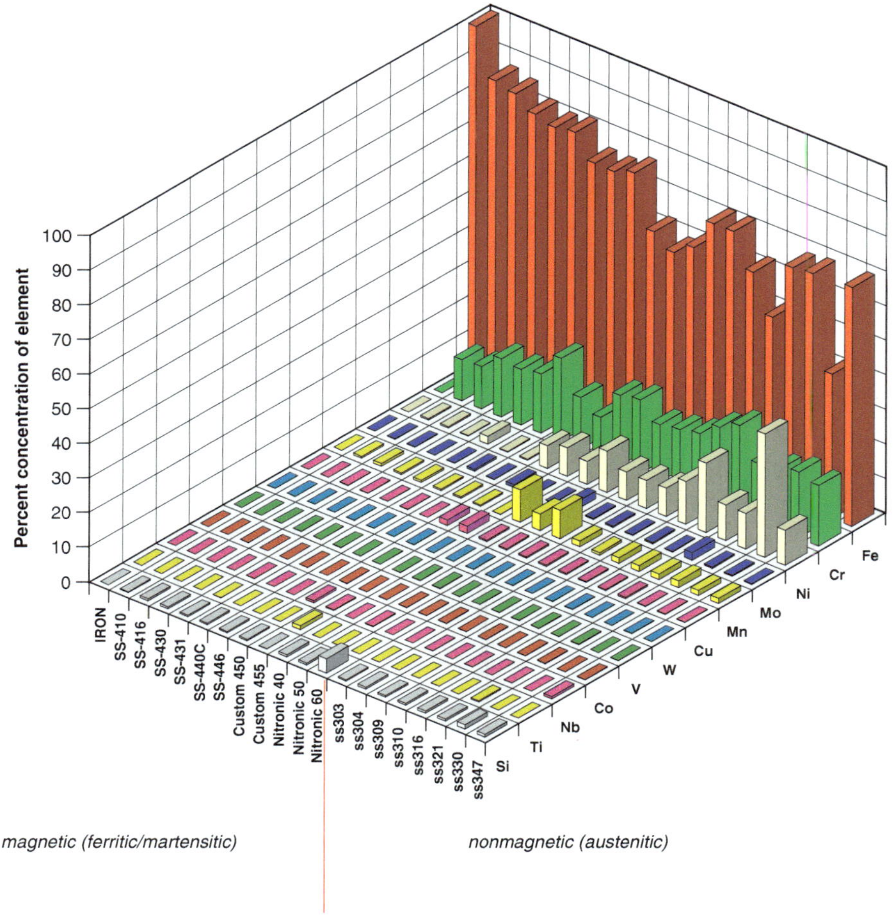

Figure 6.4 Diversity of compositions between alloys within the same family. Example of high temperature and Stainless Steels. Note the correlation between Ni content and magnetic properties of an alloy.

reference alloys. The spectral features are usually spectral intensities from individual elements in the measured sample. The multidimensional vector of intensities from the unknown alloy sample is compared for co-linearity with similar vectors generated for known alloy standards, and stored in the analyser memory (library). The alloy reference whose intensity vector is most similar to that of unknown is selected as a match.

When using the XRF analyser, identification *via* pattern recognition is much faster than *via* a grade table. Usually, the first can be completed within 5 s, while the grade ID may require up to 30 s measurement time per sample.

The advantage of speed and no need for knowledge of alloy specifications in the identification *via* pattern recognition is somewhat compromised by the absolute need of possession of physical samples (standards) of various alloys; a task

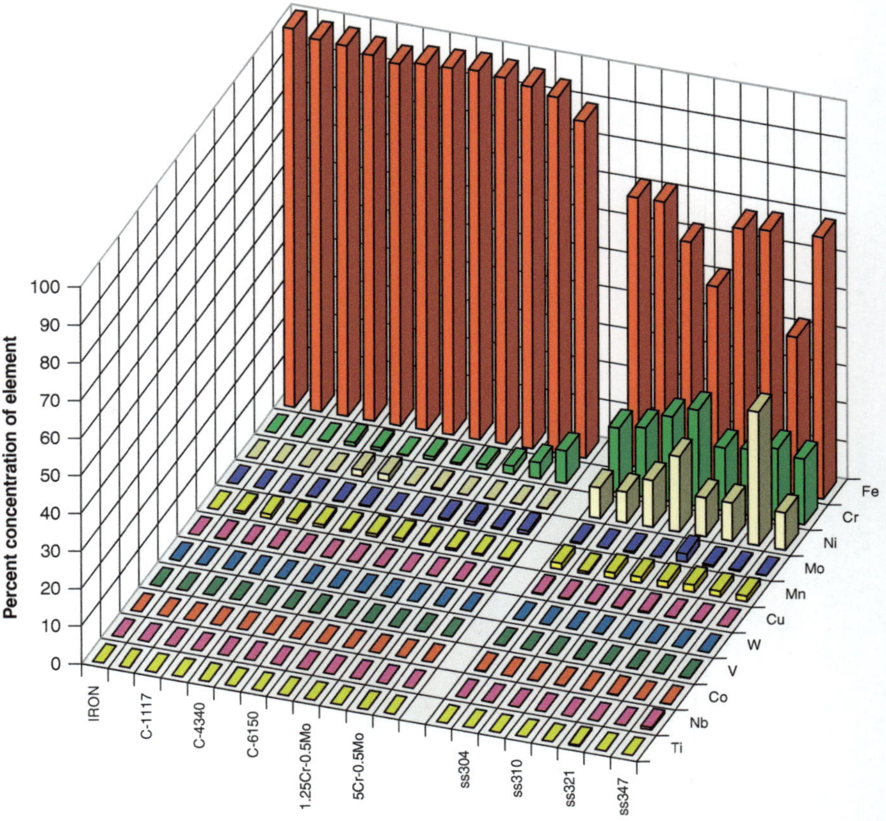

Figure 6.5 Diversity of compositions between alloys within the same family. Carbon and Cr-Moly steels. Stainless Steels of the 300 series are included for comparison.

that is not always feasible. The identification of alloys by spectral signatures is most often used by the scrap recycling industry, where the exact knowledge of alloy composition is not critical once the alloy is identified (we are looking at alloy already made to specs). Coincidentally, spectral signature mode is also faster, and the speed of alloy sorting is of primary importance in alloy recycling yards.

In the following sections we look at specifics of analytical approach to alloy analysis with portable XRF, and review practical embodiments of portable XRF analysers for alloy identification and sorting, their analytical features and performance.

6.3 Analytical Approach

XRF analysis is based on the phenomenon of the emission of X-rays by the constituent atoms of sample when excited by an external source of radiation.

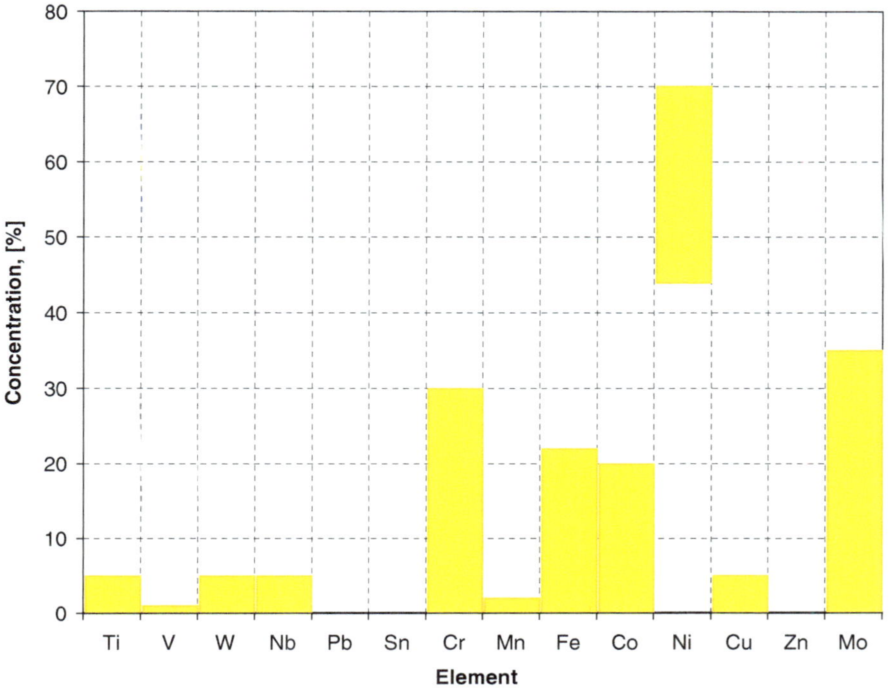

Figure 6.6 Typical concentration ranges for alloying elements in Ni-based alloys.

When a gamma- or sufficiently energetic X-ray from an isotope or the emission from an X-ray tube impinges on an atom of the sample material it may eject one of the inner-shell electrons of the atom. The vacancy created is almost instantaneously (in less than 10^{-20} s) filled by one of the electrons from the higher energy shell. The energy difference between the two energy shells involved in the process is released in the form of X-ray radiation. We call this radiation a characteristic X-ray because its energy is specific and unique to each element (atom). By being able to measure the energy and intensity of the characteristic X-rays of elements, we realize qualitative and quantitative aspects of XRF analysis, respectively. A practical embodiment of an instrument capable of performing those tasks is an X-ray Fluorescence Spectrometer or Analyser. It consists of three essential components: a source of exciting radiation (which can be an isotope or X-ray tube), a means for reproducible sample presentation, and a detector along with its multichannel analyser (MCA) and analytical software. During measurement, the spectrometer acquires an X-ray spectrum of the sample, which contains all information about sample composition. It is the information extracted from the spectrum that is then converted into qualitative and quantitative data of the elemental concentrations in the material tested. Figure 6.9 shows an example of X-ray spectra excited in samples of alloys.

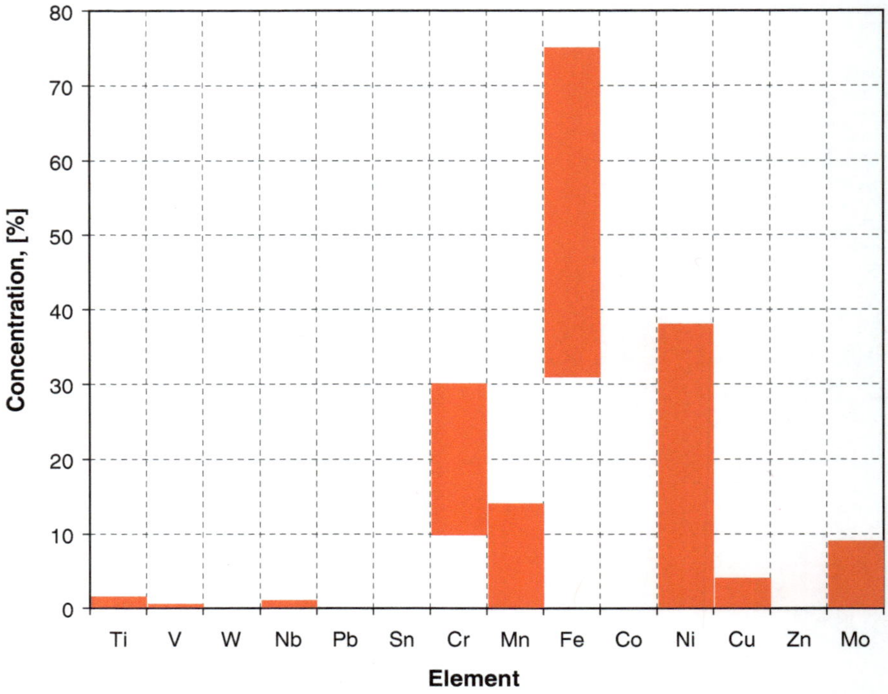

Figure 6.7 Typical concentration ranges for alloying elements in Stainless Steels.

6.3.1 Quantitative Assaying Schemes used in XRF Analysis of Alloys

The goal of any analytical method is quantification of the experimentally obtained data. Specifically, it is the conversion of the signal measured from the sample into a meaningful value/number that numerically describes the feature of the sample with which the measured signal is correlated. In our case, we measure characteristic X-ray intensities of elements in an alloy with an intent to convert them into weight fractions. The manner of converting a series of experimentally obtained X-ray intensities, I_j, into the corresponding weight fractions, w_j, is called the *calibration* of the instrument. Obviously, the quality of analysis, *i.e.* the accuracy and to some extent precision, very much depend on the quality of a calibration, that is on proper selection of mathematical model used for calibration. The calibration equation in its general form may be written as follows:

$$C_i = f(I_i) \tag{6.1}$$

where C_i is concentration of analyte, i, in sample, I_i is net intensity of analyte i, and the functional relationship $f(I_i)$ most often takes form of a polynomial.

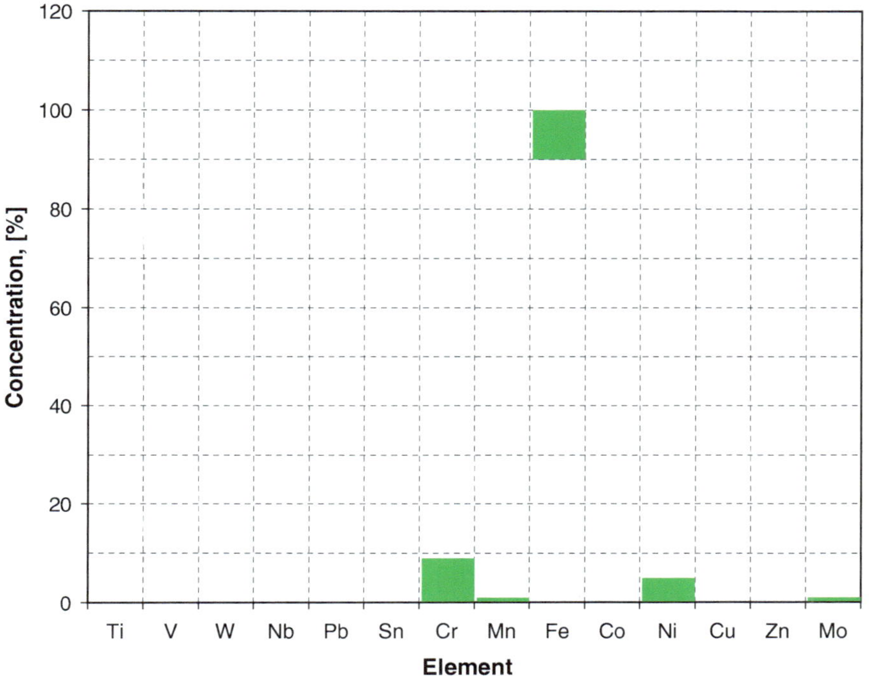

Figure 6.8 Typical concentration ranges for alloying elements in carbon and Cr/Mo steels.

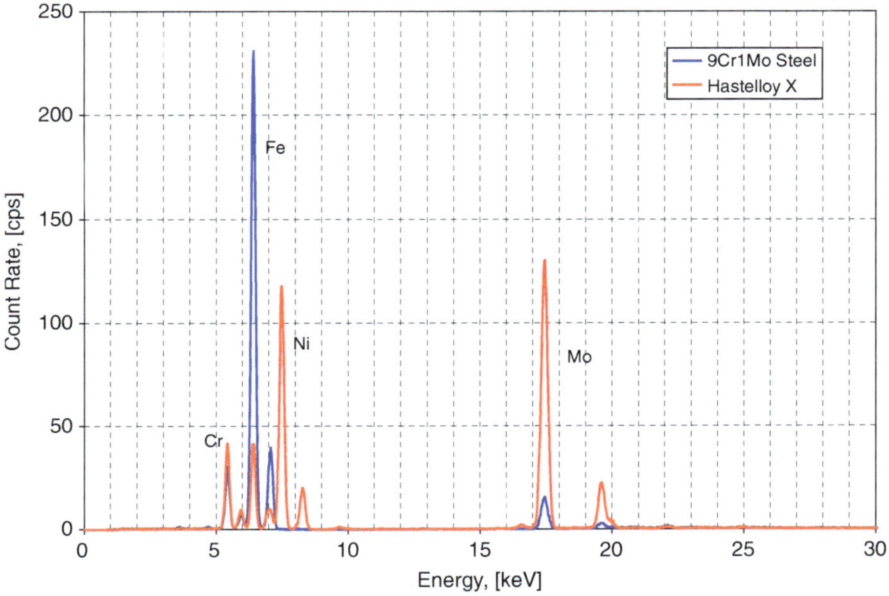

Figure 6.9 Examples of X-ray spectra of 9Cr1Mo steel and Ni-based alloy, Hastelloy X.

6.3.1.1 Empirical Calibration

As in any comparative analytical technique, the simplest and the most intuitive method of calibration in XRF is based on the empirically derived correlation between the X-ray intensities of elements and their weight fractions in samples of well-known composition. The first use of XRF to analyse material was based on empirical formula proposed in 1945 by von Hamos.[11] However, the empirical approach sometimes requires access to large numbers of well-analysed samples, a task not easily fulfilled. This is because the XRF technique is matrix dependent, *i.e.* the measured X-ray intensity of the analyte is dependent not only on its own concentration but also to a high degree on the concentrations of the other elements that make up the sample. For example, if we take two alloys, one with 90% of Ni and 10% of Fe, and another with 60% of Ni, 30% of Mo and 10% of Fe, the X-ray signal from iron from both samples will be very different, despite the fact that iron concentrations in both alloys are identical. Should we ignore this fact and try to derive iron concentration solely from the intensity of its X-rays, the results would be totally wrong. Therefore, the calibration equation for iron must somehow account for the influence of molybdenum (and nickel) on the iron results. This is also why the samples used for instrument calibration must be as similar to the unknowns in composition and morphology as possible. Notably, this requirement is very often ignored by those analysts who try to extend their experience to XRF from well-established techniques such as atomic absorption or emission in which matrix effects are not significant or are non-existent. Despite the inconveniencies of the empirical approach, it remains the most accurate one within the given calibration range.

6.3.1.2 The Standardless or Fundamental Parameters Based Calibration

The so-called "standardless" calibration techniques used in XRF analysis represent a diametrically different approach. They are based on equations that describe exactly the physical processes of interaction of X- and γ-radiation with matter. The equations contain many physical parameters that are fundamental for each element (such as mass absorption coefficients, excitation efficiencies, *etc.*) and express exactly the concentrations of elements as functions of all the X-ray intensities measured from the sample. Therefore, the "standardless" methods are often referred to as the "fundamental parameter approach". The mathematical formalism of the equations is very well known and has been verified to be accurate. However, the equations are transcendent and therefore can only be solved numerically by iteration. This is why standardless methods, while known since the late 1960s, were used mainly in laboratory spectrometers with access to mainframe computers. To circumvent this obstacle and simplify (shorten) the calculation step, X-ray spectroscopists devised clever approximations that combined theory with experiment, and also made calibration equations linear. Their efforts resulted in several different calibration models/equations, each structured in such a way that some coefficients could be calculated in advance from theory and that only a few measurements of well-characterized calibrants were necessary to calibrate the spectrometer.

All the calibration techniques proposed correction for inter-element matrix effects (absorption and enhancement), and are referred to as influence coefficient algorithms. These are divided into two groups: those that are based on the fundamental parameter approach and those that use various forms of correction equations (models). The latter group may use either X-ray intensities of elements or their concentrations in a sample as a means of correcting for matrix effects. Recently, an excellent, critical review and comparison of various matrix effect correction algorithms has been published.[12]

It is only because of the most recent progress in computer processor technology that portable analysers can now also perform complex mathematical operations in real time.

We now illustrate the two basic approaches to quantitative XRF analysis, with examples of practical realization of calibration schemes.

Example of Empirical Calibration. One of the most popular, empirical models correcting for matrix effects is the intensity based empirical coefficients model by Lucas-Tooth and Price.[13] Its general equation for the concentration C_i of analyte i has the following form:

$$C_i = r_0 + I_i \left[r_i + \sum_n r_{in} I_n \right] \tag{6.2}$$

where C_i is concentration of analyte, i, in the sample, r_0 is the empirical constant (intercept), r_i is empirical coefficient (slope) for intensity of analyte, i, r_{in} is the empirical coefficient ("slope") correcting for matrix effect of element n on analyte i, I_i is net intensity of analyte i, and I_n is net intensity of element n.

A unique property of this equation is that analyte concentration is expressed only *via* net intensities of elements in sample. In other words, to develop a calibration equation for one analyte, we need to know concentrations of only this analyte in a suite of calibration samples. Coefficients r_i, and r_{in} are determined for each analyte by a least-squares, multiple linear regression method and they are assumed to be constant for a given set of calibration samples. One needs at least $n + 1$ independent samples with varying concentrations C_i to determine a set of coefficients r_i, r_{in} for a given analyte. Equation (6.2) can be written for each sample and, therefore, a set of linear equations is generated with as many equations as the number of samples used, which is then solved only once by a simple matrix inversion, during calibration of the analyser. This is an important property of the method, making it very useful for portable instrumentation, where computing power and speed are at premium.

Being an empirical method it works satisfactorily only within the concentration ranges covered by calibration samples. The narrower the concentration range, the more accurate the results. Also, the accuracy improves with the number of calibration samples used per analyte. While the Lucas-Tooth and Price method was originally devised for routine analysis of alloys, it became popular among X-ray spectroscopists due to the already mentioned fact that the concentration of an analyte could be expressed by a combination of X-ray intensities measured from a sample, without the need for assays of the elements other than the analyte. This method, with minor modifications, was successfully implemented in the series of portable XRF analysers of the X-MET family produced since early 1980s.[10] A practical example of Equation (6.2) in action is illustrated in Figure 6.10. The data presented are for nickel in nickel-based alloys. A set of 19 Ni-based alloy reference materials (ARMI,[15]) of certified composition was measured, and for each alloy net X-ray intensities for Ni, Cr, Fe, Cu, Co, Ti, W, Nb, and Mo were recorded. To convert measured intensity of nickel K_α X rays into weight percent of nickel, one might just correlate the measured Ni-intensities with known percent concentrations of Ni in samples. The results of such simplistic calibration are shown in Figure 6.10 as blue diamonds. As we can see, the scatter of the data points around the fitted, blue straight line is very large, and nowhere is the predicted concentration of nickel in agreement with its certified value. However, if we account for the presence of other major components by way of multiple linear least-squares regression, we obtain the results marked with red circles. It is obvious which approach is

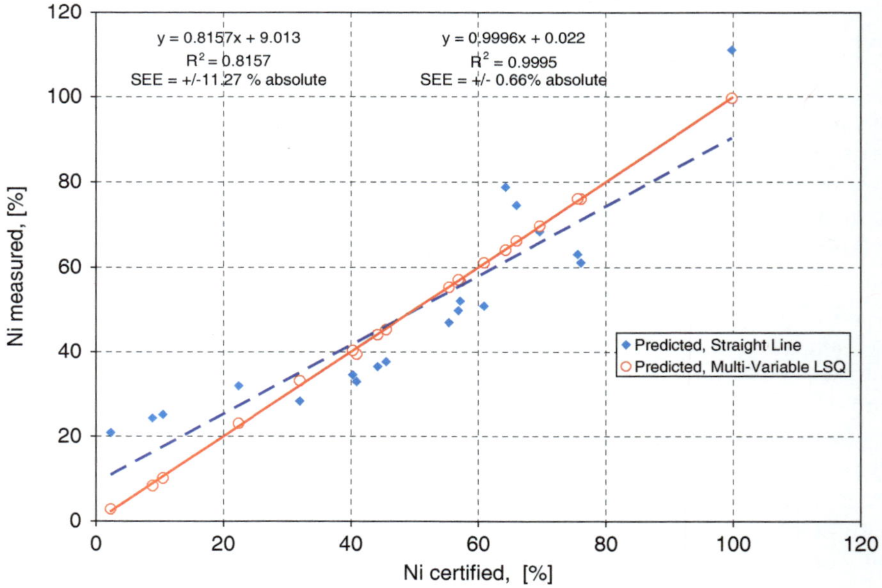

Figure 6.10 Effects of the type of empirical calibration equation on quality of results.

better. For completeness both equations converting intensities of X-rays into percent concentration of nickel are given below:

Straight line fit (diamonds on the graph):

$$C_{\mathrm{Ni}} = 18.96 + 0.0323 \cdot I_{\mathrm{Ni}} \tag{6.2a}$$

Standard error of estimate (SEE) for this equation was ±11.27% m/m Ni absolute.

Multivariable linear least-squares fit (red circles on the graph):

$$\begin{aligned} C_{\mathrm{Ni}} = {} & 16.36 + 0.029864 \cdot I_{\mathrm{Ni}} - 0.024487 \cdot I_{\mathrm{Co}} \\ & + (0.000042 \cdot I_{\mathrm{Cr}} + 0.000005 \cdot I_{\mathrm{Mo}} + 0.000018 \cdot I_{\mathrm{Fe}}) \cdot I_{\mathrm{Ni}} \\ & - 0.006800 \cdot I_{\mathrm{W}} \end{aligned} \tag{6.2b}$$

The SEE for this equation was ±0.66% m/m Ni absolute.

The concentration ranges for the elements were as follows (% m/m): Ni, 2–100; Fe, 0–45; Co, 0–59; Cr, 0–30; W, 0–15; and Mo, 0–28.

Example of Fundamental Parameters based Calibration. One of the more versatile and robust algorithms is the Broll–Tertian influence coefficient algorithm.[16] Its basic equation is shown below:

$$C_i = R_i \left[1 + \sum_j \left\{ \alpha_{ij} - \varepsilon_{ij} \frac{C_i}{R_i} \right\} C_j \right] \tag{6.3}$$

where $C_i, \ldots C_j$ are weight fractions (concentrations) of an analyte, i, and matrix elements, j, in a sample; α_{ij} is theoretical influence coefficient correcting for absorption effects of element j on analyte i. The term ε_{ij} is the theoretical influence coefficient correcting for enhancement effects of element j on analyte i. The term R_i is the intensity ratio, equal to $I_i/I_{(i)}$, where I_i is the net intensity of analyte element i in the sample and $I_{(i)}$ is the net intensity of the pure analyte element i.

As can be seen, this is a concentration based algorithm, *i.e.* elemental concentrations are present on both sides of the equation. As such, it can only be solved by iteration. Another characteristic difference between the Broll–Tertian equation and Equation (6.2) is the use of relative intensities, R_i, rather than the absolute ones, I_i. This allows the use of pure elements (such as pure metals) for calibration instead of well-characterized, multi-element alloys, which are much more difficult to obtain than any pure element calibrant (or compound). This is particularly useful in alloy analysis as almost all elements that make up alloys are metals. During calibration of the instrument all pure metals included in the model are measured and their characteristic X-ray intensities are stored to serve as references for the intensities of the analytes in any unknown material measured. No other alloy calibrants are required for instrument calibration.

Equation (6.3) is written for each analyte i and, therefore, to obtain the elemental concentrations in an unknown alloy one needs to solve a system of equations, or a matrix equation of $n \times n$ dimension. However, for the system of Equation (6.3) to converge and produce physically meaningful results it has to be complemented with a normalizing condition, expressed by Equation (6.4) below, which simply requires that all concentrations in an alloy add up to 100% m/m (or to 1.0).

$$\sum_{n} C_i = 1.0 \tag{6.4}$$

Incidentally, this normalizing condition can also significantly neutralize the negative effects of sample shape and surface imperfections (such as roughness, *etc.*) on the results of analysis. This is not so for empirical calibrations.

6.3.2 Qualitative Identification and Sorting of Alloys

When we think of identification we are asking the question: "What is the identity of an alloy out of the many possible?" In other words, if there are several hundred alloys possible, which of them is the one in which we are interested. When we think of sorting, we usually have another question to answer: "Is the unknown alloy equivalent to the one selected for comparison or not?" In other words, if the unknown alloy is not identical to the one selected as reference for comparison we do not care about its identity.

6.3.2.1 *Identification via Grade Table*

As mentioned earlier, alloys can be identified either by their chemical composition or by spectral signature. In the first instance, called identification *via* a grade table, we need to first determine as accurately as possible the chemical composition of the alloy, and then – by sifting through alloy composition data tables – decide on the identity of analysed sample. The general formula used to determine the identity of an alloy based on measured concentrations is given by Equation (6.5):

$$t_{xr} = k \cdot \sum_{i=1}^{n} (C_{xi} - C_{ri})^2 \tag{6.5}$$

where C_{xi} and C_{ri} are concentrations of the ith element in the unknown, x, and in the rth reference, respectively; n is the number of analytes (elements) measured in the sample; and k is a normalizing factor.

As can be seen, the value of the calculated comparison parameter, t_{xr}, can be interpreted as the squared, normalized distance between two points, the unknown x and the reference r, in an n-dimensional space of elemental concentrations. After the concentrations of elements in a sample are determined, the identification procedure calculates t_{xr} numbers for as many reference alloys, r, that there are in the instrument library of alloy composition tables. Next the t_{xr} numbers are sorted from the smallest to the largest, and the smallest t_{xr} is compared against the acceptance criterion for parameter t. If the t_{xr} meets the acceptance criterion, the reference alloy that yielded this t_{xr} is selected as the identity of the measured alloy. The acceptance criteria for parameter t, which are solely based on probability and statistic, can be adjusted by the operator to the required confidence level. The robustness of this mode of identification strongly depends on the accuracy of measured concentrations, which may be determined empirically or by FP-based methods.

6.3.2.2 *Identification via Spectral Signature (Fingerprint Matching)*

Another way to determine the identity of an alloy is by comparing its X-ray spectrum with the spectra of alloys stored in the instrument memory. This method is referred to as signature or fingerprint matching. Obviously, for this method to work, one must first acquire X-ray spectra of all the alloys of interest. An example of the identification formula, which has been successfully used in one of the portable XRF analysers, is given in Equation (6.6):

$$t_{xr} = K \cdot \sum_{i=1}^{n} \frac{(I_{xi} - I_{ri})^2}{(\sigma^2_{I_{xi}} + \sigma^2_{I_{ri}})} \tag{6.6}$$

where I_{xi} and I_{ri} are X-ray intensities of the ith element in the unknown, x, and in the rth reference, respectively; σI_{xi} and σI_{ri} are the standard deviations of these intensities; n is the number of intensities (elements) measured in the sample; K is a normalizing factor.

The formula is similar in its basic form to Equation (6.5) given above. As can be seen, the value of the calculated comparison parameter, t, is the function of X-ray intensities of elements rather than their percent concentrations. It also depends explicitly on measurement time as standard deviations of X-ray intensities are inversely proportional to the square root of measurement time. Once a spectrum of the sample is acquired, the identification routine calculates individual t value for each possible pair of unknown x and reference r. Then, all t values are sorted and compared against acceptance criterion. The spectrum of reference alloy that produced the passing t value is then accepted as equivalent to the spectrum of the unknown sample x, and, consecutively, the identity of the unknown sample is accepted as being the same as that of the reference alloy. The acceptance criteria for parameter t, which are based on *chi-square* statistic, can be adjusted by the operator to the required confidence level.

This method is the identification *sensu stricto*, as it takes the original spectral data and directly derives from them the identity of an alloy. Bypassing the concentration determination phase, this method is much faster than the previous one. Its obvious drawback is availability of all alloys of interest needed to create a spectral library.

6.3.3 Pass/Fail Sorting

As pointed out earlier, sorting is a procedure that typically answers the question of whether the tested alloy is identical to the one pre-selected as reference. For example, we need to select from a pile of steel scrap only those pieces that are stainless steel of 321 grade (SS321). To do that we may use any of the two identification approaches described. The difference would be that the unknown sample is compared to only one reference alloy, be it *via* specification table or spectral signature. This type of analysis is called Pass/Fail Sorting. Because comparison is performed with only one reference alloy the Pass/Fail is very fast, although given the speed of contemporary microprocessors this may be of lesser importance.

6.3.4 Type Calibration

One more quantification scheme deserves a mention when discussing alloy analysis. It is called "Type Calibration", and is a variation of the empirical method. It has been noted earlier that empirical methods require that samples used to develop calibration curves possess as similar a composition as possible to the expected unknown samples. This statement is the foundation of type calibration. Let us imagine that as a producer of a certain alloy we need to control its composition to very tight specifications. Therefore, the analytical method used to control the composition of an alloy must be very accurate, specifically for those concentrations that correspond to the expected composition of an alloy in question. To achieve this we would take a well-analysed, representative sample of the alloy and measure it with the X-ray analyser. Deviations between the measured and certified concentrations would then be introduced as additive corrections to calibration curves. This way the calibration curves would become the most accurate for the particular alloy. One must remember that this approach is justified only for very small deviations of measured concentrations from the expected ones.

Type calibration combined with spectral signature identification can provide extremely accurate assays of ready made alloys. Here we utilize the fact that any alloy, once it is produced, does not have just any composition, but a very specific one. Its concentrations fall into very tight ranges. We might express this differently by saying that ready made alloys make up discrete sets rather than a continuous range of element concentrations. For example, if by using spectral signature identification we find that the alloy under test is Inconel 700, its composition is pretty well defined within the specifications for Inconel 700. If, in addition to the X-ray spectrum, we also knew the concentrations of elements in reference alloy Inconel 700, we could calculate elemental concentrations in the unknown alloy by straight proportion as follows:

$$C_{xi} = C_{ri} \cdot \frac{I_{xi}}{I_{ri}} \tag{6.7}$$

where C_{xi} and C_{ri} are concentrations of the *i*th element in the unknown, *x*, and reference alloy, *r*, respectively; I_{xi} and I_{ri} are X-ray intensities of the *i*th element measured in the unknown, *x*, and reference, *r*, respectively.

The rationale behind this approach is that the small deviation from a specific point on the calibration curve can be approximated by a straight line, even if the calibration curve itself is nonlinear. Obviously, to utilize this method one needs not only X-ray spectra of reference alloys but also accurate assays for each one of them. Nevertheless, if applicable, this kind of type calibration can produce the most accurate results of all the methods discussed in this chapter.

In the next section we find out how the methods and techniques discussed here are implemented in practical embodiments of commercially available portable X-ray analysers for alloys.

6.5 Modern, Field-Portable XRF Analyser for Alloys

6.5.1 Hardware Considerations

Until a few years ago a portable X-ray analyser for alloy analysis still consisted of two separate components: a hand-held probe connected *via* a cable to an electronic unit. The probe contained X-ray source(s), a detector and a means of a reproducible presentation of the sample for measurement. The electronics unit accepted the signal from the probe, processed it, and displayed the results. It also housed power supplies (rechargeable batteries) and interfaces for communication with the operator and the peripheral devices. This scenario drastically changed in 1998 when the first *one-piece* X-ray analyser was offered for alloy sorting and forever redefined the concept of portable instrumentation. Surface mounted electronics, digital signal processing technology, availability of Li-ion batteries and some original creativity made it possible to integrate the whole analyser into a single, lightweight, ergonomically designed hand-held device.

A small, rugged, sealed radioisotope capsule, emitting either X-rays or low energy gamma rays, has been a preferred source of primary radiation for portable XRF analysers. Typical radioisotopes used are ^{55}Fe, ^{109}Cd and ^{241}Am, with strengths of up to 1.48, 0.74, and 1.11 GBq, respectively. Table 6.1 shows their properties.

The year 2002 signified the advent of a new generation of portable XRF analysers with a miniature X-ray tube instead of radioisotopes. These low power X-ray tubes typically work at a high voltage of 35–50 kV and up to 50 μA of current. Their low power consumption (less than 4 W) easily allows for at least 8 hours of continuous operation before the instrument's Li-ion battery needs recharging. The heart of any spectrometric, XRF analyser is its X-ray detector. In the past this could be a gas-filled proportional counter designed to offer enhanced resolution, or a semiconductor detector such as a Si(Li), liquid nitrogen cooled detector for even better energy resolution. Contemporary designs use exclusively semiconductor detectors either as silicon "p-i-n" diodes,[17,18] mercuric iodide crystals[19] or cadmium/telluride devices.[20] These detectors are

Table 6.1 Properties of radioisotope sources used in portable XRF analysers for alloy applications.

Radio-isotope	*Half-life (years)*	*X- or γ-ray energy (keV)*	*Photons per disintegration*
^{55}Fe	2.7	Mn K X-rays (5.9, 6.5)	0.28
^{244}Cm	17.8	Pu L X-rays (14–21)	0.08
^{109}Cd	1.3 (464 days)	Ag K X-rays (22.1, 25); Γ-rays, at 88.03	1.02; 0.04
^{241}Am	432.7	γ-rays, at 59.54; Np L X-rays (11.9–22.2)	0.36; 0.43

thermoelectrically cooled (utilizing the Peltier effect) to about –25 °C. Semiconductor detectors are an order of magnitude smaller than the gas filled detectors and command much better energy resolution. The better the energy resolution, the better the analytical performance of the instrument. The practical energy resolutions of 190 to 230 eV for Mn K-alpha line are typical for modern portable analysers with silicon "p-i-n" diodes.

No trivial matter for a portable instrument is its weight and size because, for obvious reasons, the portable analyser should not challenge the operator "gravitationally"! To comply with this requirement, the hand-held analyser should not weigh more than ~1 kg. There is one other very important requirement specific to analysers used in the field for alloy analysis and identification and that is the shape and size of its front end that comes in contact with the analysed object. During construction many metal parts are joined by welding them together. In such a situation it is important to know not only the identities of joined parts but also that of the weld seam. Therefore, the front end of the analyser has to be narrow enough so that it can reach into the corner weld, and the measuring aperture has to be shaped in such a way that it is possible to analyse the weld seam only. Large pipelines in chemical and petrochemical installations are often equipped with inspection ports, which can be accessed through 100 mm diameter holes. To test the alloy of the inspection port, the measuring end of the analyser must fit into such hole. The ergonomics of analyser design cannot be overstated; it is very often as important to the user as its analytical performance.

6.5.2 Software

A modern, portable XRF analyser features not only empirical calibration software, but also the so-called "standardless" calibration software, based on the fundamental parameters approach, which is usually included with the analysers featuring semiconductor detectors.[12] In fact, the "standardless" software is nowadays a "must have" feature at the expense of a more complex empirical calibration capability. This is because the analytical software based on fundamental parameters approach provides very versatile instrument calibration over a 0–100% concentration range for each analyte without user participation and without the need for an extensive set of calibrants. This feature facilitates automation of the measuring, analysis and decision making processes, making them independent of the user. We discuss analytical software in the next section.

The user interface software in portable analysers is limited in features when compared to laboratory instrumentation. There are several reasons for this. First, portable instruments are designed to perform routine rather than research type measurements under unforgiving field conditions. Second, the LCD display of the portable device is much smaller, has lower resolution and most likely cannot display colour. Consequently, the amount of information that can be output on a single screen is limited. The use of graphical symbols (icons) in menus instead of textual instructions is the preferred way of informing the user what to do next. The user software is also simplified, because more sophisticated operations are usually provided by PC resident software that runs the instrument when connected *via* a serial cable or wireless link.

Figures 6.11 and 6.12 show an earlier generation portable XRF analyser, still as a two module system – probe and main electronic unit – while Figures 6.13 and 6.14 illustrate the most advanced, one-piece, hand-held analysers, with X-ray tube or radioisotope excitation.

As one can notice, a common characteristic of all instruments is a narrow, wedge-like shape of the front end. Even the oldest model, which incorporated a gas-filled proportional detector, had the probe shaped for an easy access to corner welds. All instruments are splashproof, and are built to be rugged enough to withstand the harsh conditions of industrial environment and maintain the integrity of radioisotope sources. Extreme instances of "wear and tear" are shown in Figures 6.15 and 6.16. For isotope-based analysers it is important that the source(s) do not lose their mechanical integrity.

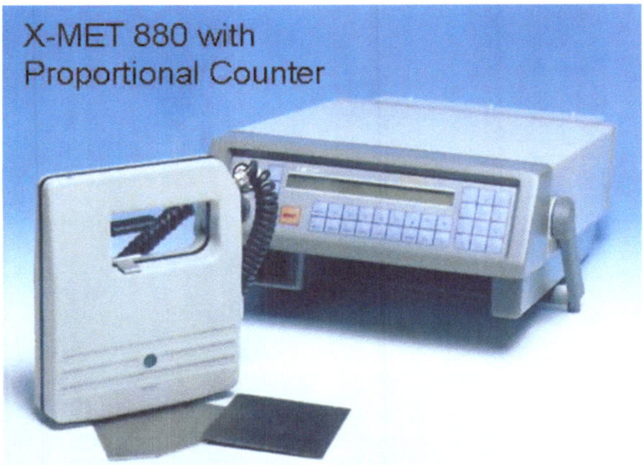

Figure 6.11 Portable XRF analyser with gas-filled proportional detector. Note the two separate modules, probe and electronic unit. The total weight of the system was about 8 kg. Introduced in 1984 by Metorex Inc., it was manufactured for over 12 years. The LCD display is just two lines by 40 characters each.

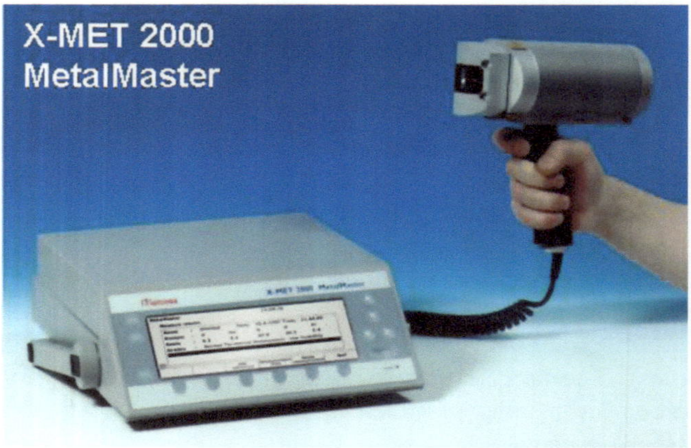

Figure 6.12 Modernized version of X-MET 880 with probe housing silicon "p-i-n" diode detector. Introduced in 1998. Better display, PC based architecture, but still a two module design.

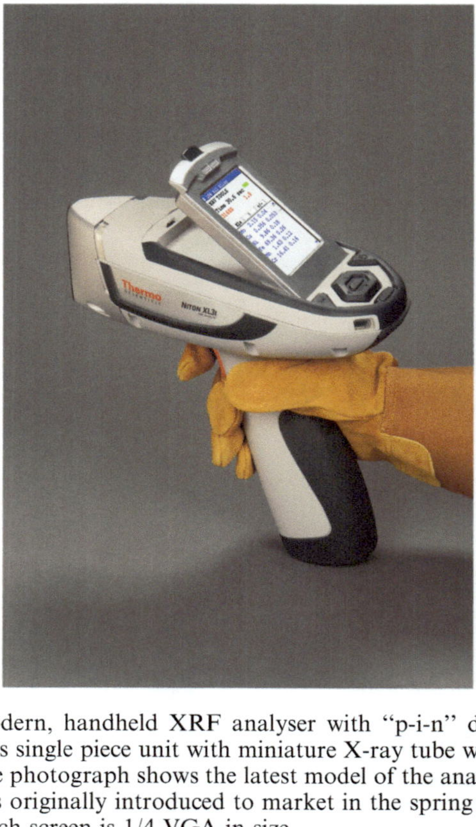

Figure 6.13 Modern, handheld XRF analyser with "p-i-n" diode silicon detector. This single piece unit with miniature X-ray tube weighs less than 1.4 kg. The photograph shows the latest model of the analyzer – its predecessor was originally introduced to market in the spring of 2002. Tilted, LCD touch screen is 1/4 VGA in size.

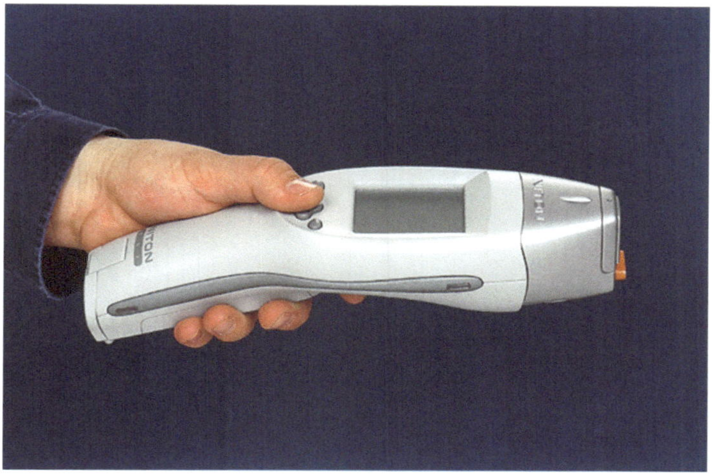

Figure 6.14 Modern, handheld XRF analyser with "p-i-n" diode silicon detector and isotope excitation. This unit can incorporate up to three different isotopes for increased elemental range.

Figure 6.15 Model XLt 800 with miniature X-ray tube after damage in fire.

Contemporary instruments are also equipped with a proximity button (an orange protrusion at the bottom of the front surface) that prevents exposure when the sample is not covering the measuring aperture.

6.5.3 Commercially Available Instruments – Feature Comparison

The comparison Tables 6.2 and 6.3 are included to provide a condensed overview of the features of commercially available XRF-based, portable alloy

Figure 6.16 An older Model XL 800 with radioisotope sources, apparently run over by some heavy equipment. The left end of the instrument houses radioisotopes, which survived the accident.

analysers available in 2005. The information in these tables has been obtained from the manufacturers' published data such as product brochures and web sites. Cost of ownership of any instrument is included for guidance purposes only as the individual price of an analyser depends on specific configuration, optional equipment, quantity discounts, and local market conditions. More detailed information as well as pictures of the equipment may be found on web sites of the manufacturers.[14,16,21]

In the next section we focus on the analytical capabilities of these instruments and their typical performance data.

6.5.4 Typical Performance Data

It is difficult to make a general statement about the analytical performance of an instrument as it depends on many instrument specific factors, such as:

- Measurement geometry of the particular instrument design (relative distances and angles between source, sample and detector);
- detector characteristics, such as energy resolution, thickness, effective surface area, and count rate throughput capability;
- type of excitation, *i.e.* isotope or X-ray tube, whether filters are used, *etc.*;
- measurement time;
- matrix of the material analysed.

Therefore, the performance data provided in the following tables should only be regarded as a guide.

Table 6.4 illustrates typical precision performance that can be expected from commercially available portable X-ray alloy analysers. Most of the precision

Table 6.2 Commercially available isotope-based XRF alloy analysers.[14,16]

Isotope-based XRF Alloy Analyzers, Features Comparison		
MODEL:	*NITON XLi 800 Series*	*Thermo Model TN Alloy Pro*[c]
FEATURES:		
Instrument Configuration	HAND-HELD ONE PIECE Inline ergonomic design	HAND-HELD, TWO-COMPONENT, ONE PIECE Pistol grip design (Analyzer + detachable iPAQ™ PDA)
Weight	1.7 lbs (0.8 kg) complete	1.7 lbs (0.8 kg)
Excitation source	**^{241}Am - 30 mCi Infiniton** or ^{109}Cd -10 mCi and/or ^{55} Fe -14 mCi and/or ^{241}Am -14 mCi (Up to three sources in one instrument)	^{109}Cd - 4 mCi and ^{55}Fe - 20 mCi (two sources in one instrument)
Radioactive Emission (direct in-beam dose rate in millirem per hour at 5 cm from instrument aperture)	From 0.45 to 315 mR/hr dependant upon source type and activity level	
Safety Features	Password protected operation (4) shutter open LED indicators Auto close/lock mechanism (activates on power failure) Sample proximity sensor (opt. activation in U.S.) Two-handed safety interlock (opt. activation in U.S.)	
System Electronics and Operating System	Hitachi SH-4 CPU ASICS high-speed DSP (digital signal processor) 4096-channel MCA (multi channel analyzer) Dedicated operating system with export and update functionality	400 MHz Intel™ Xscale processor (in HP iPAQ Pocket PC™) Performs all data collection and analysis functions Windows™ CE™ operating system
Power Source	Rechargeable Li-Ion Battery (not subject to memory effects) AC Adapter	14.8 V Rechargeable Li-Ion Battery (not subject to memory effects)

Table 6.2 (*Continued*).

Isotope-based XRF Alloy Analyzers, Features Comparison		
MODEL:	*NITON XLi 800 Series*	*Thermo Model TN Alloy Pro*[c]
Spare Battery Pack	Standard	Optional
Battery Life per Charge	6–8 hours dependant upon duty cycle	over 6 hours
Detector	High resolution Si-PiN, 220 to 230 eV energy resolution typical	Si-PiN detector with 250 eV energy resolution
Calibration	Factory Fundamental Parameters	Factory Fundamental Parameters
Periodic Standardization	Automatic standardization using internal microsource	
Standard Sorting Library	More than 300 Common Alloys Editable by user	Over 230 alloys
User Programmable Library	> 500 Alloys	Unlimited, with Type II Flash Cards
Data Input	Integrated Barcode Reader Virtual Keyboard Drop-down list of user-preset values	iPAQ virtual keyboard
Ease-of-Use	Very easy - minimal user expertise, training and/or technical knowledge necessary	Moderately complex - requires basic knowledge of Windows CE™ operating system, requires use of iPAQ stylus for certain software functions. Display can be pivoted to any angle.
PERFORMANCE:		
Alloy Testing-Protocol	Automatically determines alloy base. Auto-grade identification and verification-quality chemical composition. Automatically normalizes for analysis of small or odd-shaped samples	Automatically determines alloy base. Auto-grade identification and verification-quality chemical composition. Automatically normalizes for analysis of small or odd-shaped samples
Sample Preparation	Minimal - only non-representative material (paint, coating, etc) must be removed	Minimal - only non-representative material (paint, coating, etc) must be removed

Table 6.2 (*Continued*).

Isotope-based XRF Alloy Analyzers, Features Comparison		
MODEL:	*NITON XLi 800 Series*	*Thermo Model TN Alloy Pro*[c]
Fingerprint Matching	1–2 clock seconds[a] Performs spectral fingerprint match from up to 500 user-stored alloy fingerprints Results updated continuously on screen	several seconds
Grade Matching (with verification chemistry)	2–5 clock seconds[a] Results updated continously on screen	several seconds
Precise Alloy Chemistry	5–10 clock seconds[a] Results updated continously on screen	10–30 clock seconds
ECONOMY:		
U.S. List Price	Average system \$35,000 (all alloy bases)	Average system about \$32,000 (all alloy bases)
Upkeep and Expendables costs	^{109}Cd approx. once every two years[b] ^{55}Fe approx. once every five years[b] ^{241}Am does not require replacement	^{109}Cd approx. once every two years[b] ^{55}Fe approx. once every five years[b]

[a]With fresh radioisotope sources or Infiniton source. Cd and Fe sources will decay.
[b]Based on published price lists. Half-life of ^{109}Cd source = 15 months; Half-life of ^{55}Fe source = 32.4 months; ^{241}Am = 432 years.
[c]Thermo Model TN Alloy Pro has been discontinued in January of 2006.
No entry in a cell means information was not readily available.

data are shown as a range, with the lower value typical for X-ray tube excitation and the higher value typical for radioisotope excitation. The X-ray tube based analyser used to collect data for Table 6.4 was equipped with a miniature X-ray tube with the transmission end-window, silver anode, operated at 35 kV and 5 μA of anode current. The isotope-based analyser was equipped with a 1.48 Gbq (40 mCi) ^{109}Cd and a 1.48 Gbq (40 mCi) ^{55}Fe source. Both types of analysers were equipped with silicon p-i-n diode detector. Data quoted in Table 6.4 apply also to older type instruments with gas-filled proportional detectors, such as Metorex Model X-MET 880.

Table 6.5 shows the performance of portable XRF analysers in alloy identification *via* spectral signature. The data listed were obtained by testing each individual alloy at least ten times in sequence against a library of spectra of up to 130 alloys. The ratio of the number of positive identifications to the total number of attempted tests represents the success ratio expressed in percent, and these results are listed under the heading "Identification results".

Table 6.3 Commercially available X-ray tube-based XRF alloy analysers.[14,16,21]

	X-Ray Tube-based XRF Alloy Analyzers, Features Comparison		
MODEL:	*NITON XLt and XL3t 800 Series*	*Metorex X-MET 3000T*	*Innov-X Systems 200 Series*
FEATURES:			
Instrument Configuration	HAND-HELD ONE PIECE Pistol grip design	HAND-HELD,TWO-COMPONENT, ONE PIECE Pistol grip design (Analyzer + detachable NEC®PDA)	HAND-HELD, TWO-COMPONENT, ONE PIECE Pistol grip design (Analyzer + detachable iPAQ™ PDA)
Weight	3.0 lbs (1.4 kg) complete	5.5 lbs (2.5 kg) complete (with PDA and battery)	5.5 lbs (2.5 kg) complete (with iPAQ and battery) Innov-X spec of ~4 lbs is a misrepresentation
Sources	50 kV x-ray tube Typical power output of 2 W	40 kV x-ray tube Typical power output of 2 W	35 kV x-ray tube Typical power output of 4 W
Radioactive Emission (direct in-beam dose rate in milli-rem per hour at 5 cm from instrument aperture)	1200 mR/hr	> 500 mR/hr	3300 mR/hr
Safety Features	Password protected operation (3) shutter open LED indicators Auto shutdown (activates on power failure) Tube ramp-up on measurement (no X-rays produced between tests) Sample proximity sensor (opt. activation in U.S.) Two-handed safety interlock (opt. activation in U.S.)	System lock key (key must be inserted while system is activated) Tube ramp-up on measurement Optical proximity sensor	Trigger lock button Constant tube ramping (X-rays always on) Optional auto ramp-down after user-specified idle period
System Electronics and Operating System	Hitachi SH-4 CPU ASICS high-speed DSP (digital signal processor)	Intel StrongARM™ (in NEC PDA) Performs all data collection and analysis functions.	Intel StrongARM™ (in iPAQ Pocket PC™) Performs all data collection and analysis functions.

	4096-channel MCA (multi channel analyzer)	Microsoftx® PocketPC 2002 operating system	Windows™ CE™ operating system
	Dedicated operating system with export and update functionality		
Ruggedness/Environmental Issues	Self-contained unit, Sealed against moisture, dust, and UV exposure. Can be used effectively in rain or dusty work environments.	Two-component system. NEC PDA specs state to avoid exposure to moisture, dust and UV (sunlight) and that it cannot be used in rain or dusty environments.	Two-component system. PDA specs state to avoid exposure to moisture, dust and UV (sunlight) and that it cannot be used in rain or dusty environments.
Power Source	Rechargeable Li-Ion Battery (not subject to memory effects)	Rechargeable Li-Ion Battery (not subject to memory effects)	Rechargeable Li-Ion Battery (not subject to memory effects)
	AC Adapter	AC Adapter	
Spare Battery Pack	Standard	Standard	Standard
Battery Life per Charge	6–8 hours dependent upon duty cycle	$\sim$4 hours	2–4 hours dependent upon duty cycle
Detector	High resolution Si-PiN	High resolution Si-PiN	High resolution Si-PiN
Calibration	Factory Fundamental Parameters, User calibrations	Empirical Analysis-only.	Empirical Analysis-only on base models.
	NEVER necessary	Effective only for specific element ranges for which it has been calibrated	Factory Fundamental Parameters on high-end models.
Periodic Standardization	Automatic standardization using internal reference sample		Standardization must be performed on initial startup priol to testing and after every four hours during use. External standardization accessory must be clipped over the nose of the unit during standardization procedure.
Standard Sorting Library	More than 300 Common Alloys	Editable by user	250 Common Alloys
	Editable by user		Editable by user
User Programmable Library	> 500 Alloys	> 500 Alloys	> 500 Alloys
Data Input	Integrated Barcode Reader	Windows™ CE	iPAQ virtual keyboard
	Virtual Keyboard	Virtual keyboard	Graffiti pad
	Drop-down list of user-preset values	Graffiti pad	
Ease-of-Use	Very easy-minimal user expertise, training and/or technical knowledge necessary	Moderately complex - requires basic knowledge of Microsoft® PocketPC 2002 OS, requires use of stylus for user input functions and software navigation	Moderately complex - requires basic knowledge of Windows CE™ operating system requires use of iPAQ stylus for use input functions and software navigation

Table 6.3 (*Continued*).

	X-Ray Tube-based XRF Alloy Analyzers, Features Comparison		
MODEL:	*NITON XLt and XL3t 800 Series*	*Metorex X-MET 3000T*	*Innov-X Systems 200 Series*
PERFORMANCE:			
Alloy Testing-Protocol	Automatically determines alloy base. Auto-grade identification and verification-quality chemical composition. Automatically normalizes for analysis of small or odd-shaped samples	Automatically determines alloy base. Auto-grade identification and verification-quality chemical composition. Automatically normalizes for analysis of small or odd-shaped samples	Automatically determines alloy base. Auto-grade identification and verification-quality chemical composition. Automatically normalizes for analysis of small or odd-shaped samples
Sample Preparation	Minimal - only non-representative material (paint, coating, etc.) must be removed	Minimal - only non-representative material (paint, coating, etc.) must be removed	Minimal - only non-representative material (paint, coating, etc.) must be removed
Fingerprint Matching	1–2 clock seconds Performs spectral fingerprint match from up to 500 user-stored alloy fingerprints Results updated continuously on screen	2–5 clock seconds Performs spectral fingerprint match from up to 500 user-stored alloy fingerprints	5–10 clock seconds Performs spectral fingerprint match from up to 500 user stored alloy fingerprints
Grade Matching (with verification chemistry)	2–5 clock seconds Results updated continously on screen	2–5 clock seconds	10–15 clock seconds (calculation time extends actual measurement by 3–12 seconds)
Precise Alloy Chemistry	5–10 clock seconds Results updated continously on screen	10–30 clock seconds	18–30 clock seconds (calculation time extends actual measurement by 3–12 seconds)
ECONOMY:			
U.S.List Price	Base System: $35,000 (all alloy bases) Average system $35.000 (all alloy bases)	Base System: $35,000 (all alloy bases) Average system $35.000 (all alloy bases)	Base System: $22,750 (single base - addl $1500ea.) Average system $35.000 (all alloy bases)
Upkeep and Expendables costs	Tube replacement est at 3–5 years: $4500	Tube replacement est at 3–5 years: (unpublished)	Tube replacement est. at 3–5 years: $4500

No entry in a cell means information was not readily available.
Reader is advised that due to constant improvements and new designs the data in table may change.

Table 6.4 Typical precision data for a contemporary, commercially available portable XRF alloy analyser. Values quoted are in percent absolute for measurement time of 30 s per sample.

Alloy group	*Ti*	*Cr*	*Mn*	*Fe*	*Co*	*Ni*	*Cu*	*Zn*	*Nb/Mo*	*Sn*	*Pb*
Low alloy steels	0.005	0.02	0.05	0.15	0.15	0.05	0.03	0.10	0.003	0.10	0.10
	0.01	0.05							0.005		
			0.1	0.25		0.15	0.09				
Stainless, Hi-Temp	0.010	0.10	0.05	0.20	0.20	0.20	0.01	0.06	0.02	0.30	0.05
steels	0.020	0.20							0.05		0.20
			0.10	0.90		0.30	0.05				
Ni/Co alloys	0.10	0.20	0.10	0.12	0.10	0.20	0.05	0.30	0.02	0.30	0.15
		0.50							0.08		
			0.30	0.80	0.70	0.80	0.30				
Cu alloys, brass/	0.06	0.10	0.02	0.02	0.10	0.10	0.15	0.07	0.02	0.01	0.20
bronze										0.20	0.30
			0.10	0.10		0.20	0.80	0.50			
Aluminium alloys	0.03	0.05	0.10	0.05	0.05	0.05	0.05	0.06	0.003	0.01	0.01
		0.20							0.005	0.20	0.02
				0.10							
Titanium alloys	0.50	0.10	0.10	0.06	0.05	0.05	0.02	0.02	0.008	0.01	0.01

Table 6.5 Performance of a portable X-ray analyser in alloy identification *via* spectral signature match. *Note*: if two results are given, the first refers to gas-filled proportional detector while the second is to a solid state, silicon p-i-n diode detector.

Alloy group	*Measured elements*	*Identification results (% feasible)*
Nickel and cobalt alloys	Ti, Cr, Fe, Co, Ni, Cu, Nb, Mo, W, Hf, Ta	100
Copper alloys	Mn, Fe, Ni, Cu, Zn, Pb, Sn	90–100; 98–100
Stainless steels and high-temperature alloys	Ti, Cr, Mn, Fe, Co, Ni, Cu, Nb, Mo	90–100; 100
Cr/Mo steels	Cr, Fe, Ni, Mo	95–100; 100
Low alloy steels	Cr, Mn, Fe, Ni, Cu, Nb, Mo	65–80; 90–100
Titanium alloys	Ti, V, Mn, Cr, Zr, Mo, Sn	95–100
Aluminium alloys	Mn, Fe, Cr, Cu, Zn	90–100; 95–100

As can be seen, the great majority of alloys can be identified 100% of the time. The difficulties arise from the fact that some non-ferrous alloys, such as copper and aluminium based, contain significant amounts of Si, Al, Mg, which are not measurable directly *via* this XRF method. One may also notice that some low alloy steels also pose difficulties. However, the nickel and cobalt group are always identified with 100% success. This is because these two alloy groups contain elements that are readily analysed by PXRF and the differences in compositions between alloys of the same family are relatively large (typically several percent) when compared to other groups such as low alloy steels (typically less than 1–2%).

The analytical accuracy of an XRF analyser with miniature X-ray tube is documented by examples listed in Tables 6.6 and 6.7. Measured data are listed along with certified concentrations of measured alloy reference materials obtained from Brammer Standard Co.,[22] NIST,[23] and ARMI.[15] All data were obtained at 35 kV, 5 μA current, and up to 20 s measurement time per sample. Data are shown for 300 series stainless steels and for some typical Ni-based

Table 6.6 Examples of performance of X-ray tube-based analyser in analysis of stainless steels. (NITON Model XLt 800.)

Analytical Accuracy and Precision for 300 Series Stainless Steels

SAMPLE:	**RM-BS81E**			**Portable Analyzer Results after:**				
AISI 304	Cert	Error	2 Sec	+/−2σ	5 Sec	+/−2σ	20 Sec	+/−2σ
Grade ID →			**304/321**		**304**		**304**	
Cr	18.31	0.13	18.33	0.83	18.42	0.63	18.29	0.32
Ni	9.52	0.06	9.56	0.87	9.45	0.66	9.70	0.34
Mn	1.73	0.05	2.14	0.51	1.94	0.40	1.99	0.21
Mo	0.38	0.02	0.40	0.10	0.42	0.08	0.38	0.05
V	0.14	0.02	nd		nd		<0.30	0.20
Ti	0.00	n/a	nd		nd		<0.30	0.20
Fe	bal.	n/a	67.51	1.73	67.82	1.33	68.05	0.68
SAMPLE:	**RM - BS321A**			**Portable Analyzer Results after:**				
AISI 321	Cert	Error	2 Sec	+/−2σ	5 Sec	+/−2σ	20 Sec	+/−2σ
Grade ID →			**321**		**321**		**321**	
Cr	17.20	0.06	17.54	0.82	17.09	0.60	17.18	0.31
Ni	9.38	0.08	9.31	0.87	9.61	0.67	9.62	0.34
Mo	0.20	0.01	nd		nd		0.22	0.05
Mn	1.22	0.03	1.41	0.50	1.51	0.38	1.56	0.19
Cu	0.28	0.02	nd		nd		0.29	0.12
Ti	0.51	0.02	nd		nd		0.57	0.20
Fe	bal.		69.38	1.79	69.70	1.34	69.26	0.68
SAMPLE:	**RM - BS CA31A316 - 1**			**Portable Analyzer Results after:**				
AISI 316	Cert	Error	2 Sec	+/−2σ	5 Sec	+/−2σ	20 Sec	+/−2σ
Grade ID →			**316**		**316**		**316**	
Cr	17.44	0.05	17.81	0.85	17.82	0.65	17.55	0.33
Ni	11.21	0.10	11.30	0.93	11.27	0.69	11.30	0.37
Mo	2.08	0.05	2.03	0.20	2.18	0.16	2.16	0.08
Mn	1.54	0.04	1.94	0.51	1.72	0.40	1.83	0.20
Fe	sbal.		64.47	1.71	64.79	1.33	65.31	0.68
SAMPLE:	**RM - IARM 8B**			**Portable Analyzer Results after:**				
AISI 347	Cert	Error	2 Sec	+/−2σ	5 Sec	+/−2σ	20 Sec	+/−2σ
Grade ID →			**347**		**347**		**347**	
Cr	17.61	0.06	18.07	0.86	17.96	0.64	17.85	0.31
Ni	9.06	0.04	8.69	0.86	8.92	0.66	9.14	0.33
Mn	1.43	0.01	1.29	0.51	1.55	0.38	1.53	0.19
Nb	0.630	0.02	0.61	0.10	0.64	0.07	0.66	0.05
Mo	0.50	0.01	0.44	0.12	0.53	0.09	0.50	0.05
Cu	0.26	0.01	nd		nd		0.32	0.12
Fe	bal.		68.89	1.84	69.02	1.36	69.07	0.67

Table 6.7 Examples of performance of an X-ray tube-based analyser in the analysis of Ni-based alloys. (NITON Model XLt 800.)

Analytical Accuracy and Precision for Nickel Based Alloys

SAMPLE:	RM - IARM 201A			Portable Analyzer Results after:				
Inconel 690	Cert	Error	2 Sec	+/−2σ	5 Sec	+/−2σ	20 Sec	+/−2σ
Grade ID →			**690**		**690**		**690**	
Cr	29.90	0.08	29.19	1.08	29.12	0.73	29.39	0.45
Fe	9.09	0.04	8.72	0.61	8.78	0.41	8.74	0.25
Ni	59.90	0.06	60.41	1.36	60.19	0.93	60.13	0.57
Ti	0.30	0.01	nd		nd		0.42	0.21
SAMPLE:	**RM - IARM 56C**			**Portable Analyzer Results after:**				
Inconel 718	Cert	Error	2 Sec	+/−2σ	5 Sec	+/−2σ	20 Sec	+/−2σ
Grade ID →			**718**		**718**		**718**	
Cr	18.21	0.09	18.47	1.07	18.35	0.65	18.46	0.39
Fe	18.27	0.23	18.12	0.90	18.14	0.55	17.98	0.33
Nb	5.19	0.03	5.02	0.28	5.26	0.18	5.26	0.10
Mo	2.94	0.04	2.73	0.27	2.81	..17	2.76	0.10
Ti	1.01	0.01	1.31	0.80	1.07	0.48	1.06	0.28
Ni	53.36	0.62	52.83	1.48	52.93	0.89	53.03	0.53
SAMPLE:	**RM-BS 750A**			**Portable Analyzer Results after:**				
Inconel X750	Cert	Error	2 Sec	+/−2σ	5 Sec	+/−2σ	20 Sec	+/−2σ
Grade ID →			**750**		**750**		**750**	
Cr	15.68	0.20	15.22	0.82	15.61	0.54	15.39	0.34
Fe	7.07	0.07	6.75	0.46	6.71	0.30	6.92	0.19
Ti	2.60	0.07	2.72	0.75	2.85	0.50	2.76	0.31
Nb	1.07	0.03	1.13	0.13	1.11	0.09	1.13	0.05
Mo	0.22	0.02	nd		nd		0.20	0.05
Ni	71.90	0.25	72.38	1.38	72.24	0.91	72.42	0.58
SAMPLE:	**RM - NIST 1246**			**Portable Analyzer Results after:**				
Incoloy 800	Cert	Error	2 Sec	+/−2σ	5 Sec	+/−2σ	20 Sec	+/−2σ
Grade ID →			**801/800**		**800**		**800**	
Ni	30.80	0.10	30.83	1.29	31.48	0.84	31.21	0.53
Cr	20.10	0.10	20.24	0.93	20.36	0.59	20.41	0.38
Fe	46.20	0.10	45.15	1.40	44.68	0.89	44.84	0.57
Mn	0.910	0.02	1.13	0.46	1.07	0.28	1.01	0.18
Cu	0.49	0.02	0.56	0.39	0.49	0.25	0.55	0.16
Mo	0.36	0.02	0.35	0.11	0.38	0.07	0.39	0.05
Ti	0.32	0.02	nd		nd		0.53	0.22

alloys. Measurement errors are quoted as two sigma precision data. Similar or better performance may be expected for other alloy groups, such as Co-based, Ti-based, or aluminium alloys. Note how a precision of analysis improves with measurement time. The instrument does not display element concentration as long as it is below limit of detection, which – being the function of measurement error – improves with measurement time. This is why, for example, titanium, copper and molybdenum data in the AISI 321 sample were not displayed until the measurement time reached close to 20 s. In general, we can say that for the

Table 6.8 Estimate of limit of detection in copper- and iron-based alloys.

	Cu, Copper Base			*Fe, Iron Base*	
Element	*X-ray tube 35 kV, 5 uA*	*Isotope 40 mCi ^{109}Cd*	*Element*	*X-ray tube 35 kV, 5 uA*	*Isotope 40 mCi ^{109}Cd*
Zn	0.30	0.30	**Ti**	0.33	0.34
Sn	0.18	0.08	**V**	0.17	0.25
Pb	0.04	0.03	**Cr**	0.13	0.17
Ni	0.10	0.32	**Mn**	0.29	0.35
Fe	0.04	0.12	**Co**	0.41	0.45
Mn	0.05	0.18	**Ni**	0.43	0.42
Cr	0.10	0.30	**Cu**	0.14	0.20
Bi	0.07	0.05	**Nb**	0.01	0.01
Zr	0.10	0.03	**Mo**	0.01	0.02
			W	0.11	0.28

major components of alloys one sigma error of measurement is about 1% relative, and about 10% relative for minor components.

Table 6.8 sheds light on limits of detection achievable with a portable X-ray analyser. Until recently, this has not been an important issue in alloy analysis since – as explained before – ready made alloys can be distinguished and sorted with the help of minor and major constituents. However, the ever growing environmental concerns of society have resulted in new restrictions on use of metals known to be harmful to humans. The most recent effort in this field was initiated by the European Union, which enacted a ban on using lead.[24] The consequence of this is a major switch taking place within electronic industry to lead-free solders. Therefore, we now have to monitor lead content in lead-free tin solders at levels way below 1000 mg kg^{-1}.

The limit of detection for each element listed was determined by measuring at least ten times an alloy sample that essentially did not contain the element of interest, yet the element was quantified. The standard deviation of the series of such ten measurements, multiplied by a factor of 3, is a measure of the Limit of Detection as quoted in Table 6.8.

6.6 Practical Issues to Consider

6.6.1 Radioisotope or X-ray Tube Excitation?

This question is probably the one most often asked by potential users of XRF analysers. As it turns out, the answer depends on more than one factor. Overall, however, the X-ray tube system is in most instances the preferred choice. The benefits of using an X-ray tube rather than radioisotope are many. For example:

- An X-ray tube allows for maximum optimization of excitation that cannot be afforded by radioisotope based devices. This in turn translates into the best possible analytical sensitivity and limits of detection.

- An X-ray tube generates a flux of photons that is 2–3 orders of magnitude more intense than that from the comparable radioisotope source. This allows for better precision and shorter measurement times.
- An X-ray tube system does not "slow down" with time; tubes do not suffer from natural radioactive decay as radioisotope sources do.
- An X-ray tube-based analyser does not emit radiation when turned off and not in use.
- Tube-based instruments are much easier to transport from one worksite to another due to less restrictive regulations.
- Unlike isotope-based instruments, tube-based analysers are not a source of radioactive waste in the form of decayed isotopes that have to be properly disposed off.

On the other hand tube-based analysers:

- require more power to operate,
- are electronically more complex than radioisotope counterparts,
- are somewhat larger and heavier,
- are not welcome in environments susceptible to fire or explosion hazards.

Therefore, to make a proper decision, one should consider all factors, and evaluate the benefits of performance along with logistics issues of proper training and security (isotope-based instruments must always be operated by specifically trained personnel, and securely stored when not in use). We now illustrate the advantages of X-ray tube excitation over radioisotope excitation with real life examples taken from the alloy scrap industry.

One of the notorious problems is the separation of stainless steel 304 from 321. The main difference between the two grades comes from titanium content, which in SS304 is typically present at 0.02% or less and at less than 0.40% in SS321. The difficulty arises from the fact that the relatively very small peak from the K_{α} line of Ti is superimposed on a high background tail of intense iron and chromium peaks, and to make the matter even worse it overlaps with the silicon escape peak due to the iron K_{α} peak. Figure 6.17 illustrates the point in case.

The spectra are shown on a logarithmic scale to highlight the details of the Ti region. Curves consisting of blue symbols show spectra collected at a standard setting of 35 kV, 5 μA current, and with a 1.5 mm thick Al primary filter. At this setting all elements up to molybdenum are excited. As can be seen, the difference between SS321 and SS304 in the Ti region is very small, although it corresponds to about 0.51% m/m Ti content. Also, the peak structure on SS304 spectrum in the Ti region is actually an escape peak of silicon caused by a major peak of the iron $K\alpha$ line. However, when the high voltage on the tube is reduced to 7 kV, which is below the absorption edge of iron at 7.111 keV, and the primary filter of Al removed, the situation drastically changes in favour of the Ti $K\alpha$ line. Now, iron is not excited at all, while titanium is excited with much better efficiency. This is shown by the red curves. Additionally, because of the absence of the imposing $K\alpha$ peak of iron, there is no silicon escape peak in the Ti region. Overall, by such a change of excitation conditions the titanium signal

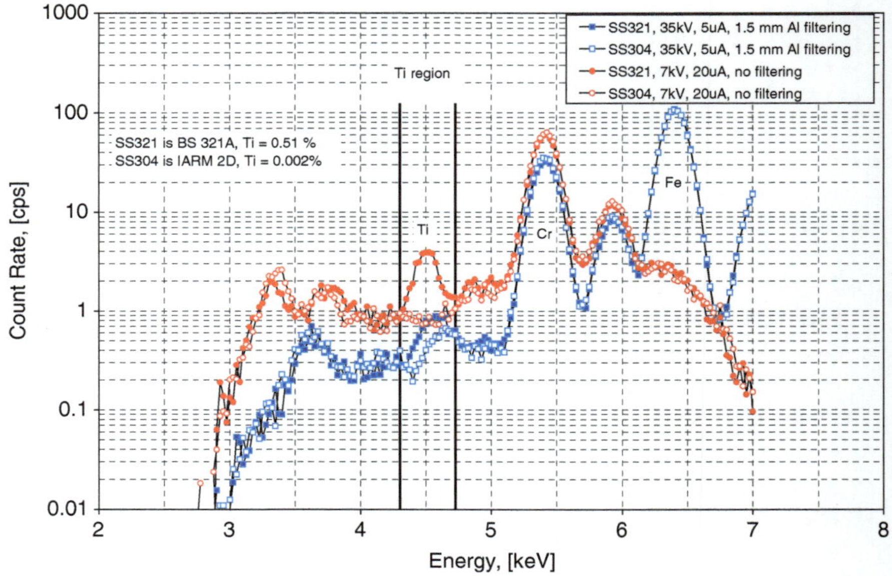

Figure 6.17 Example of optimized excitation for analysis of Ti in stainless steels.

is improved by a factor of 7. This kind of optimization can be achieved in many ways. For example, a high voltage of 20 kV and primary filter of chromium would also result in improved titanium analysis. Table 6.9 shows the results that can be obtained with such a setup.

Note that each measurement was taken in two steps. To be specific, the first measurement was performed for 2 s at 35 kV, followed by a second measurement for 3 s at a lower, optimized for Ti and V analysis high voltage. When we compare the results for the first two samples from Table 6.9 with the results for the same two samples listed in Table 6.6 we will see that the results in Table 6.9 are much, much better; the error of measurement for Ti reported in Table 6.9 is an order of magnitude smaller than that in Table 6.6.

Another of the many examples of difficult alloy identification is a pair of Zr-based alloys used in the nuclear power industry. These are known under trade names of Zr-2 and Zr-4 (or Zircaloy-2 and Zircaloy-4, respectively). They are used for the production of fuel "rods" for nuclear reactor cores. The fuel rod has a form of a tube capped at both ends, containing sintered pellets of nuclear fuel such as enriched UO_2. The tube of the fuel rod is made of one of the Zircaloys. Zirconium is used because of its very low absorption cross-section for thermalized neutrons. In addition, it has good corrosion resistance in superheated water. The essential, typical elements these two alloys are composed of are listed in Table 6.10.

The exact specifications can be found in UNS.[7] There, we will find that Ni in Zircaloy-2 has an allowable range of 0.05–0.08% m/m, while Zircaloy-4 is allowed to have maximum of only 0.0070% m/m Ni. The key functional difference between the alloys is due to nickel, which tends to absorb hydrogen

Table 6.9 Performance of XRF analyser with X-ray tube for alloys with low concentrations of Ti and V. (Data courtesy of Volker Thomsen, Niton LLC, NITON Model XLt 800 series.)

Analytical Performance for Some Low Ti and Low V Alloys

SAMPLE:	RM - BS81E		Portable Analyzer Results after:					
AISI 304	Cert	Error	2 + 3 Sec	+/−2σ	5 + 5 Sec	+/−2σ	10+10Sec	+−2σ
Grade ID →			**304**		**304**		**304**	
Cr	18.31	0.13	18.08	0.47	18.36	0.32	18.27	0.23
Ni	9.52	0.06	9.67	0.82	9.27	0.55	9.59	0.40
Mn	1.73	0.05	1.71	0.53	2.04	0.37	1.85	0.26
Mo	0.38	0.02	0.40	0.05	0.40	0.04	0.41	0.03
V	0.14	0.02	nd		0.19	0.04	0.14	0.03
Ti	0.00	n/a	0.01	0.04	0.00	0.04	0.01	0.02
SAMPLE:	**RM - BS321A**		**Portable Analyzer Results after:**					
AISI 321	Cert	Error	2 + 3 Sec	+/−2σ	5 + 5 Sec	+/−2σ	10 + 10 Sec	+/−2σ
Grade ID →			**321**		**321**		**321**	
Cr	17.20	0.10	17.29	0.42	17.31	0.30	17.42	0.22
Ni	9.38	0.08	9.77	0.83	9.35	0.54	9.48	0.40
Mn	1.22	0.03	nd		1.45	0.34	1.32	0.25
Cu	0.28	0.02	nd		0.37	0.16	0.37	0.12
Mo	0.20	0.01	0.22	0.04	0.21	0.03	0.21	0.02
Ti	0.51	0.02	0.50	0.07	0.53	0.07	0.53	0.05
SAMPLE:	**RM - IARM 52B**		**Portable Analyzer Results after:**					
K Monel	Cert	Error	2 + 3 Sec	+/−2σ	5 + 5 Sec	+/−2σ	10 + 10 Sec	+/−2σ
Grade ID →			**K500**		**K500**		**K500**	
Cu	30.21	0.05	29.84	0.74	29.97	0.53	30.11	0.39
Fe	0.77	0.05	0.83	0.14	0.81	0.10	0.80	0.08
Mn	0.77	0.04	0.85	0.18	0.78	0.13	0.77	0.09
Ti	0.49	0.03	0.52	0.08	0.51	0.06	0.54	0.05
P91/F91	**RM - IARM 238A**		**Portable Analyzer Results after:**					
P91/F91(9Cr + V)	Cert	Error	2 + 3 Sec	+/−2σ	5 + 5 Sec	+/−2σ	10 + 10 Sec	+/−2σ
Grade ID →			**P91 + V**		**P91 + V**		**P91 + V**	
Cr	8.24	0.11	8.29	0.25	8.07	0.17	8.19	0.13
Mo	0.94	0.03	0.97	0.09	1.00	0.05	1.01	0.04
Mn	0.40	0.02	nd		nd		0.50	0.19
Nb	0.086	0.01	0.09	0.03	0.08	0.02	0.09	0.01
V	0.22	0.03	0.24	0.05	0.26	0.04	0.23	0.03

so that Zircaloy-4 absorbs less hydrogen than Zircaloy-2 when subjected to high temperature water corrosion.

From an analytical point of view, there is no chance to reliably and repeatedly measure the difference between 500 mg kg^{-1} nickel and no nickel using standard excitation conditions of 35 kV, which generates huge Zr peaks. However, again, by reducing the tube high voltage to 17 kV, below the excitation edge of zirconium at 17.998 keV, we create preferential excitation conditions for nickel. This is illustrated in Figure 6.18.

These two examples clearly illustrate the advantage of X-ray tube over isotope source instruments. This kind of optimization of excitation conditions to

Table 6.10 Elemental composition of the alloys Zircaloys 2 (R60802) and 4 (R60804).

	Zircaloy-2 (R60802)	*Zircaloy-4 (R60804)*
Sn	1.5% m/m	1.5% m/m
Fe	0.12%	0.18%
Cr	0.01%	0.01%
Ni	**0.05%**	**none**
Zr	rem.	rem.

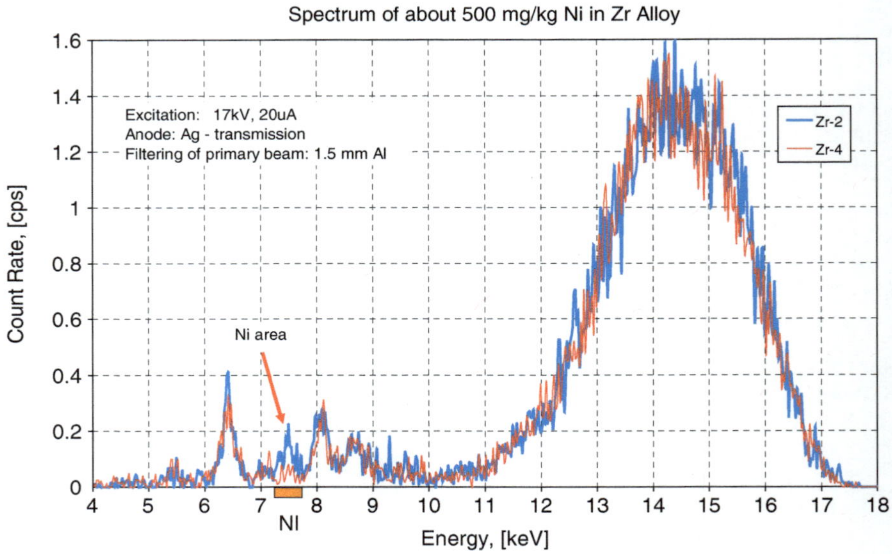

Figure 6.18 Identification of Zr-2 and Zr-4 alloys by preferential excitation of nickel, using an X-ray tube analyser.

particular analytical problems is what makes the X-ray tube so much more desirable than radioisotope sources which cannot change the energy of emitted radiation and are intrinsically limited by self-absorption in their output intensity.

6.6.2 Sample Condition

The penetration range of X-rays in metals is relatively very short, specifically in steels, copper and nickel-based alloys. Figure 6.19 shows, for the five most common alloy matrices, the maximum depth from which the characteristic X-rays of a particular element can reach the surface of metal and be detected. As we can see, in alloys based on iron, nickel or copper, we reach at best only the first 500 μm of thickness when analysing for elements from Al through Sb, and only 70 μm when looking for heavier elements that are analysed *via* their L-series lines. In other words, for practical reasons, XRF analysis of alloys is a "*surface*" technique. For example, there are many copper and titanium alloys

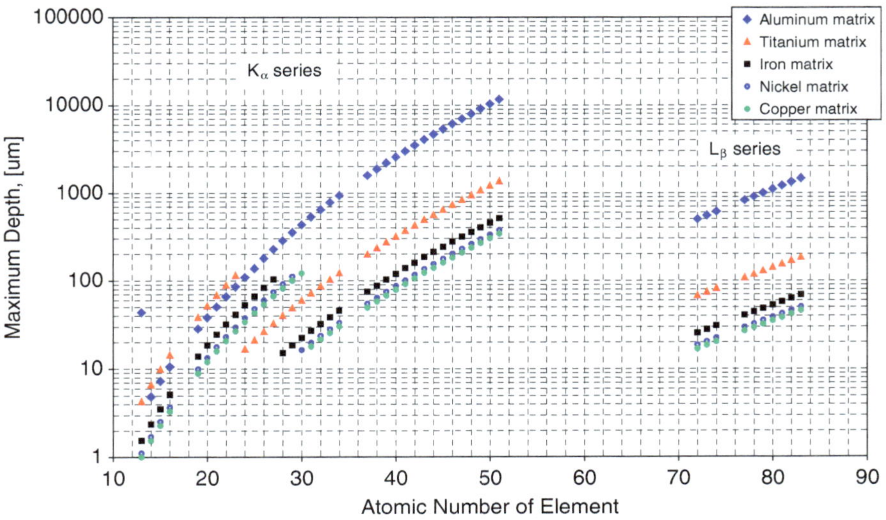

Figure 6.19 Maximum depth from which characteristic X rays of elements can reach the surface of an alloy from the five most common alloy matrices.

that contain up to 10–15% m/m of Si and/or Al. Figure 6.19 tells us that for these analytes we are collecting a signal from just a few micrometers off the surface. This implies at least two conclusions:

1. The homogeneity of the alloy is critical as the layer thickness from which we collect the signal is comparable with the grain size of the alloy under investigation.
2. The surface preparation technique may be source of diffraction of X-rays of elements which can distort the results of the analysis.

The surface of an analysed alloy must be kept free of debris such as scale, alloy dust (such as from grinding, welding), paint, oil residue. When analysing alloys with XRF one must be aware of the fact that alloys may be plated or clad with another metal.

Surface preparation technique is extremely important when analysing leaded brasses and bronzes. This is because lead, being much softer than copper, tends to smear over the alloy surface when rubbed, thus creating a film of lead overlay, which obviously will result in an inflated lead concentration.

Another important factor is the thickness of the analysed sample. The data in Figure 6.19 indicate that if we would analyse the composition of an alloy from samples thinner than the maximum depth shown in the graph, the results would depend not only on actual composition but also on sample thickness, a very undesirable effect. To prevent this effect, one needs to stack thin samples to create a thickness of material that exceeds the critical penetration depth. Similarly, if we are faced with analysis of small objects such as screws, pins or thin wires, we should take and place a number of them in a special sample cup and only then analyse them. Analytical programs developed for portable XRF analysers usually have a built-in degree of tolerance for sample size and/or shape that diverges from the ideal, large, flat and "infinitely" thick plate. Typically, it is possible to analyse reliably non-flat shaped objects such as rods, screws, down to about 3–5 mm in diameter. Similarly, if the surface is not flat, but the deviation from flatness is not greater than 3–4 mm the results will not be adversely affected. Manufacturers of portable XRF alloy analysers usually provide accessories that facilitate analysis of such non-typical samples.

6.7 Summary and Conclusions

Despite the fact that first commercially available portable XRF instruments were designed about 35 years ago, they did not gain much ground until the end of the last century. The real explosion in their popularity has its roots in the successful integration of the analyser into a single, one-piece, light-weight package. This change and the possibility of replacing inconvenient isotopes with miniature X-ray tubes increased the attractiveness of the portable XRF method and in many ways redefined the market by making it ten times bigger than just few years ago. Alloy analysis and identification became the flagship application for portable XRF analysers. Finally, industry has realized and appreciated the intrinsic attributes of the XRF method, that it is truly non-destructive in character, its speed and instantaneous results, as well as its ruggedness and simplicity. This positive feedback has led to even wider use of XRF in field applications. Field-portable XRF is an example of an excellent balance between portability, ruggedness, speed and versatility, yet with uncompromised analytical performance often matching or exceeding that of laboratory systems.

Future improvements in field-portable XRF will be directed at turning the analysers more into problem solving devices that make decisions rather than producing numbers. They are already being equipped with wireless technology (Bluetooth) that links a portable device with a PC, and with GPS capabilities for linking the measurement results with its physical location. One may be assured that any future progress in electronics and detector technology will find a way into portable devices, XRF analysers included.

Acknowledgements

The author thanks the many scientists and practitioners who generously shared the information and data used in this chapter.

References

The list of references is deliberately short, simply because it would take a whole chapter to give a full listing. Readers interested in the intricacies of XRF analysis as such will benefit from the excellent book *Handbook of X-ray Spectrometry*, ed.

R. VanGrieken and A. Markowicz, Marcel Dekker, New York, 2002. A multitude of various papers on alloy analysis *via* XRF techniques may be found in *Advances in X-ray Analysis*, which is a series of the proceedings of the annual "Denver X-ray Conference", which started in 1951. The proceedings on searchable CDs are available from the International Centre for Diffraction Data, 12 Campus Boulevard Newtown Square, PA 19073-3273 U.S.A. Some of the most recent ones may be viewed on the web site of the Center at: http://www.dxcicdd.com/. The references listed below are those that the author of this chapter considers as the most relevant for the development of portable XRF alloy Analysers.

1. ISO 9000 www.iso.org.
2. A. Holmes, *Rapid Spot Testing of Metals and Alloys*, ASM International, Materials Park, OH 44073, 2002, (ISBN 0-904477-25-8).
3. R.D. Brown, W.D. Riley and C.A. Zieba, *Rapid Identification of Stainless Steel and Superalloy Scrap*, U.S. Bureau of Mines Report, RI 8858, 1984.
4. F.X. Spiegel and E. Horowitz, *Instruments for the Sorting and Identification of Scrap Metal*, The John Hopkins University Center for Material Research, Oct. 15, 1981.
5. R. Newell, R.E. Brown, D.M. Soboroff and H.V. Makar, *A Review of Methods for Identifying Scrap Metals*, U.S. Bureau of Mines Report, IC 8902, 1982.
6. *Woldman's Engineering Alloys*, 7th edn, ed. John P. Frick, ASM International, Materials Park, OH 44073, 1990 (ISBN 0-87170-408-0).
7. *Metals and Alloys in the Unified Numbering System*, 8th edn, (SAE HS-1086 JAN99, ASTM DS-56 G), Society of Automotive Engineers, INC., 400 Commonwealth Drive, Warrendale, PA 15096-0001, USA, 1999; (ISBN 0-7680-0407-1, later editions are available too).
8. *Stahlschlüssel*, ed. C.W. Wegst, Verlag Stahlschlüssel Wegst GmbH, D-7142 Marbach, 1989; (ISBN 3-922599-07-9, later editions are available too).
9. http://www.key-to-metals.com.
10. S. Piorek and J.R. Rhodes, Application of a microprocessor based portable analyser to rapid, nondestructive alloy identification, Proceedings of International Conference and Exhibit, ISA 86 Houston, Texas, October 13–16, 1986, Instrument Society of America 1986.
11. L. Von Hamos, *Ark. Mat. Astron. Fys.*, 1945, **31a**, 1.
12. J.P. Willis and G.R. Lachance, Comparison between some common influence coefficient algorithms, *X-Ray Spectrom.*, 2004, **33**, 181–188.
13. H.J. Lucas-Tooth and B.J. Price, *Metallurgia*, 1961, **64**, p. 149.
14. Niton, L.L.C, Billerica, M.A., http://www.niton.com, January 2005.
15. Analytical Reference Materials International, 700 Corporate Circle, Suite A, Golden, CO 80401, USA, http://www.armi.com, January 2005.
16. Thermo Electron Corporation, http://www.thermo.com/com/cda/product/detail/1,1055,22789,00.html, January 2005.
17. Amptek Inc., 6 De Angelo Drive, Bedford, MA. 01730 U.S.A. http://www.amptek.com/5x5x500.html.

18. MOXTEK™, Inc., 452 West 1260 North, Orem, Utah 84057 http://www.moxtek.com.
19. P.F. Berry and G.R. Voots, On-site verification of alloy materials with a new field-portable XRF analyser based on a high-resolution HgI2 semiconductor detector, in *Non-Destructive Testing*, Proceedngs 12th World Conference on Non-Destructive Testing, Amsterdam, The Netherlands, April 23-28, 1989, ed. J. Boogaard and G.M. van Dijk, Elsevier, Amsterdam, 1989, pp. 737–742.
20. *Semiconductors for Room Temperature Nuclear Detector Applications*, Vol. 43 in Semiconductor and Semimetals Series, ed. T.E. Schlesinger and Ralph B. James, Academic Press, San Diego, 1995.
21. InnovX Systems, Woburn, MA, http://www.innov-xsys.com/, January 2005.
22. Brammer Standard Company, Inc. 14603 Benfer Rd. Houston, TX 77069-2895 USA http://www.brammerstandard.com/, January 2005.
23. NIST, 100 Bureau Drive, Stop 1070, Gaithersburg, MD 20899-3460. http://www.nist.gov/, January 2005.
24. European Parliament and Council Directive 2002/95/EC of 27 January 2003 on the restriction of the use of certain hazardous substances in electrical and electronic equipment (RoHS). *Official Journal* L 37, 13/2/2003 P. 0019–0023.

CHAPTER 7
Geochemical Prospecting

GE LIANGQUAN

Dept. Nuclear Technology Engineering, Chengdu University of Technology, No 1 Erxiangiao Dongsan Road, Chengdu, Sichuan 610059, China

7.1 Introduction

X-Ray fluorescence spectrometry was one of the first spectroscopic techniques that could be applied in the field for *in situ* analysis of rocks, soils and sediments. Applications of portable XRF analyzers in mining and mineral exploration were first reported in the mid-1960s.[1] After the introduction of on-board memory and microprocessors into the portable XRF units and with the development of room-temperature, high-resolution X-ray detectors, modern portable XRF analyzers have found wider acceptance and applications in this area. Field portable X-ray fluorescence (PXRF) analyzers can be used in every stage of mineral exploration and exploitation. They can be used for *in situ* analysis, where the PXRF probe is directly placed on the surface of natural rock, soil or sediment. They can also be used in a field mobile laboratory for the analysis of fully or partially prepared soil or rock samples. Special forms of portable XRF analyzer can be utilized as a logging probe for *in situ* use in a drill hole for the determination of elemental concentration or ore grade.

The significant features of PXRF analysis are simplicity, speed of operation, flexible requirement for sample preparation, *in situ* measurement capability, immediate availability of analytical results and the non-destructive nature of determinations.

This chapter discusses the capabilities and the relevant techniques of PXRF in geochemical prospecting. Applications of PXRF technique in mining and mineral processing are also included.

Portable X-ray Fluorescence Spectrometry: Capabilities for *In Situ* Analysis
Edited by Philip J Potts and Margaret West

7.2 *In Situ* PXRF Analysis

In geochemical prospecting, soil, rock and sediments are mostly targeted as sampling media. The concentration of elements associated with mineralization in soil, rock and sediments is important information for finding ore bodies. A field portable XRF analyzer can be used for the *in situ* determination of elements in such media. A large analytical error can usually be tolerated, and in some cases the relative concentration of elements or even the intensity of characteristic X-rays of target elements may be acceptable. The significant advantages of the PXRF technique over traditional laboratory-based methods are not only a saving in the expense of labor for sampling, transportation of samples, grinding and sample preparation, but also a saving the time, so that geochemical anomalies can be identified, located and evaluated in the field.

7.2.1 Natural Soil

Portable XRF instrumentation has been used to analyze natural soil *in situ* for elements such as potassium, calcium, titanium, chromium, manganese, iron, nickel, cobalt, copper, zinc, lead, arsenic, strontium, molybdenum, tin, silver, antimony, barium, *etc*. In these cases, portable XRF analysis is usually used at the reconnaissance geological survey stage or when a detailed geological study is being undertaken to look for possible mineralized bodies, where the background concentration of a target element in soil is sufficiently high to be detected by PXRF analysis.

The main factors that affect the results are moisture, density and the mineral grain size of the soil. The density of soil is usually controlled by means of pressing the area of soil to be analyzed with a spade. Changes in soil moisture can cause up to 10–20% relative errors.[2]

Many successful applications in geochemical prospecting have been obtained, even when using instrumentation that incorporates X-ray detectors with relatively poor energy resolution, such as a NaI(Tl) scintillation counter or proportional counter.[3–7]

Zhou *et al.*[3] have reported the successful application of PXRF for defining copper mineralization in a copper prospecting area in China. The portable XRF analyzer incorporated a NaI(Tl) scintillation counter with an energy-balanced filter, a ^{238}Pu isotope excitation source and a double-channel pulse height analyzer. The results were recorded as the counts under two characteristic peaks in a given energy range or the ratio of the counts in the first channel to those in the second. Since the detection limit of copper in soil is about 50 $\mu g\,g^{-1}$, which is greater than the background concentration of copper (about 20–50 $\mu g\,g^{-1}$), it is difficult to locate a copper mineralized area by determining the copper concentration directly in soil. As a result of geological investigations, some occurrences of copper mineralization were found to be associated with an ore-bearing zone formed typically by medium-high temperature hydrothermal processes. Galena, sphalerite and pyrite are then found to associate with and accompany chalcopyrite. The PXRF instrument incorporated a Co/Se energy-balanced filter with an energy range from 8.00 to 12.00 keV, permitting the detection of the total intensity of the Cu Kα (8.01 keV), Zn Kα (8.64 keV) and Pb Lα (10.56 keV) characteristic X-rays. Figure 7.1 shows the transmission characteristics of a Co/Se filter and typical XRF spectra. The difference between the area of the peaks recorded with the Se filter and with the Co filter in a given channels is a measure of the total intensities of characteristic X-rays of the target elements. To overcome the absorption of iron on the Cu Kα, Zn Kα and Pb Lα characteristic X-rays, the ratio of the total intensities of characteristic X-rays

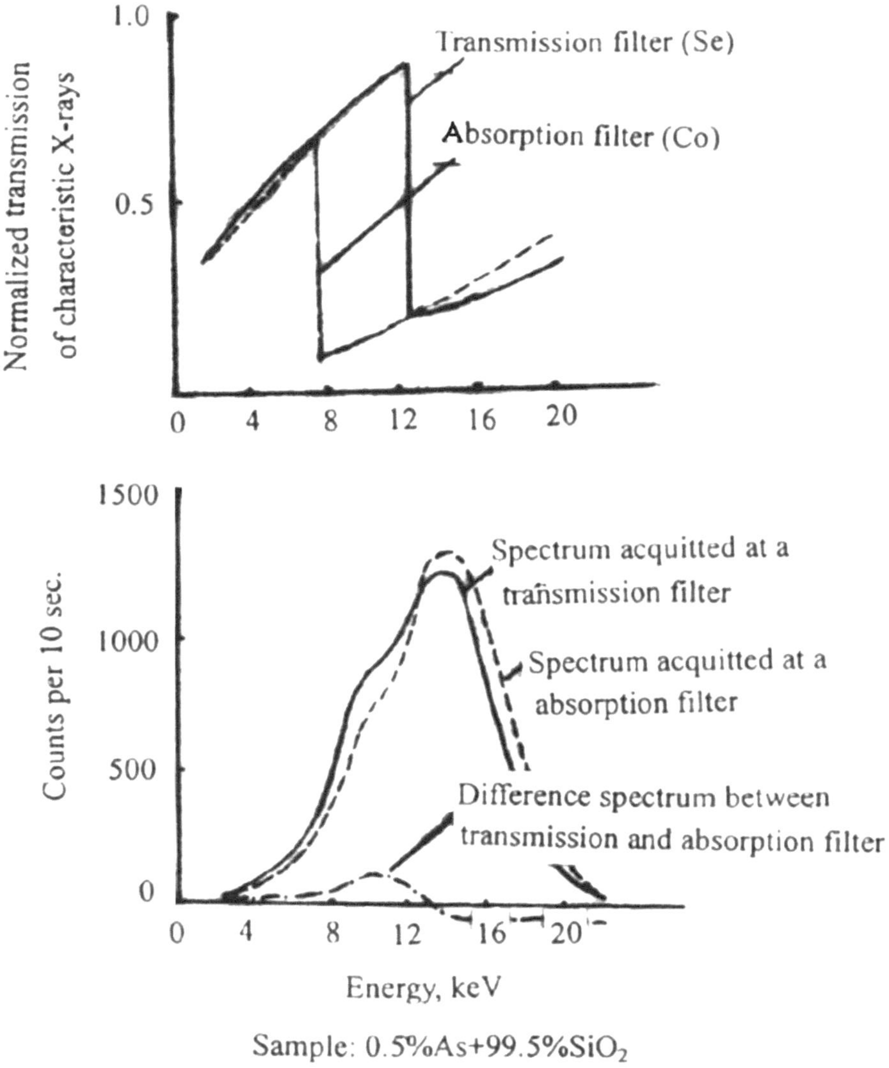

Figure 7.1 Transmission characteristics of Co/Se filter and typical XRF spectra.

Table 7.1 Statistical results of *in situ* PXRF measurements in soil at a copper prospecting area in northern China.

	Ratio of total intensities of Cu K, Zn K, Pb L characteristic X-rays to the intensities of scattered radiation		
Section number	*Range of this ratio in the background area*	*Average of this ratio in the mineralized area*	*Description*
1	0.9–3.0	6.4	The average background value: 1.8
2	0.6–2.8	8.2	The threshold value for
3	0.7–2.7	12.5	an anomaly: 3.0

(in the first channel) to that of scattered X-rays in the energy range 13–17 keV from ^{238}Pu primary radiations (in the second channel) was taken as the measurement parameter. Results in the prospecting area showed that the average background ratio was 1.8, with the measured range of 0.6–3.0. Three anomalies were mapped by contoring at a threshold ratio of 3.0. Table 7.1 shows the statistics of the results on these three anomalies in the copper prospecting area. As can be seen, the ratio of the anomalies is three to seven times that of the background.

Zhang *et al.*[4] applied a portable XRF analyzer with a sodium iodide [NaI(Tl)] scintillation counter, a ^{241}Am source and a Ag/Cd energy-balanced filter whose energy transmission band is from 25.6 to 26.8 keV to measure *in situ* the intensity of antimony characteristic X-rays in soil at a depth of 0.3 m to locate potential antimony deposits at the Guang Yang antimony deposit, Guangxi province, China. Zhai[5] used a portable analyzer for the *in situ* determination of barium and tin in soil at a tin and celestite prospecting area in Yunnan province, China. Liu[6] reported two successful examples of the application of PXRF at a tin exploration area in Sichuan province, China.

Ge *et al.* applied the IED-2000P PXRF analyzer (Figure 7.2) to the *in situ* determination of elements in soil and sediments in three copper prospecting areas in Yunnan, China. This instrument was fitted with a ^{238}Pu sealed source (4.44×10^8 Bq), a Si-PIN X-ray detector and a 2048-channel MCA unit. Nine elements, potassium, calcium, manganese, iron, copper, zinc, arsenic, lead and strontium, were analyzed simultaneously using a count time of 200 s. The detection limit of this instrument for some mineralized elements was about 5–10 $\mu g\, g^{-1}$ (Table 7.2). Significant differences were found between *in situ* elemental concentrations in soil measured before and after precipitation with the IED-2000P PXRF analyzer (Figure 7.3). The concentration of copper and zinc measured after rain decreased by about 20% relative, due to the absorption and scattering of X-rays by water in the soil.

The PXRF technique has also been used in gold mineral prospecting.[6,7,9–11] Although the gold abundance in rock or soil (about 1 $ng\, g^{-1}$) is far less than the detection limit of PXRF, some sulfophile elements closely geochemically

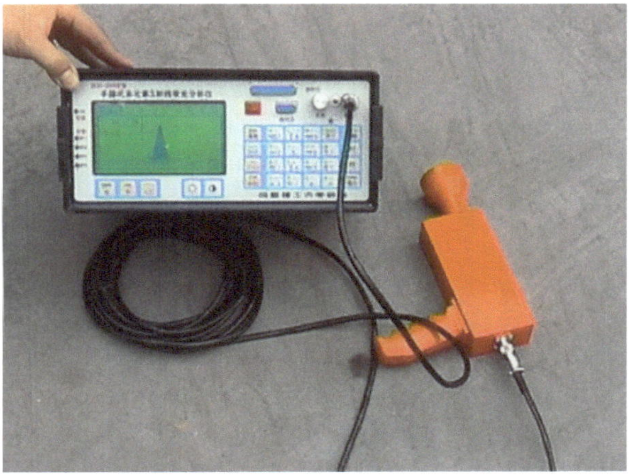

Figure 7.2 IED-2000P PXRF analyzer which incorporates a Si-PIN detector

associated with gold mineralization, such as arsenic, copper, zinc and lead, can be analyzed in the field by PXRF instruments. In the 1980s and 1990s, more than ten large-scale gold deposits in China were directly or indirectly found or evaluated by such a technique.

Depending on the characteristics of mineralization and the behavior of associated elements in soils and any additional requirements of the geochemical prospecting program, the sampling methodology for *in situ* PXRF analysis should include consideration of the following:[12]

(1) The optimum depth of measurement: Soil can be sub-divided from the surface of the soil to the bedrock into four layers, marked as A, B, C and D (Figure 7.4).

The "A" layer is known as the humic soil layer. This layer is heavily influenced by the activities of animals, insects and plants. It is generally less than 400 mm thick and is mostly black or gray-black.

The "B" layer is called the leached layer or passage layer.

The "C" layer is known as the weathered layer. It consists of residual material and pieces of broken rock.

Finally, the "D" layer is the bedrock.

One finds that these four layers are not fully developed everywhere. In some areas, the A layer is absent, and in other areas the C or B layers are absent. The optimum layer for *in situ* PXRF analysis for geochemical prospecting is the C layer. In some places, it is difficult to access the C layer with a simple stainless-steel trowel since the A and B layers could be more than 1 meter thick. If this is the case, the B layer should be selected. Measurement of the A layer is always avoided, since some elements in the A layer readily migrate and are easily absorbed by plants, animals or are affected by other factors thereby disturbing the primary distribution of the elements of interest.

Table 7.2 Detection limits (μg g^{-1}) of IED-2000P PXRF analyzer and typical abundance in some rocks. (Abundance data of elements are from Liu Yinjun *et al.*, 1984.[8])

Element	*Cr*	*Mn*	*Fe*	*Cu*	*Zn*	*As*	*Sr*	*Ag*	*Sb*	*Sn*	*Ba*	*Pb*
Detection limits	30–100	30–100	30–100	10–20	10–20	10–20	50–100	10–30	10–30	10–30	100	10–20
Clarke value of the crust	100	950	5.6	55	70	1.8	375	0.07	0.2	2	425	12.5
Magmatic rock	1–36000	500–3000	14 000–94 000	27–59	48–100	10–24	141–270	50–100		<1–600	3–4000	0.01–>600
Sedimentary rock				30–35	0.1–150	5–22	245–2075	0.19–2.2		0.1–40	10–5000	0.01–400
Metamorphic rock				30–35			315–388	0.3–1.0		0.7–21	90–1500	0.1–100

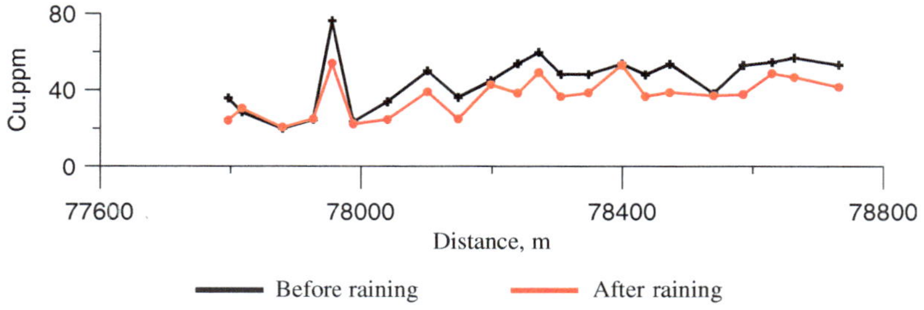

Figure 7.3 Concentration of copper in soil measured *in situ* before and after raining using a IED-2000P PXRF analyzer.

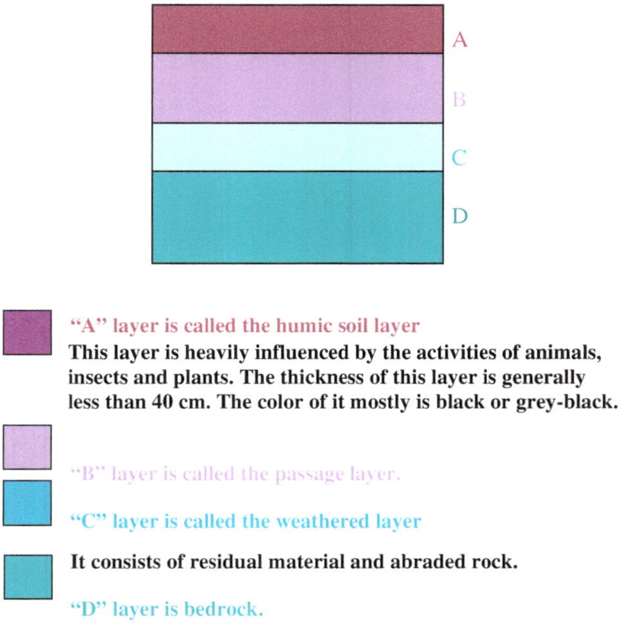

Figure 7.4 Layers of soil from the soil surface to the bedrock.

Therefore, the first step for *in situ* PXRF measurement in soil is to remove the A layer soil to a depth of up to 400 mm with a stainless-steel trowel to expose the B or C layer soil.

(2) The second step is to remove any large or non-representative debris from the surface of the B or C layer soil, such as rock, pebbles, leaves, roots *etc.*
(3) The third step is to make the soil surface as smooth as possible so that the probe window will make effective contact with the surface of soil.
(4) The fourth step is to tamp the soil with the stainless-steel trowel to increase soil density and compactness for better repeatability and representativity of the analysis.

(5) The last step is the *in situ* measurement with the portable XRF instrument following the manufacture's protocols.
(6) *In situ* measurements should be undertaken at least twice per sample location with the final results being the mean of two or three measurements.
(7) Another requirement is that the soil must not be saturated with water, especially during or after rainy days.

7.2.2 Natural Rock

Rapid assessment of the mineral concentration in a rock is highly desirable in geochemical prospecting, since the distribution of mineralized elements that form the so-called primary geochemical dispersion halo can used to predict the potential mineral body or concealed mineral body. Obviously, the value of a direct reading instrument is greatest when there are no obvious visible indications of the presence of minerals. In fact, the unprepared rock surface represents the most challenging situation for measurements by PXRF analysis. The irregularity of the rock surface, heterogeneous distribution of minerals and variations in grain size affect, to a great extent, the measurement results. These effects may be partially overcome when determining high Z elements ($Z >$ ca.50) by the excitation and detection of K line X-rays. Under these conditions grain size effects are likely to be least, larger surface irregularity effects can be tolerated and the greater penetration depth of characteristic X-rays enables a more representative sample to be examined. In most reported cases the reliability of results should be improved by increasing the number of measurements.[1]

A few studies and discussions on the surface irregularity effects have been reported.[13–15] In fact, the effects of surface irregularity are difficult to correct and the variations in surface morphology always exist since the surface of the ore cannot be polished before measuring. Potts *et al.*[15] investigated discrepancies that arose from surface irregularity in the field analysis of geological and archaeological rock samples by PXRF spectrometry. To overcome these discrepancies, the measured intensity was normalized by multiplying it by the ratio of the scatter peak intensity from a compositionally similar flat reference sample to the scatter peak intensity measured from the sample itself. The scatter peak data were obtained from ^{55}Fe, ^{109}Cd and ^{241}Am sources incorporated in the instrument used for this investigation and the ^{55}Fe scatter peak intensity was favored for this correction (Figure 7.5). Under controlled conditions, this correction proved to be successful in compensating for effective air gaps of up to 3 mm in the analysis of the K lines of higher atomic number elements (Rb, Sr, Y, Zr, Nb, Ba) and up to 1 mm for the Fe K line. Principal limitations to the application of this method to larger air gaps were (a) the change in scatter angle and, therefore, relative scatter intensity as the air gap was increased and (b) the increasing contribution from scatter in air, particularly to the measured ^{55}Fe scatter peak at large air gaps between sample and analyzer.

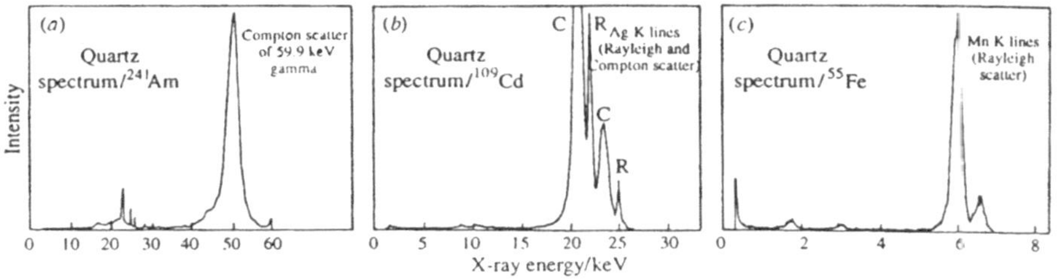

Figure 7.5 Spectrum of quartz excited with (a) ^{241}Am, (b) ^{109}Cd and (c) ^{55}Fe excitation sources, showing the respective Compton and Rayleigh scatter peaks.

Potts *et al.*[16] also investigated the effect of mineral grain size in rocks. Based on replicate measurements on five rock slabs/blocks (dolerite, quartz andesite, micro-granite, medium-granite and coarse-granite), the results were evaluated to determine the number of individual measurements that must be averaged to achieve relative standard deviations of the mean of 2, 5, 10 and 20% (Table 7.3). The data in Table 7.3 show that determinations can be made with sampling precisions (expressed as the relative standard deviation of the mean) of better than 10% (almost 5% in the case of the dolerite) on the four fine- to medium-grained samples listed. If five determinations are averaged, the 5% criteria are met for these samples (almost 2% for the Whin Sill dolerite). However, an average of five determinations on the coarse-grained Shap granite will only offer an average sampling precision of 20%. These data illustrate the limitations of sampling precision that affect *in situ* PXRF measurements on crystalline silicate rocks and, no doubt, demonstrate the importance of undertaking similar measurements on representative samples before the technique is used in new applications.

To obtained good performance when undertaking PXRF analysis on natural rock surfaces, the following points should be adopted in the analytical protocol:[12]

1. The most important point to keep in mind for *in situ* measurement by PXRF on primary rocks is that the measurement surface of rock must be "fresh" since the effective detection depth of PXRF analysis is only about 1 mm or less. Fractured surfaces of rock should be avoided as the measurement surface may be covered with oxide minerals and secondary concentrations of elements of interest.
2. The second requirement is that the surface of rock should be made as flat and smooth as possible with a hammer so that the probe window will make as effective a contact as possible with the surface.
3. The last requirement is to undertake three to five measurement points at a distance of 0.5 to 1 m apart in the area of interest so as to obtain a representative result.

There are many successful applications of the PXRF technique for outlining geochemical primary haloes or zones of elements associated with an

Table 7.3 Number of determinations that must be averaged to estimate the bulk composition to a specified precision (in terms of relative standard error of the mean).

Number of determinations required to achieve a precision of:	*Dolerite (Whin Sill)*		*Quartz andesite (Imperial porphyry)*		*Micro-granite (Threlkeld)*		*Medium-grained granite*		*Coarse-grained granite*	
	Mean	*Range*	*Mean*	*Range*	*Mean*	*Range*	*Mean*	*Range*	*Mean*	*Range*
20% S.D. of mean	0.1	0.01–0.2	0.2	0.1–0.4	0.2	0.1–0.4	0.2	0.1–0.3	5.1	0.7–21.7
10% S.D. of mean	0.3	0.05–0.8	0.8	0.3–1.5	0.9	0.5–1.5	0.7	0.3–1.2	20.5	2.7–86.6
50% S.D. of mean	1.2	0.2–3.1	3.1	1.0–6.1	3.6	2.0–5.9	2.7	1.0–4.9	81.9	11.0–347
2% S.D. of mean	7.3	1.2–19.5	19.6	6.3–37.9	22.5	12.8–36.7	16.7	6.4–30.9	512.0	68.7–2166

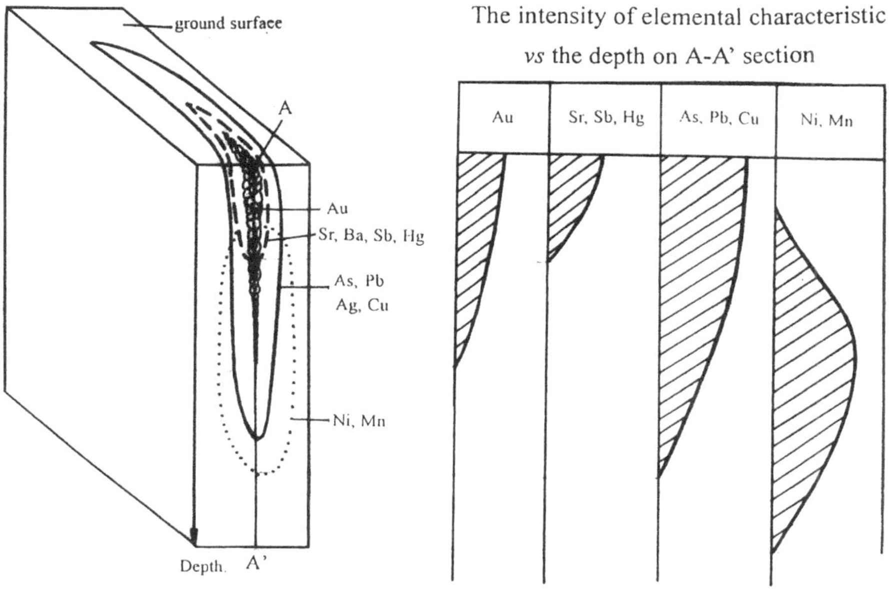

Figure 7.6 Distribution of elements around a gold body, based on the characteristic intensities of elements on the walls of tunnel and trenches measured *in situ* by a PXRF analyzer.

ore-body boundary.[6,7,17,18] In some cases, the operators only measured the relative intensities of characteristic X-rays from elements of interest. Figure 7.6 shows the distribution of elements around a gold body in Sichuan, China, which is based on the characteristic intensities of elements on the walls of tunnels and trenches measured with a PXRF instrument fitted with a proportional counter.[6] As can been seen, arsenic, lead, silver and copper are closely related and almost coincide with the gold body; strontium, barium, antimony and mercury are also associated with the gold body, but they appeared in front of the gold mineralization; conversely, manganese and nickel appear at the rear of the gold mineralization. Clearly, the position of the gold body can be predicted based on the distribution of the above elements.

Figure 7.7 illustrates the determination of the extent of a gold body identified by measuring the As characteristic K line with an PXRF analyzer, incorporated with a NaI(Tl) scintillation counter.[17] The results represent the wall of a trench at a gold deposit in Hebei province, China. Although the fracture zone with limonite was identified as the gold-bearing zone, the intensity of the As K line allowed the operators to locate the highest concentration of Au within this fracture zone, which was subsequently verified by the channel-chemical sampling method.

Figure 7.8 shows an evaluation of the extent of alteration of multi-element-bearing hydrothermal solution by *in situ* determination of the ratio of the

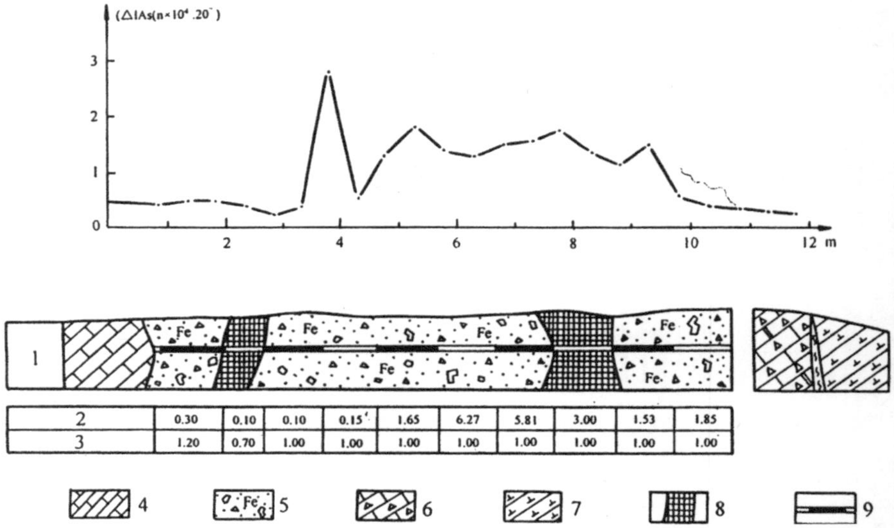

Figure 7.7 Determination of the extent of a gold body by measuring *in situ* the As K line with a PXRF analyzer. (1) Geological section; (2) gold grade, g $tonne^{-1}$; (3) length over which the sample was composited; (4) limestone; (5) fracture zone with limonite; (6) cataclastic limestone; (7) andesite; (8) gold body; (9) position of sample.

intensity of As K + Pb L + Zn K + Cu K characteristic lines to scattered radiation with a PXRF analyzer incorporating a NaI(Tl) scintillation counter and a Co/Se energy-balanced filter on the outcrop of a quartz vein at Dongga glod exploration area in Tibet.[18]

7.2.3 Drill Core and Borehole Logging

The PXRF analyzer offers an attractive analysis opportunity for the *in situ* determination of elemental concentrations on drill cores. As drilling is normally a continuous and expensive process, there is an obvious merit in having a rapid indication of the mineral grade in an ore-body so as to guide the drilling pattern in real time in the most effective manner.

In contrast to the measurement of unprepared rock surfaces on the ground, the geometric arrangement of the cylindrical drill core can be fixed in relation to the hand-probe (source and detector), noting that and the surface of the core is relatively smooth. Hence, errors due to surface irregularity are negligible. However, problems due to heterogeneity of the mineral deposit, variations in grain size and changes in matrix composition remain. Nevertheless, the accuracy of measurements on drill-cores should be higher than is possible on rock surfaces.

Some typical elements analyzed by PXRF in drill cores include iron, copper, lead, zinc, molybdenum, tin, antimony, barium, strontium and gold.[6,7] The relative error between laboratory-based and *in situ* PXRF analysis was

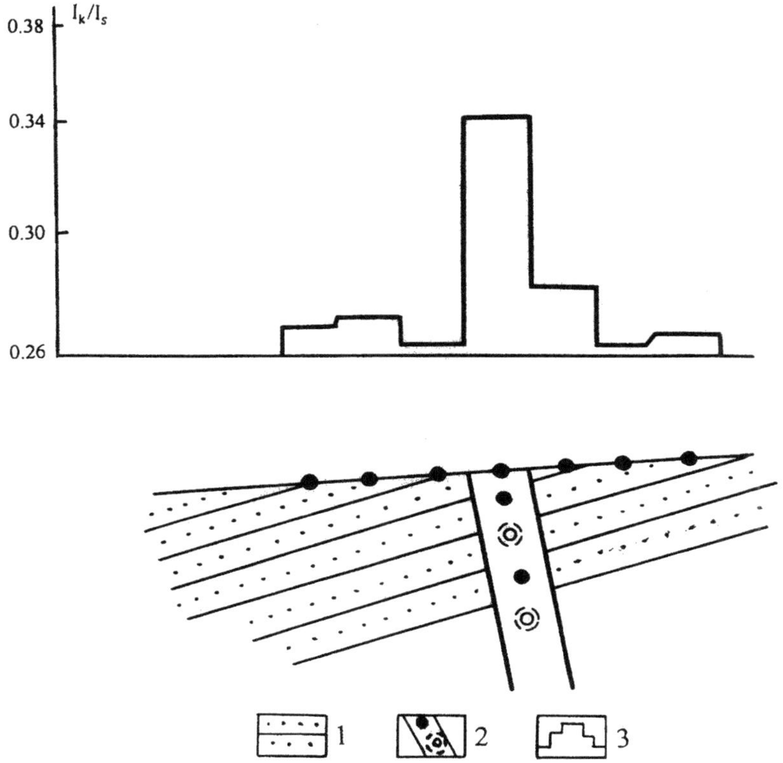

Figure 7.8 Evaluation of the extent of alteration by a multi-element bearing hydrothermal solution at the outcrop of a quartz vein by *in situ* PXRF technique. (1) Sand rock; (2) mineral-bearing quartz vein; (3) curves of the ratio of the intensities of As K + Pb L + Zn K + Cu K characteristic lines ratioed *versus* the scattered radiation (I_k/I_s).

within 20–25% in the range 3–0.3% m/m Cu and the average relative error was within 10%.

An alternative to the analysis of drill core for obtaining mineral concentrations is to use radioisotope X-ray fluorescence logging, by which an XRF borehole probe is placed directly down the borehole to determine *in situ* the mineral concentration on the wall of the borehole.

Earlier XRF probes incorporated a sodium iodide [NaI(Tl)] scintillation counter with an energy-balanced filter.[7,19,20] This probe was usually used to determine one elemental concentration per measurement. Later XRF probes incorporated a proportional counter filled with lower-pressure gas, which can determine the concentration of 2 to 4 elements per measurement.[21] XRF probes incorporating a Si(Li) semiconductor detector with cryogenic cooling have been developed and have a multi-element determination capability.

The main restriction on detection by XRF logging in a natural borehole is the presence of drill fluid (mud or water) between the wall of the borehole and the detection window of XRF probe. The mud or water, on the one hand, absorbs characteristic X-rays from primary radiation from the source and the elements of interest. On the other hand, it strongly scatters characteristic X-rays and primary radiation so as to increase the background counts in the detected spectrum. So, XRF logging appears practicable for elements of $Z > 50$ in wet boreholes.

An effective way of overcoming the influence of drill fluid is to ensure that the detection window of the probe is as close as possible in contact with the wall of the hole. Ge[22] reported a specially designed XRF logging probe. The detector and sources were placed in a short probe (about 450 mm) that was connected by two support arms to the main part of the XRF logging probe. The short probe with the detection window could be automatically positioned in close contact with the wall of the borehole. Figure 7.9 is a schematic diagram of the mechanical structure of this instrument. Experiments showed that the distance between the detection window and the wall of the borehole was less that 5 mm, provided that the wall did not collapse (diameter $<$ 250 mm) and was not broken over a length of $<$45 mm. If the wall of the borehole was smooth, the distance was generally less than 2 mm.

Ge[22] also proposed a technique for correcting the absorption and scatter of mud and water in XRF logging, using double scattered radiation with different energies. This procedure was tested by carrying out a series of measurements on the wall of an imitation borehole. Table 7.4 shows the mean of three independent determinations on different thicknesses of a water layer (0 and 6 mm thick). The standard deviations (SD) listed in this table were calculated for a 68% level of confidence. It was also found that the detection limits of XRF logging instruments for lower atomic number elements, such as, Fe, Cu, Zn, Ni, Cr and V, were heavily restricted in wet boreholes. For example, when the water layer was 3 mm thick, the detection limits of the XRF logging instrument for Pb or Zn was about 0.7% m/m; but when it was 5 mm thick, the detection limit was 3% m/m.

Some operations in natural boreholes were reported by Zhang[23] for tin, antimony and barium, Liu[24] for tin and Ge[21] for strontium and barium. Figure 7.10 is the result of XRF logging at the Xinglong strontium deposit in Sichuan province, China. As can been seen, there is a good correspondence in elemental concentrations between the *in situ* XRF logging method (column

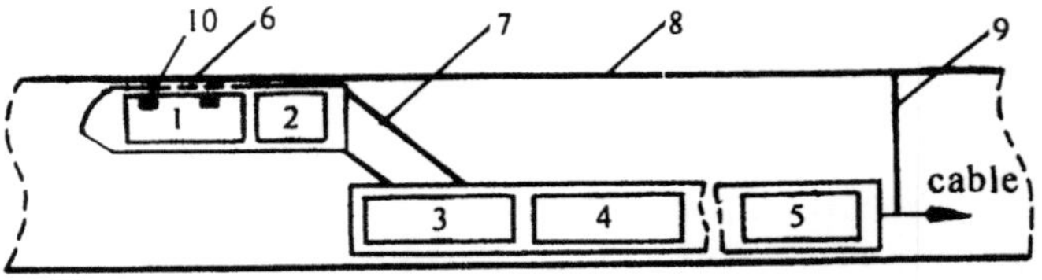

Figure 7.9 Schematic diagram of XRF logging probe. (1) X-ray detector; (2) charge-sensitive amplifier; (3) electric thruster; (4) system for amplifying and transmitting signal; (5) power supply; (6) detection window; (7) supporting arms; (8) borehole wall; (9) fixed arm; and (10) source.

Table 7.4 Results of correction for the influence of drilling fluids.

Lead-zinc model	*Zn (%) m/m* *Standard value*	*Measured value* *Mean*	*SD*	*Pb (%) m/m* *Standard value*	*Measured value* *Mean*	*SD*
PZ1	0.5	0.73	1.26	0.11	0.38	0.67
PZ2	0.1	0.39	0.78	0.5	0.99	0.71
PZ3	0.7	1.16	0.68	0.9	1.15	0.81
PZ4	1.1	1.23	1.25	1.5	1.07	0.93
PZ5	1.5	1.63	0.42	2.1	2.77	0.84
PZ6	2.4	3.23	1.80	3.0	3.25	0.81
PZ7	4.7	4.77	0.39	5.0	4.80	0.45
PZ8	7.5	6.59	0.08	8.0	7.78	0.95
PZ9	15.0	14.93	0.94	14.0	14.23	0.51
PZ10	1.1	1.99	0.76	10.0	10.27	0.85
PZ11	9.0	8.77	0.26	1.5	1.61	0.80

No. 5 in Figure 7.10) and a conventional core-sampling chemical analysis method (column No.4 in Figure 7.10).[21] The analytical results from XRF logging were obtained in real time in the field and that from the core-sampling chemical analysis were provided one month later.

7.2.4 Sediments

Measurements on natural stream sediments with presently available portable XRF analyzers are not satisfactory for most elements, especially measurements on the mapping scale of 1/200 000 or less. These usually require high accuracy, high precision, low detection limits (up to the order of 1 ng g^{-1}) with a requirement for the determination of over 70 elements. However, stream sediments in some mineralized regions or prospecting areas can be measured where the abundance of an element is high enough to be measured directly by PXRF analysis. The main factors affecting the measurement results are basically the same as those for natural soil measurements, *i.e.*, moisture, grain size, and surface irregularity effects. In most cases, stream sediments are partially or fully prepared for analysis, including drying, sieving with 60–80 mesh sieves and weighting so as to obtain repeatable measurements. Figure 7.11 compares the analytical results of steam sediment samples from a multi-metallic mineral prospecting area in Guangxi province, China, where data was obtained using both the PXRF analyzer with a NaI(Tl) scintillation counter and a conventional laboratory XRF method.[17]

An ability to measure quickly the elemental concentrations of marine sediments on the seabed or on board a survey vessel is another important requirement for marine mineral prospecting. A capability for obtaining these objectives has been achieved by the PXRF technique.

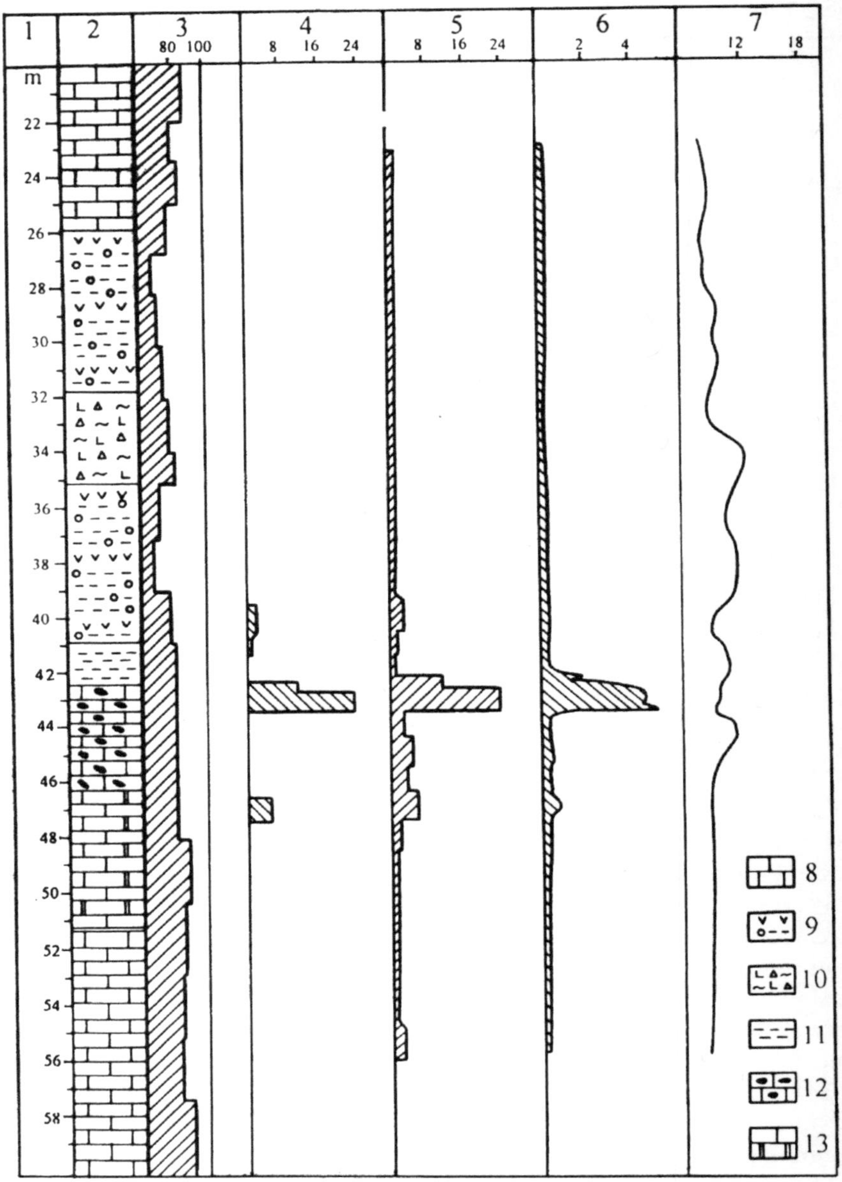

Figure 7.10 Comprehensive section of XRF logging at Xinglong strontium deposit in Sichuan, China. (1) Depth of borehole; (2) geological column; (3) percentage of core from the borehole; (4) concentration of strontium oxide by core-sampling chemical analysis; (5) concentration of strontium oxide by XRF logging method; (6) curve of the ratio of strontium characteristic line intensity to scattered radiation; (7) curve of borehole diameter; (8) limestone; (9) slurried material; (10) breccia; (11) mud rock; (12) dolomite with strontium oxide mineral; (13) dolomite.

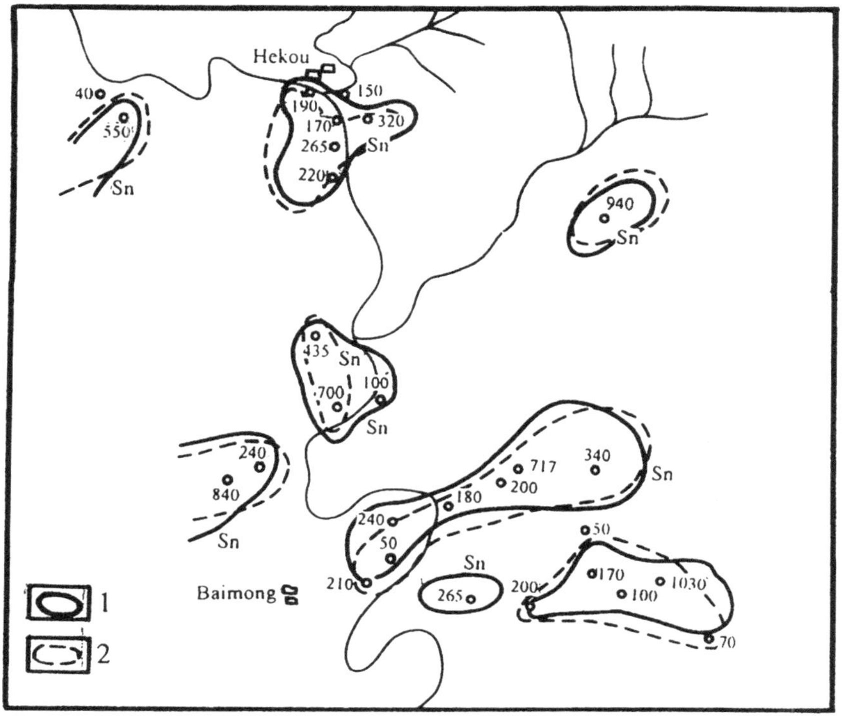

Figure 7.11 Comparison of the contour lines of tin concentration ($\mu g\, g^{-1}$) from steam sediment samples provided by a PXRF analyzer and a laboratory-based XRF method in a tin prospecting area. (1) Contour lines of tin concentration from PXRF analysis; (2) contour line of tin concentration from laboratory-based XRF method.

Ge[25] reported an XRF probe that could be used for the *in situ* determination of elements on the seabed. The probe incorporated a thermoelectrically cooled Si-PIN detector, a ^{238}Pu 222 MBq) or ^{241}Am (37 MBq) source, and a 1024 channel analyzer (Figure 7.12). The detection window of the probe was made of a 1.5 mm thick Be layer. The X-ray spectra formed by the analyzer were transferred to a portable computer on board the survey vessel through a 1000 m long cable. The probe was tested in the South China Sea for the determination of Ti, Mn, Fe, Cu, Zn, As, Pb and Sr in natural seabed sediments. Detection limits for these elements are about 20–100 $\mu g\, g^{-1}$.

7.3 Prepared Soil and Rock Samples

In mineral prospecting, a portable PXRF analyzer is usually used as a base station instrument in the field, operated in a mobile analytical laboratory. By undertaking appropriate sample preparation, we can obtain better performance for PXRF analysis than its use for *in situ* measurements.

Figure 7.12 Photograph of an ANTG-2000M PXRF probe for *in situ* determination of elemental concentrations in sediments on the seabed.

Since prepared soil and rock samples have been dried, ground and may have been passed through a 180 mesh sieve, the influence of mineral heterogeneity on analytical results should be insignificant. The main factors that influence analytical precision are likely to be matrix effects. In some cases, grain size effects may be another influencing factor, especially for roughly prepared soil or stream sediment samples, where the sample has only been passed through 30–80 mesh sieves.

Following the development of room-temperature, high-resolution X-ray detectors and the incorporation into electronic units of on-board memory and microprocessors, the modern PXRF analyzer can perform the complex data manipulations required for matrix correction methods that were previously only available in laboratory-based XRF instruments.[26] Among the different matrix correction methods, the Fundamental Parameters (FP) method is widely applied in modern PXRF analyzers that use HgI_2 and Si-PIN X-ray detectors.[2] Other matrix correction methods adopted include the Compton Scatter Normalization method (CSN), and the Influence Coefficient method (IC).[6] See Chapter 2.

Figure 7.13 shows the comparison between the analytical results of stream sediments samples from the Laba copper prospecting area in Yunnan province, China by PXRF analyzer and a conventional laboratory-based XRF method. As can be seen, good agreement of elemental concentration anomalies between analytical methods was obtained. The time taken for PXRF measurement was about 2 days (about 275 samples) and the analytical results were obtained at the field station. The analytical results by the laboratory-based XRF method

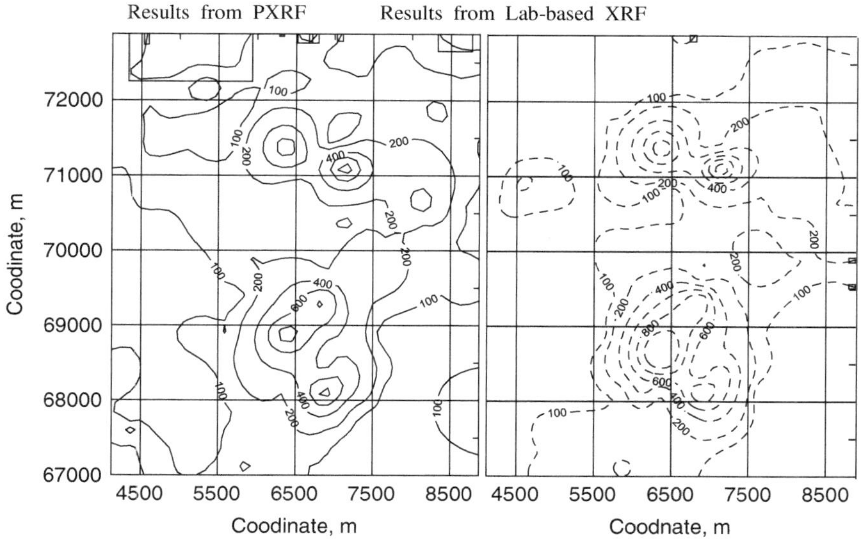

Figure 7.13 Comparison of the contour lines of copper concentration in steam sediment samples provided by a PXRF analyzer and a laboratory-based XRF method at a copper prospecting area (unit, $10^{-6}\,\mu g\,g^{-1}$).

were obtained after a delay of two months. The PXRF analyzer employed used a Si-PIN semiconductor detector and X-ray tube with a molybdenum anode. The stream sediments were dried, ground and passed through 80 mesh sieves. The concentrations of iron, manganese, zinc, arsenic and lead in samples were obtained within the same measurement. Based on these preliminary analytical results by PXRF measurement and the results of a geophysical induced-polarization measurement, nine survey lines in the northern-east and the southern-west of the prospecting area were laid out. *In situ* PXRF measurements on the soil surface with the IED-2000P PXRF analyzer incorporating a ^{238}Pu isotope source were performed immediately. The detailed distribution of copper in soil was revealed and two new significant geochemical anomalies were found. The time that elapsed between base camp analysis and *in situ* measurements was about one week.

PXRF analysis was applied successfully to many other geochemical samples in China in the 1980–1990s.[6,27,28] It is estimated that more than 50 geological survey teams or companies employed a PXRF analyzer for geochemical spectrometry during this period.

7.4 Applications in Mining

Applications of XRF in mining date back almost to the origins of portable X-ray analysis. They were the first applications for which the rugged PXRF analyzer was used. Basically, mining applications do not differ significantly in procedure from those used for geochemical exploration applications. In fact, mining applications may be even easier because the concentration of mineralized elements is higher (in the range of 0.1–30% m/m for most elements) and the matrix is simpler, and well suited to PXRF analysis. However, analytical accuracy and precision must be higher to meet the needs of mining production. Typical mining applications involve determining the ore grade and identifying the boundary of ore bodies at a mining site to ensure recovery of the appropriate ore grade material.

Surface irregularity, mineral heterogeneity and matrix effects are the major sources of error in quantitative PXRF analysis. The effects of surface irregularity are difficult to correct and variations in surface texture always exist since the surface of the ore cannot be polished before measurement. From the point of view of *in situ* PXRF analysis of ore, the geometric rock surface structure may be divided into four types (Figure 7.14):[13] (a) convex, (b) concave, (c) planar, and (d–f) undulating, with respect to the size of the analyzing probe. The surface contours (d), (e) and (f) in Figure 7.14 show that variations in surface texture cause different source–sample distances, such as the largest distance for (d), the smallest for (e) and the average for (f). The surface irregularity causes the following effects:

1. Changes to the effective ranges of primary and secondary radiation in air;
2. Changes to the effective detection area of the probe;
3. Changes in the absorption of radiation in the air gap.

Although the roughness can be very pronounced over a relatively large measurement area, as a result of general observations of the rock outcrops, it is normally the case that the amplitude between wave peak and wave valley on a typical rock section is generally restricted to 10 mm, *i.e.*, within the effective

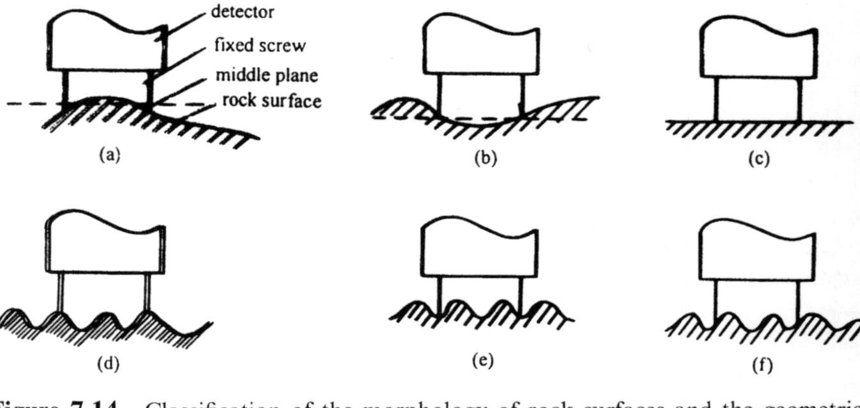

Figure 7.14 Classification of the morphology of rock surfaces and the geometrical arrangement between PXRF probe and rock surface. (a) Convex; (b) concave; (c) planar; (d)–(f) undulating.

detection area of a probe, which is usually about 20–30 cm^2 in terms of the solid angle of a detector or a source. Therefore, Ge[13] proposed a procedure to account for and correct the effects of changes in surface contours by assuming a frequency of irregularity of the order of 10 mm peak–valley amplitudes when assaying element concentrations with a PXRF analyzer. They investigated this procedure for correcting the effects of surface irregularity by *in situ* X-ray fluorescence elemental analysis of ore.[13] This effect can be estimated quantitatively by three parameters: the source–sample distance, the surface peak–valley amplitude and the frequency estimated by the number of convex or concave surfaces within the effective detection area of the probe. Based on theoretical and experimental results, three measures can be used to reduce the surface irregularity effect to a minimum:

1. Taking the ratio of the intensity of characteristic-to-scattered radiation as a basic measurement parameter and making the energies of characteristic and scattered radiations as similar as possible.
2. Taking the average ratio of the intensities of characteristic-to-scattered radiation.
3. Avoiding convex or concave morphology within the effective detection area of the probe when measuring, *i.e.*, (a) and (b) in Figure 7.14.

Under the above conditions, the relative standard deviation in the intensity of characteristic-to-scattered radiation measured on planar and undulating surfaces of a man-made model representing lead-zinc mineralization in a rock was is less than 20%.

To reduce errors caused by the surface irregularity effect to a minimum, Ge and Zhang[29] introduced an optimal source–sample distance. Based on the experimental results, these optimal source–sample distances for *in situ* determination of Sn, Pb or Zn, Mo and Fe concentration on rock or ore surface are 23, 16, 20 and 14 mm, respectively.

Mineral heterogeneity is another source of error in *in situ* XRF analysis, especially when analyzing ores containing minerals with a coarse grain size or when minerals exist in veins. Zhou[30] investigated the influence of tin mineral veins on *in situ* PXRF analysis in tin mining. The distribution of elemental concentrations at a point or in a small volume of ore can be described by a normal distribution (N):

$$c \sim N(\mu,\sigma^2) \tag{7.1}$$

where c is the ore grade at a point or in a small volume of ore, the typical unit of which is % m/m; μ represents the true grade of body; and σ is the standard deviation. The distribution of average ore grade in an area of interest corresponds to:

$$\bar{c} \sim N(\mu, \sigma^2/\sqrt{n}) \tag{7.2}$$

The relative error (η) between the average elemental concentration and the true concentration of the ore body is given by Equation (7.3):

$$\eta = \frac{\bar{c} - \mu}{\mu} = \pm \frac{\sigma/\sqrt{n}}{\mu} \tag{7.3}$$

where, n is the number of points measured in the area of interest.

From the above equation, we can see that the larger the number of measurement points, the lower the relative error, assuming that none of the measurements overlap. Therefore, the best way to minimize the influence of mineral heterogeneity is to arrange an appropriate measurement array. That is, as many measurement points as possible should be selected and there should no overlap in the measurement area under the effective detection area of the probe between any of these points (Figure 7.15c). In some cases, the measurement grid shown in Figure 7.15(a) is also adopted when the minerals in ores are homogenous and the distance between two measurement points is dependent on the mineral homogeneity. In any case, the example in Figure 7.15(b) should be avoided in *in situ* PXRF measurement.

The influence of moisture in ores on the intensity of characteristic X-ray lines should be considered for those elements with an atomic number of less than 26. Experiments have shown that a 10% moisture content in iron ores will produce a 4% relative error for the iron ore grade (Figure 7.16).[6] This influence can be

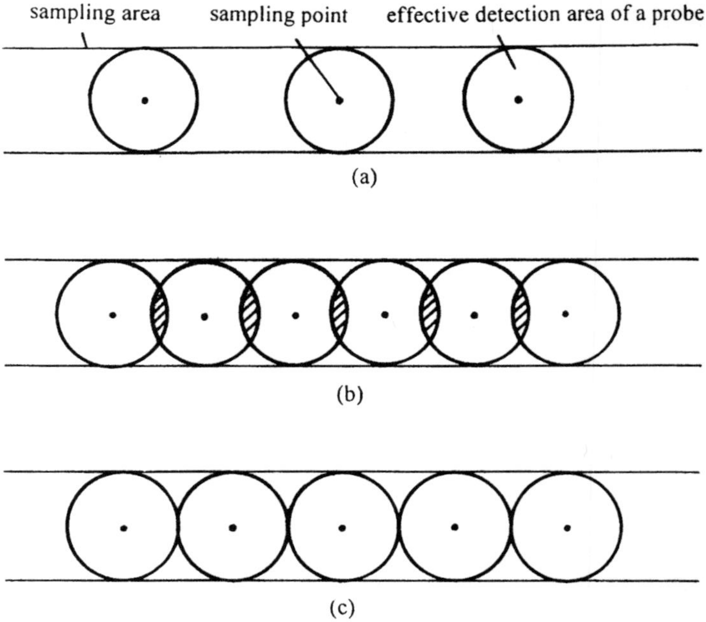

Figure 7.15 Arrangement of measurement points. (a) Initial arrangement; (b) overlapped arrangement; (c) correct arrangement.

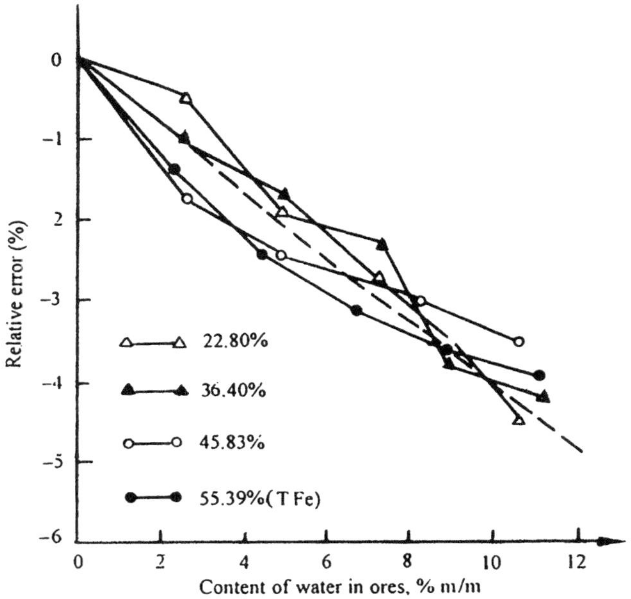

Figure 7.16 Experimental curves showing the relative error in iron concentration by PXRF analysis *versus* the increasing concentration of water in iron ores.

effectively corrected noting that the intensity of scattered radiation is directly proportional to the content of water in ores (Figure 7.17).

Because *in situ* PXRF is used to analyze natural ore and rock surfaces, whereas laboratory techniques are used to analyze well-prepared samples collected from the field, there are significant problems in demonstrating the comparability in performance of these two methods when characterizing samples from exploration sites. Ge *et al.*[6] investigated this problem in the assessment of lead–zinc, molybdenum, and copper deposits and proposed the idea of a Maximum Allowable Random Error Limit (MAREL) as a method of assessing whether the performance of *in situ* PXRF or a conventional method of sampling and analysis was satisfactory. It was assumed that both methods involved analyzing samples from a channel dug at the site for *in situ* or laboratory techniques.

To take into account the heterogeneous distribution of elements of mineralization found at most deposits, the method of assessment involved analyzing a representative number of *in situ* PXRF and laboratory samples first from ore channel and then from a duplicate channel dug immediately below the first (Figure 7.18). The average concentration of both the first (C') and the second (C'') channel was then calculated. This procedure was then repeated for a representative number of channels ($n \geq 30$) across the deposit. Differences in the mean composition of each pair of channels are due to analytical uncertainty and sampling uncertainty, the latter caused by the inhomogeneous distribution of the elements of interest across the site.

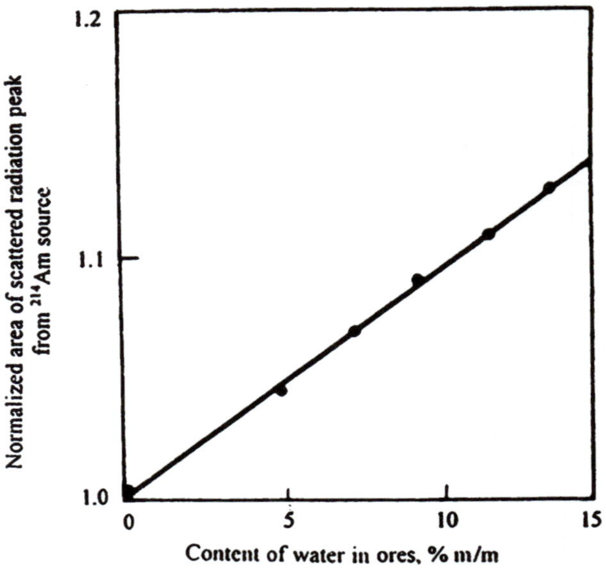

Figure 7.17 Experimental curves of the intensity of scattered radiation from source *versus* the concentration of water in iron ores.

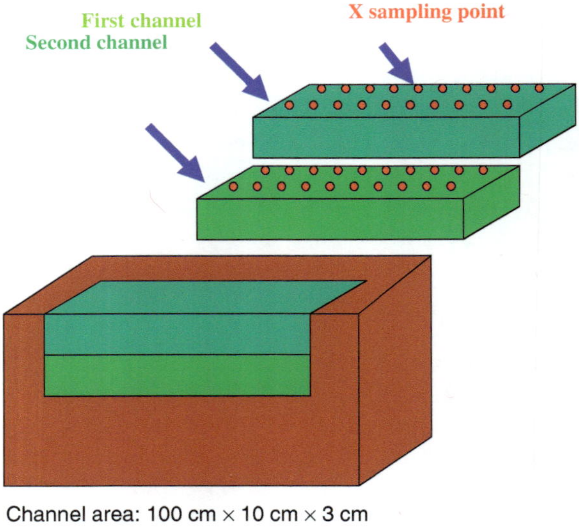

Figure 7.18 Arrangement of duplicate channels and measurement points for PXRF analysis.

Thus, the mean differences between all pairs of channels analyzed is given by:

$$\Delta\bar{C} = \frac{1}{n}\sum_{i=1}^{n} |\Delta C_i| = \frac{1}{n}\sum_{i=1}^{n} |C'_i - C''_i| \tag{7.4}$$

and the standard deviation of these data by:

$$S = \sqrt{\frac{\sum_{i=1} (\Delta C_i - \Delta\bar{C})^2}{n-1}} \tag{7.5}$$

In fact, the major contribution to the mean difference ΔC between pairs of channels can be attributed to the heterogeneity of mineralization. To set a standard of performance for *in situ* PXRF or laboratory techniques it is necessary to identify the additional contribution to uncertainty over and above heterogeneity effects, leading to the concept of Maximum Allowable Random Error (ε) where:

$$\varepsilon_x = \Delta\bar{C} + 1.65S \tag{7.6}$$

From an appropriate statistical analysis (Figure 7.19) it is expected that, for the 95% confidence level, results will fall within the MAREL. The advantage of this approach is that it sets a standard of performance, which takes into account sample heterogeneity effects. Analytical bias in the technique is not considered.

To evaluate this approach, results are listed in Table 7.5 of the MAREL analysis of several copper, tin, lead–zinc and molybdenum deposits, for which values of ΔC, S and ε have been calculated.

A comparison of performance between *in situ* PXRF and laboratory analysis is shown in Table 7.6 for a similar range of deposits. Data in this table lists the number and proportion of measurements that comply with the MAREL principal when results from duplicate channels are compared. These results show that PXRF gives the same or a slightly higher proportion of results that fall below the maximum allowable random error limit, indicating that this technique is as effective as sampling and laboratory analysis procedures in assessing

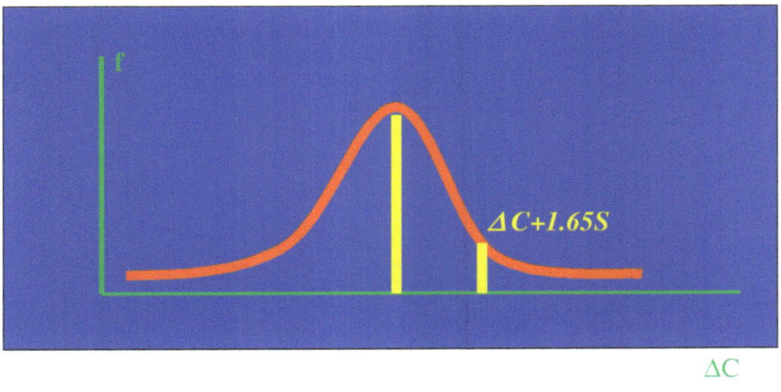

Figure 7.19 Determination of maximum allowable random error at the 95% confidence level.

Table 7.5 Maximum allowable random error level (MAREL) for *in situ* PXRF analysis for copper, tin, lead–zinc and molybdenum mineral deposits.

Types of ores	*Range of ore grades (% m/m)*	ΔC^a	S^b	ε (%)[c]	*Notes*[d]
Cu	>1.0	14.6	9.27	30	RE
	0.4–1.0	20.59	12.67	40	RE
	0.2–0.4	21.54	19.40	50	RE
	<0.2			0.09	AE
Sn	>5	14	9	28	RE/19
	2–5	15	11	30	RE/50
	1–2	20	16	45	RE/42
	0.3–1	29	20	60	RE/61
	0.1–0.3	32	26	75	RE/26
	<0.1			0.1	AE/13
Pb	1.0–20	8.09	7.28	20	RE/13
Zn	1.0–25	9.59	5.93	19	RE/13
Mo	>0.2	14.76	7.83	28	RE/36

[a]ΔC is the difference in concentration between measurements in each pair of channels.
[b]S is the standard deviation associated with ΔC when all paired measurements are taken into account.
[c]ε is the maximum allowable random error.
[d]Number of statistical samples; RE stands for the relative standard error between the analytical results of *in situ* PXRF method and laboratory technique; AE stands for the absolute error between the analytical results of *in situ* PXRF method and laboratory technique.

mineral deposits. This approach has already been accepted in some regions in China, and is commended for consideration in other exploration areas.

Generally, *in situ* PXRF analysis has a higher precision than the conventional channel-chemical analysis method. The relative errors of the ore grades repeatedly determined with *in situ* PXRF is far less than MAREL. Table 7.7 shows some results of repeated *in situ* PXRF analysis in a tin mine in Guangxi province, China.

7.5 Applications in Mineral Processing

The first on-stream XRF analyzer applications were at metal concentrators. Modified laboratory analyzers were installed in flotation concentrators in the late 1960s. In the early 1970s, the first on-stream analyzers specially designed for the industry came onto the market.[31] To date, many successes have been reported and EDXRF equipment for analyzing Ag, Ba, Ca, Co, Cu, Fe, Mo, Ni, Pb, Sn and Zn is now installed in a range of processing plants worldwide.[1,31,32] Devices (XRF probes) have been fitted to the input conveyor belt of feedstock (raw material) to control the removal of waste or the concentration of associated minerals. XRF probes have also been placed down stream in plants to measure the solid weight fraction and the concentration of elements in the slurry or placed on the output belt to control the quality of concentrate products.

Table 7.6 Comparison between the statistical analysis of the results of *in situ* PXRF and laboratory analysis at tin, copper, lead–zinc and molybdenum mines.

Types of ores	*Type of sampling*	*Number of samples*	*Numbers less than the MAREL[a] limit*	*Proportion of samples below the MAREL[a] limit (%)*
Cu	X-radiation sampling	133	104	78.2
	Second channel sampling	101	72	71.3
Mo	X-radiation sampling	56	45	80.4
	Second channel sampling	36	26	72.2
Sn (vein ores)	X-radiation sampling	105	78	74.3
	Second channel sampling	88	62	70.5
Sn (disseminated ores)	X-radiation sampling	100	84	84
	Second channel sampling	100	28	72
Pb	X-radiation sampling	15	11	73.3
	Second channel sampling	13	10	76.9
Zn	X-radiation sampling	15	13	86.7
	Second channel sampling	13	11	84.6

[a]MAREL is the maximum allowable random error limit.

The XRF analyzers in use are mainly based on radioisotope X-ray source and scintillation or proportional counter detectors. NaI (Tl) detectors do not resolve K line X-rays from Fe, Ni, Cu and Zn and this problem is overcome by use of energy-balanced filters and by selecting the optimum energy of the exciting K line X-rays.[1] By using these techniques, along with preferential absorption methods, Fe, Ni, Cu, Zn, Sn and Pb can be detected at all but the lowest concentrations normally encountered in mineral processing.

Semiconductor detector probes are used, but only when justified by the need to measure very low concentrations such as those that occur in porphyry copper residues, where determinations at ±0.005% m/m Cu are required in slurries.[1]

Low-power X-ray tubes are also being used in on-stream XRF analyzers. Figure 7.20 illustrates the principle of the COURIER 40 diffraction analyzer.[32] The excitation source is an X-ray tube and the X-rays from it are diffracted into a five-wire detector for XRD measurement, with proportional counters used for supplementary XRF measurements.

Table 7.7 Results of repeated *in situ* PXRF analysis at a tin mine in Guangxi province, China.

Number of channel	*First in situ PXRF result; Sn (%m/m)*	*Second in situ PXRF result; Sn (%m/m)*	*Relative error (%)*	*MAREL limit (%)*
9921A	0.97	0.95	2	60
9923A	1.17	1.18	1	45
9924A	3.03	3.03	0	30
9935A	0.61	0.75	23	60
9936A	1.93	1.49	23	30
9937A	1.72	1.38	20	45
9938A	0.33	0.44	33	60
9940A	0.64	0.67	5	60
9941A	0.63	0.55	13	60
9951A	2.26	2.67	18	30
9960A	0.51	0.52	4	60
9966A	0.22	0.24	9	60
9973A	1.19	1.13	5	45
9974A	0.14	0.10	29	75
9975A	0.16	0.17	6	75
9976A	1.59	1.48	7	45
9977A	3.13	3.34	7	30
9978A	1.93	1.77	8	45
9980A	1.45	1.57	8	45
9984A	2.05	2.18	6	30
9902A	0.64	0.84	31	60
Mean	1.252	1.260	12.30	

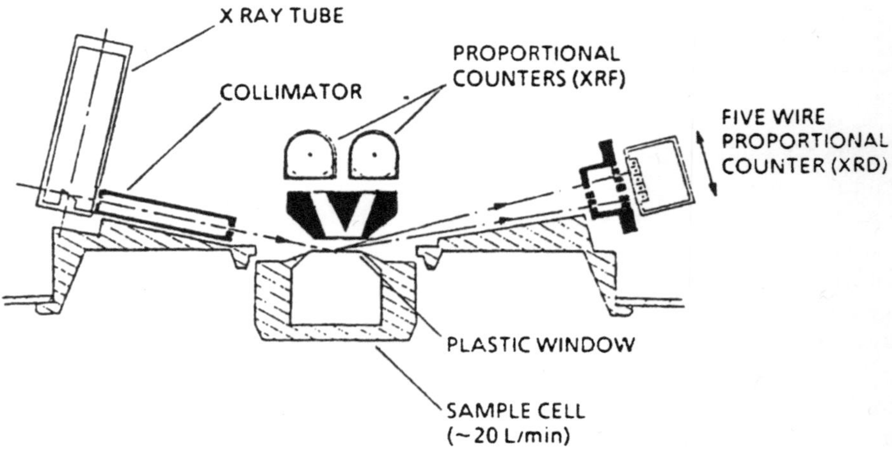

Figure 7.20 Principle of the COURIER 40 diffraction analyzer.

In one system, information from various detector devices is processed and routed to a central microprocessor, which combines the signals from the various probes in the stream to derive the solids weight fraction and the concentrations of elements in the slurry solids. In small concentrators, probes may

Table 7.8 Plant installations of AMDEL system for the on-stream analysis of slurries in mineral concentrators in the Asia and Pacific regions.

Country	*Company/mine*	*Location of plant*	*Elements analyzed*	*Number of streams analyzed*	*Date of installation*
Myanmar	Number one mining	Bawdwin	Pb, Zn, Cu	8	1987
China	Yunna Tin Corporation	Ta Tun Mines	Sn, S	6	1985
	Hong Tou Shan	Liaoning	Cu, Zn	3	1985
	Dachang	Guanxi	Sn, Pb, Zn, As	7	1986
	Anshan Ming Co.	Anshan, Liaoning	Fe	4	1988
India	Hindustan Zinc Ltd				
	Zawar mine	Udaipur	Pb, Zn, Fe	5	1988
	Rampura Agucha	Udaipur	Pb, Zn, Fe	8	1989
	Hindustan Copper Ltd.	Malanjkhand	Cu, Fe	5	1988
Malaysia	Mamut	Sabah	Cu, Fe	4	1986
Papua New Guinea	Ok Tedi	Tabubil	Cu, Fe	11	1987
Philippines	Philes Mining Co.	Baguio	Cu	7	1983

be installed in from three to six process streams. In large concentrators, as many as 14 installations have been made.

From 1982 to 1990 there were 11 AMDEL radioisotope and at least five Outokumpu X-ray tube on-stream analysis systems installed in the Asia and Pacific region (excluding Australia).[32] Table 7.8 summarizes the AMDEL installations.

At Zawar mine's Balaria concentrator near Udaipur, India, the system controls the flotation process by varying the rates of addition of various reagents, pH and the pulp level in the scavenger bank of flotation cells, based on continuous monitoring of these variables and the on-stream analysis. The improvements in plant operation can be obtained at Zawar mine's Balaria concentrator before and after installation of systems for on-stream analysis and process control. The recoveries of lead and zinc increased by about 2 and 1.2% m/m, respectively. The corresponding increase in the tonnage of lead and zinc are 131 and 338 t/a, respectively which, at world market prices for the concentrate would be worth, respectively, about US $80 000 and $30 000 a year. The increase in silver in the two concentrates, assuming that the increases in concentrates increase the silver in proportion to the lead and zinc, is 193 kg, valued at $30 000.

Figure 7.21 is a schematic of an on-line XRF system, manufactured by Chengdu University of Technology. The system has been operated at Guaziping concentrator, Pangang Group, near Chengdu, China. The XRF analyzer incorporates a scintillation detector and two radioisotope sources, ^{238}Pu and ^{241}Am. The former is used to excite Fe K shell X-rays and the latter to produce Compton scattered radiation for the correction of the moisture

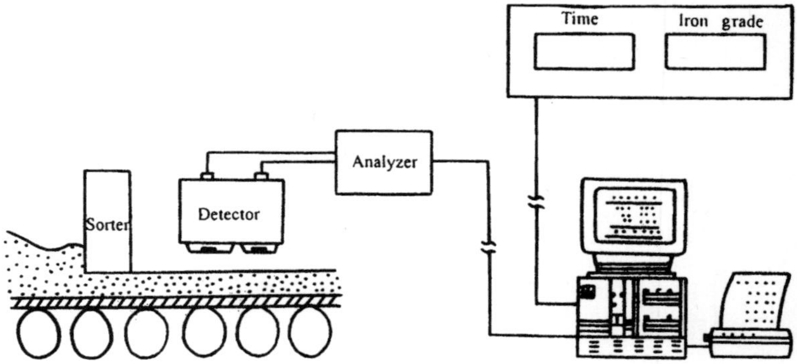

Figure 7.21 Schematic of an on-line XRF system.

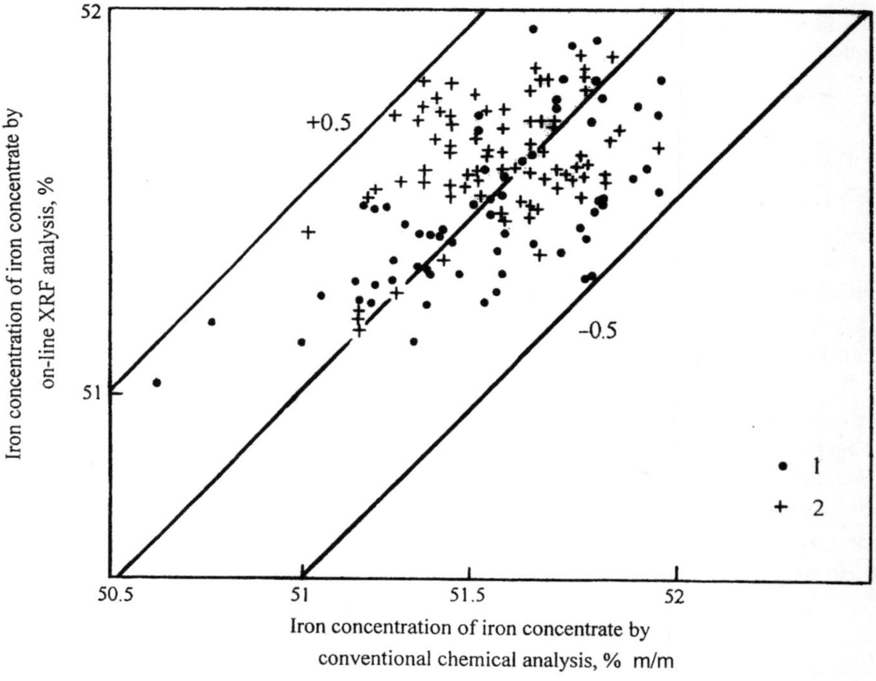

Figure 7.22 Comparison of the iron concentration by the on-line XRF system and laboratory-based technique. Results (1) prior to on-line XRF system; (2) after installation of on-line XRF system.

content of the mineral concentrate. Figure 7.22 illustrates comparisons of the concentration of iron concentrate by conventional chemical analysis and on-line XRF analysis. As can been seen, the difference between two analytical methods was less than 0.5% m/m. It was also found that the range of iron concentrations in the iron concentrate was smaller in the range 51–52% m/m after operating on-line XRF analysis than prior to it. It is estimated that the iron concentrate production has increased by about 8% since installation of the system.[33] (Figure 7.23)

Figure 7.23 Map of China.

In summary, therefore, the technique of PXRF fluorescence has opened up new vistas for geochemical prospecting, mining, and mineral processing. Different designs of PXRF analyzer are needed to meet the requirements of different parts of these areas.

References

1. C.G. Clayton *et al.*, IAEA-SM-308/21, pp. 107–134.
2. P.J. Potts, *Proceedings of Consultants Meeting on in-situ Applications of X-ray Fluorescence Technique*, IAEA, 13–17, Sept., Vienna, 1999.
3. S.C. Zhou, *Nucl. Technol. (Chinese)*, 1992, **15**, 1.
4. Y. Zhang, *et al.*, *Geol. Exploration (Chinese)*, 1985, **5**, 5.
5. Y.L. Zhai, *Symposium on the Examples of the Applications of Nuclear Geophysical Exploration Techniques*, Geology Publish House, Beijing, 1989, pp. 192–198.
6. L.Q. Ge, *et al.*, *X-Radiation Sampling*, Sichuan Science & Technology Publish House, Chengdu, 1997.
7. Y. Zhang, *et al.*, *X-ray Fluorescence Technique in Mineral Exploration*, Geology Publish House, Beijing, 1984.
8. Y.J. Liu, *et al.*, *Elemental Geochemistry*, Science & Technology Publish House, Beijing, 1987.
9. Y. Zhang, *et al.*, *Comput. Techniques Geophys. Geochem. Exploration*, 1986, **8**, 4.
10. Z. Qi, *The 4th Symposium on the Geophysical and Geochemical Exploration*, Geology Publish House, Beijing, 1986, pp. 215–220.
11. L.W. He, *The 4th Symposium on the Geophysical and Geochemical Exploration*, Geology Publish House, Beijing, 1986, 176–180.
12. L.Q. Ge, *Proceedings of Consultants Meeting on In-Situ Applications of X-ray Fluorescence Technique*, IAEA, 13–17, Sept., Vienna, 1999.
13. L.Q. Ge, *et al.*, *Appl. Radiat. Isot.*, 1998, **49**(12), 1713–1720.
14. L.Q. Ge, *et al.*, *Nuclear Technol. (Chinese)*, 1995, **18**, 6.
15. P.J. Potts, *et al.*, *J. Anal. Atomic Spectrom.*, 1997, **12**, 769–776.
16. P.J. Potts, *et al.*, *J. Geostandards Geoanal.*, 1997, **21**(1), 29–41.
17. L.Q. Ge, *et al.*, *Uranium Geol. (Chinese)*, 1995, **11**, 6.
18. L.Q. Ge, *et al.*, *Geol. Exploration (Chinese)*, 1996, **16**, 3.
19. J.R. Rhodes, *et al.*, *Nuclear Techniques and Mineral Resources*, IAEA, Vienna, 1969, pp. 353–362.
20. Y. Zhang, *et al.*, *Nuclear Instrum. Methods (Chinese)*, 1981, **1**, 1–9.
21. L.Q. Ge, *et al.*, *J. Chengdu University Technol. (Chinese)*, 1997, **24**(1), 103–107.
22. L.Q. Ge, *et al.*, *X-ray Spectrom.*, 1997, **26**, 303–308.
23. Y. Zhang, *et al.*, *Nuclear Technol. (Chinese)*, 1985, **8**(9), 9–12.
24. H.J. Liu, *The 4th Symposium on the Geophysical and Geochemical Exploration*, Geology Publish House, Beijing, 1986, 193–198.
25. L.Q. Ge, *et al.*, *J. Chengdu University Technol. (Chinese)*, 2001, **28**(1), 80–85.

26. S. Perick, *Proceedings of Consultants Meeting on In-situ Applications of X-ray Fluorescence Technique*, IAEA, 13–17, Sept., Vienna, 1999, pp.16–19.
27. Y. Zhang, *et al.*, *Nucl. Technol. (Chinese)*, 1990, **13**(6), 371–376.
28. Y. Zhang, *et al.*, *J. Geophys.*, 1989, **32**, 4.
29. Y. Zhang and L.Q. Ge, *Nucl. Geophys.*, 1994, **8**(2), 195–200.
30. S.C. Zhou, *et al.*, *Nucl. Electron. Detecting Techniques (Chinese)*, 1991, **11**(1), 42–45.
31. K. Varmola, IAEA-SM-308/23, pp. 135–149.
32. J.S. Watt, IAEA-SM-308/9, pp. 215–225.
33. L. Lei, *et al.*, *Chengdu University Technol.*, 1995, **21**, 2.

CHAPTER 8

The Application of Portable X-Ray Fluorescence Analysis to Archaeological Lithic Provenancing

OLWEN WILLIAMS-THORPE

Department of Earth Sciences, The Open University, Milton Keynes MK7 6AA, UK; Present address, High Gable House, Guisborough TS14 7BD, UK

8.1 Introduction

This chapter describes applications of portable X-ray fluorescence (PXRF) analysis to a range of lithic artefacts, within several case studies in archaeological provenancing. The work has been carried out largely at the Open University (OU), UK, and has employed a radioisotope/solid state mercury(II) iodide detector system, which is described in detail elsewhere in this volume.

Archaeological provenancing studies have, until comparatively recently, lacked a truly portable and non-destructive multi-element analytical system capable of determining high quality data for lithic and other artefacts. The development of modern PXRF has, to some degree, filled this lacuna, meeting most of the exacting requirements of archaeologists. In particular, the ease of field use of modern PXRF systems, combined with the capability of analysing objects non-destructively, has opened up new opportunities to extend the range and scope of provenancing studies. In the important field of prehistoric stone axe-head studies, for example, the application of PXRF has already greatly increased both the numbers and types of implements analysed.

Notwithstanding the real contribution that PXRF can make, and indeed already has made, to archaeological provenancing, the method is not without

Portable X-ray Fluorescence Spectrometry: Capabilities for *In Situ* Analysis
Edited by Philip J Potts and Margaret West

its limitations. It is, essentially, a near-surface analytical technique, and the analysed material is restricted to an area of typically only 25 mm across and a thickness of generally less than 10 mm. The determination of bulk rock or artefact composition, so often required for provenancing applications, may be complicated by weathered surface layers, or by a coarse mineral grain size. The PXRF system used in the applications described here operates in air, so that elements lower in atomic number than K cannot be effectively detected in silicate rocks; identifying rock types in the absence of elements such as Si, Na, and Mg presents a particular challenge.

The applications described here illustrate both the successes and the limitations of this PXRF system in the analysis of lithic materials. Accounts of the applications are prefaced by sections outlining some instrumental, practical and analytical considerations which are specific to, or of particular relevance in, archaeological use of PXRF.

8.1.1 Background and Early Applications of PXRF in Archaeology

The challenge of adapting and developing portable, non-destructive XRF methods for archaeology was taken up remarkably early in the history of the X-ray analysis techniques. Mobile instruments, employing both radioisotope and X-ray tube excitation together with liquid-nitrogen-cooled semiconductor (silicon) detectors (requiring vehicle transport), were developed at Oxford by the early 1970s by E. T. Hall and colleagues,[1] in pioneering archaeological applications of this technology. At around the same time, researchers at Edinburgh also used a radioisotope/semiconductor detection system to analyse various artefacts, including faience and a wide range of metal artefacts.[2] Their system, again requiring liquid nitrogen cooling and, therefore, transportation by road vehicle, was widely used in museums, and their location map reveals an enviable number of European and Mediterranean venues. Cesareo and Marabelli, meanwhile, were using radioisotope excitation portable XRF (with gas proportional detection) to measure, *in situ*, the bronze doors of Italian churches.[3] During the later 1970s, Stanley Warren at Bradford University initiated portable EDXRF analysis of glass,[4] though still the requirement for liquid nitrogen placed limitations on field use of this system.

During this period, portable XRF analysis had required a trade off, in terms of either maintaining good portability but with relatively poor detector resolution (using gas proportional detectors) or prejudicing portability (by requirements for liquid nitrogen) in favour of better detector resolution (using semiconductor detectors). The non-destructive requirement could be met both by portable systems, and by less portable or laboratory-based EDXRF, though often with limitations on the size of objects. In the 1980s and 1990s, laboratory-based applications of quantitative and semi-quantitative, non-destructive EDXRF in archaeology became widespread, encompassing various lithic and related materials such as obsidian,[5] felsite,[6] basalt,[7] shale and jet[8] and marble.[9]

A combination of true portability, non-destructive analysis, and good detector resolution was initially achieved by the development of radioisotope excitation instruments employing solid-state HgI detectors that require no external cooling. Instrumentation based on such portable analyser units also have the advantage that there is no limit on the size of objects that can be measured. The potential of this type of system for archaeology was first recognized in 1992 by Cesareo and colleagues,[10] and the method was adopted at the OU in the mid-1990s for archaeological lithic applications. Concurrently, other advances in XRF technology, including thermoelectrically cooled semiconductor detectors (*e.g.* Si PIN diodes and Si drift detectors) and miniaturized low power X-ray tubes, were being exploited for further archaeological applications, including metals, paintings, and pottery (Lutz and Pernicka;[11] Cesareo *et al.*[12]).

8.2 Instrumental, Practical and Analytical Considerations in Field and Museum Applications of PXRF

8.2.1 Instrumentation and its Suitability for Silicate Lithic Analysis

The instrument used for the work described in this chapter is a Spectrace TN9000, manufactured by Tracor Northern Inc. (TN Technologies, Round Rock, Texas, USA), incorporating three radioactive sources (^{55}Fe, ^{109}Cd, and ^{241}Am), and employing a solid-state mercury(II) iodide detector that requires no external cooling. Two different instruments have been used, both of this model, and both hired from Thermo FI, Crawley, Sussex, UK.

The potential of the instrument for successful lithic characterization depends on the range of elements routinely analysed, and on the analytical performance in terms of lower limits of detection, precision and accuracy, especially for those elements most useful in geochemical characterization of rocks. The elements normally analysed are K, Ca, Ti, Mn, Fe, Rb, Sr, Y, Zr, Nb, and Ba, with limits of detection of between 6 and 21 $\mu g\, g^{-1}$ for trace elements with higher counting sensitivities, and with precision and accuracy mainly between 1 and 10% rsd, and 4 and 13%, respectively[13] (and *cf.* Chapter 1 this book). Thus, the capability of analysing the geochemically important trace elements Zr, Y, Nb, and Sr at concentrations as low as 6–14 $\mu g\, g^{-1}$, and with good precision and accuracy, means that the TN9000 offers a realistic opportunity to characterize, and discriminate between, a wide range of silicate rocks.

Further trace elements, Co, V, Cr, Ni, Cu, Zn, Pb, Ga, La, and Ce, are also routinely measured, but can only be determined when present at concentrations of > 100 ppm/$\mu g\, g^{-1}$ (or > 1000 ppm/$\mu g\, g^{-1}$ for Cr). This means that they are at or below the detection limits of the instrument in many silicate rocks, so that some potentially useful geochemical discrimination diagrams, *e.g.* those involving V, Cr or Ni,[14] may be excluded from data interpretation.

The exclusion of elements below K in the periodic table results in an important limitation on data interpretation. The absence of data for Si, Al, Mg, and Na makes the geochemical identification of rock type difficult, and classifying a rock within the commonly used Total Alkali Silica diagram is not possible. In practice, the most effective procedure is to make an assessment of the likely, general, rock type in hand specimen, and to consider whether the elements that are available from PXRF are consistent with this assessment.

8.2.2 Practical and Analytical Considerations Important for Field and Museum Applications of PXRF

8.2.2.1 Radiation Safety

Since the TN9000 PXRF contains radioactive sources its use is governed by national regulations, which require appropriate registration (involving some cost and restrictions on use) for owners or keepers of the instrument. However, the TN9000 can, under current regulations, be hired by (unregistered) users in England and Wales, provided simple training is undertaken, and the instrument is under the management control of the registered owners or keepers. Thus, it may be made available to archaeologists without extensive previous experience or training in the use of the instrument.

Radiation dose rates from a TN9000 instrument measured at the OU[15] indicate a dose rate of $0.25\,\mu Gy\,h^{-1}$ when the instrument is not in use (*i.e.*, the dose measured at the instrument surface with the sources retracted and the safety cover in place), rising to a maximum of $0.6\,\mu Gy\,h^{-1}$ when the sources are exposed behind the cover. Thus, when the instrument is held against a flat surface of a silicate rock or artefact that is at least 10 mm thick, the operator receives no significant dose. Maximum dose rates at 100 mm distance from the analyser aperture are $267\,\mu Gy\,h^{-1}$ when the (Cd) source is exposed and no safety cover or sample is in place, reducing to $0.7\,\mu Gy\,h^{-1}$ at a distance of 2 m from the analyser unit. While this situation (a source exposed with no cover or sample in place) should not arise in normal use of the instrument, the measurements indicate that a person standing 2 m from the analyser, even in a direct line of sight of the aperture, is unlikely to receive a significant dose.

Use of the TN9000 in museums or other collections, and at archaeological sites, all of which may be open to the public, requires several conditions to be met. Important among these is maintaining the 2 m minimum distance (appropriate to the TN9000 that we used) for all except the person holding the instrument. Local Rules developed at the OU on the use of the TN9000 also exclude lone working with the instrument. These conditions are readily met, and have not proved a problem in any of our studies so far.

Importantly, these dose rates are specific to the TN9000 used in our studies and will vary with instrument type, and with individual instrument. Simpler regulatory requirements apply to analysers fitted with low-power X-ray tubes, where the source of radiation can be switched off when the instrument is not in use.

8.2.2.2 Portability and Setting up

The TN9000 analyser (probe) unit weighs 1.9 kg, and the spectrum acquisition and data processing unit weighs 6.7 kg. The instrument is, therefore, readily portable, and in its carrying harness may be quite easily transported over rough terrain. Battery endurance is 4–5 h.

Setting up the instrument requires five to ten minutes, plus a further ten with the instrument switched on to allow the detector to reach temperature. Preliminary work at the start of an analysis session, including checking the instrument's energy calibration (to ensure that the energy registration of spectra has not drifted) and measuring reference materials (to confirm analytical trueness in instrument response), is generally necessary, so that an overall set-up time of about 40 min is realistic.

8.2.2.3 Counting Times and Number of Analyses Required for Each Artefact

Counting times chosen will vary, depending on the particular application, considering especially the elements required, and the precision and detection limits that are acceptable. Counting times for our work on silicate rock artefacts are generally between 100–200 s for the Fe source, 50–100 s for the Cd source and 20–40 s for the Am source. The counting times selected also depend on the age (and activity) of the sources, and the manufacturers' guidelines on increasing counting times to compensate for the radioactive decay of the sources should be followed.

The number of measurements required to estimate the bulk composition of silicate rocks using PXRF (to compensate for the mineralogical heterogeneity generally exhibited by such rocks) has been assessed by Potts *et al.*[16] and is discussed in detail elsewhere in this book (Section 2.4.2). As a general guide, three measurements of a fine-grained rock, when averaged, will give an estimate of bulk composition (including key trace elements such as Sr, Zr, and Ba) with a precision of 5% relative standard deviation of the mean. The number of measurements required rises to five to six for medium-grained rocks, but may be reduced if not all the elements considered by Potts *et al.* are necessary for the study. Coarse-grained rocks require much larger numbers of analyses (11 to >100) to estimate bulk composition to the same (5%) precision, but a more modest five measurements result in a typical precision of 20% for several of the elements routinely analysed.

Since the number of measurements required depends on the rock type (in particular on the grain size), and on the element, as well as the required precision of estimate, it will, like counting times, be dependent on the particular application.

We have compared the standard deviations of the means obtained for PXRF analyses of stone axes with those obtained by Potts *et al.*[16] for rocks of similar types and/or grain sizes, and good agreement was observed in most cases. Some small discrepancies between expected and observed standard deviations of the means for about 20 dolerite axes[17] were attributed, in the main, to small differences in analytical conditions (source activity, counting times). However, a slightly larger average standard deviation observed for Fe in the axes (3.2%, as opposed to 2.0% predicted by Potts *et al.*) may relate partly to the slightly weathered, and perhaps more compositionally variable, volumes of rock analysed for the axes, as opposed to the fresh samples used by Potts *et al.*

8.2.2.4 Rock and Artefact Weathering, and Identification of Elements that Best Reflect Bulk Rock Composition

Weathering of rocks and artefacts can result in chemical alteration, especially of the near-surface layers. Since most PXRF measurements are made on the surface of lithic artefacts in applications where the aim is to estimate the bulk composition of the sample, a knowledge of the chemical effects of weathering is very important.

The effects of weathering on PXRF analysis of some dolerite, rhyolite, and microdiorite artefacts have been assessed by Williams-Thorpe *et al.*[15] By comparing PXRF analyses of weathered surfaces of artefacts with WDXRF analyses of crushed and homogenized fresh rock from the same artefacts (representing bulk composition), the authors were able to identify which elements determined by PXRF appeared to be most affected by weathering of the samples, and which elements were most representative of the bulk composition of the rock. The nine artefacts included in the study had thin weathered layers, of about 0.5 mm and less, except for one sample which had weathering up to 5 mm deep.

The results, some of which are illustrated in Figure 8.1, indicated that the PXRF surface measurements of the trace elements Sr, Zr, and Ba showed the closest comparison to bulk sample compositions, with the data straddling 1 : 1 lines of equality. Strontium was significantly depleted only in sample M10 (see Figure 8.1), probably reflecting the thicker weathered layer (5 mm) on this sample. Rubidium apparently showed a good approximation to bulk concentration, but Rb proved to be below PXRF detection limits for some samples, and so it did not give as complete a picture as the datasets for Sr, Zr, and Ba. Yttrium showed some depletion compared with bulk compositions.

The major elements determined by PXRF showed much greater discrepancies from bulk sample compositions. Iron was greatly depleted or enriched in three samples (Figure 8.1a), and Ti was depleted in several samples (Figure 8.1g). Both K and Ca (not illustrated because few data were available for these elements) also showed large differences from bulk compositions; K was depleted in two weathered surfaces, and Ca enriched in one case.

These results reflect partly the generally greater mobility of the major elements in weathering processes,[14] and also their relatively small X-ray critical penetration depths in silicate rocks (between about 0.03 and 0.17 mm). The elements Sr, Rb, Zr and Ba have larger critical penetration depths (0.63–10.8 mm), allowing analysis of a larger volume of sample that is likely to include a greater proportion of fresh, unaltered, rock.

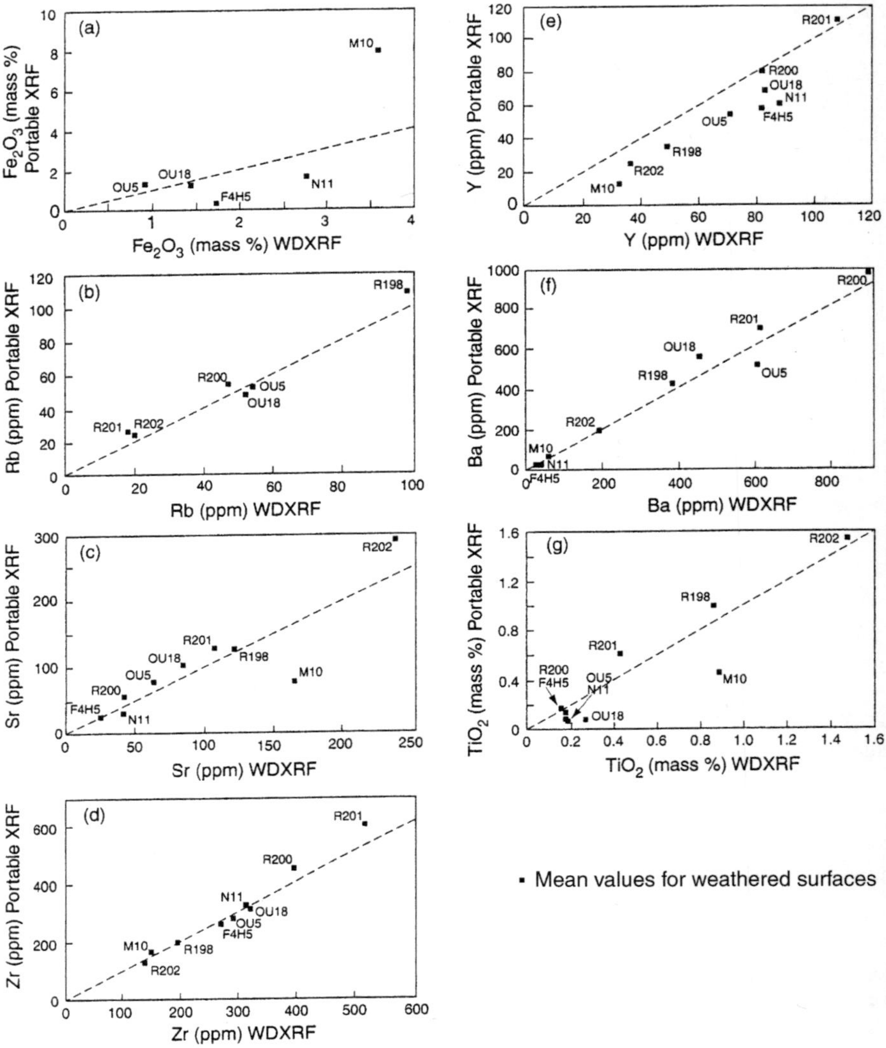

Figure 8.1 Graphs comparing analyses of weathered surfaces of artefacts of dolerite, rhyolite and microdiorite (PXRF), with bulk analyses of fresh material of the same artefacts (WDXRF). Elements shown are (a) Fe_2O_3, (b) Rb, (c) Sr, (d) Zr, (e) Y, (f) Ba and (g) TiO_2. R202 is a dolerite, R198 is a microdiorite, and the rest are rhyolites. Figure taken from *J. Archaeol. Sci.*, **26**, 225, with permission.

Notably, the discrepancies between weathered surface PXRF and bulk analysis WDXRF for Ti are difficult to interpret, because a comparison between PXRF of fresh artefact surfaces and WDXRF also showed lower PXRF Ti values. The reason for this is uncertain, but the possibility that it may relate to density differences between Ti-bearing minerals and rock matrix has been suggested.[15]

In summary, then, the trace elements Zr, Sr, Ba, and Rb are likely to be amongst the most reliable guides to the bulk rock composition in PXRF of dolerites and rhyolites that have thin (< 1 mm) weathered layers. Yttrium showed some depletion but much less than that seen for the major elements. The elements K, Ca, and Fe may be unreliable guides to bulk composition even in thinly weathered samples. Titanium data comparing weathered surface and bulk analyses are difficult to interpret because of unresolved analytical problems in Ti determinations by PXRF.

These results apply, strictly, only to the rock types studied by Williams-Thorpe *et al.*[15] Weathering effects are specific to rock types and the environment of weathering, and are therefore very difficult to predict.

8.2.2.5 Summary of Constraints on Sample Selection for Lithics

Objects measured by PXRF should be large enough to cover the analysis aperture, which is a circular area 25 mm in diameter. However, experiment has shown that samples that are very slightly smaller than this, and leave a small (*ca.* 1 mm or less) gap at the edge of the aperture, do not give lower than expected concentrations,[15] probably because the efficiency of excitation and detection of X-rays is lower near to the edge of the aperture on the instrument used for this study.[16] The object should be at least 10–11 mm thick to satisfy infinite thickness criteria for the fluorescence lines (in particular, the Ba $K\alpha$ line) routinely measured in PXRF.

Measured surfaces should either be fresh and unweathered, or have only very thin weathered layers, ideally much less than 1 mm. For slightly weathered samples, only selected elements, as discussed in the previous section, are likely to be reliable estimators of bulk composition.

Very coarse-grained samples are not suitable for PXRF analysis if an estimate of bulk composition is required to a high precision. However, PXRF could potentially be used for the characterization of individual crystals in such rocks or artefacts.

Surface relief over the area to be measured should be a maximum of 3 mm, in terms of the difference between the highest peak and the lowest pit. Samples with greater relief will not give reliable results even after the application of the relief correction recommended by Potts *et al.*[18]

The application of the relief correction developed by Potts *et al.*[18] depends on measuring the intensity of scatter peaks from either a flat surface within a particular assemblage of artefacts of similar rock type or a reference material of similar composition. It is important that such a surface or reference material is available, otherwise a PXRF analysis from a non-planar surface will not be fully quantifiable.

This chapter deals only with silicate rocks, but PXRF could be applied to other rock types such as limestones or marbles, given element concentrations that meet the lower limits of detection established by Potts *et al.*[13]

8.3 Applications

In each of the applications that follow, PXRF has provided an opportunity to investigate an archaeological problem, by allowing non-destructive and *in situ* analysis of artefacts. In each case, restrictions on destructive sampling of, or moving, artefacts would have precluded, or severely limited, the work without the availability of PXRF analysis.

8.3.1 British Neolithic and Bronze Age Stone Axes

Stone axes were used during the Neolithic and Early Bronze Ages in the UK (between about 4500 and 1500 BC) mainly for clearing forests and woodland for agricultural purposes. Thus, they played a crucial role in prehistoric survival strategies, and many thousands have been found during excavation and in recent agricultural and construction work. In archaeological terms, "stone axes" implies polished implements (as opposed to those that have chipped or flaked surfaces), and they are differentiated from axes made of flint (with the rather odd implication that flint is not, at least in this context, stone). Closely related polished stone implements (often included for convenience under the general heading of "stone axes") include battle-axes and axe-hammers, perforated to allow the insertion of a handle or haft, and mace heads which are often interpreted as ceremonial or status objects. Adzes, probably used for trimming timber, look similar to axes but were hafted with the plane of the blade edge at right angles to the shaft. The range of styles and sizes of stone axes is very large, and they vary from simple cutting tools a few cm in size, to large axes more than 30 cm long, and battle-axes with complex profiles and highly polished surfaces.

Stone axe-heads were frequently manufactured from particular rock types, and were sometimes transported by people over long distances within the British Isles. Identification of the rock types, and if possible the source outcrops, of these implements can, therefore, yield information about trade or other contact between human groups in Britain during the Neolithic and Bronze Age. Such information, bearing as it does on the more intangible, social, aspects of early human culture, is often difficult to elicit from the archaeological record, so it is unsurprising that stone axes have been the subject of intensive study for over 60 years.

The traditional method of identifying the rock type and source of a stone axe has been to remove a sample by sawing or, more recently, by coring, and to make a petrological thin section that is then examined through a microscope in transmitted light. In this way, a large and still invaluable database of stone implement rock types and provenance has been built up,[19] under the auspices of the British Implement Petrology Committee (IPC), previously a sub-committee of the Council for British Archaeology and now the Implement Petrology Group (IPG). The IPC defined 34 axe groups (identified as Groups I–XXXIV, some containing sub-groups), each one encompassing implements of a particular petrographic type, and linked to a likely geographical source area (in a few cases axe production sites, or "axe factories" have been identified through archaeological investigations).

There are, however, some disadvantages in provenancing stone axes using thin section petrography. Some potential outcrop sources of axes have very similar mineralogy, so petrography alone is not always adequate for unambiguous provenancing.[20] In addition, the method is, unavoidably, partially destructive of the implements. Early work by Davis[21] and, more recently, an extensive program of research in Ireland[22,23] introduced geochemical characterization to stone axe studies, but those projects required destructive sampling of artefacts.

The advent of new PXRF systems appeared to offer a solution to many of the problems in stone axe provenancing. The key capabilities of the method – portability, non-destructive analysis, and quantitative determination of a wide range of elements down to the $\mu g\,g^{-1}$ level – dovetailed exactly with the requirements for stone axe studies. Implements could either be placed on the probe, or the probe could be held against them, in the manner shown in Figure 8.2.

A significant amount of geochemical data on the known and potential source areas is available in the published geological literature, providing a reasonable comparative database. Therefore, a program of non-destructive stone implement analysis was initiated at the OU, in collaboration with colleagues from the IPC/IPG and from the Universities of Birmingham, Dublin and Belfast, to validate PXRF for stone axe studies, and to apply the method to a range of implements.

Figure 8.2 Portable XRF being used to measure a stone axe in the Ashmolean Museum, UK. The axe is too large and heavy to place on the instrument aperture, so the probe is held against one side of the axe. Reproduced with permission from the Ashmolean Museum.

8.3.1.1 Welsh Axes of Groups VIII and XIII

A small number of axes, axe-hammers and a mace-head, all of which had already been examined petrographically by the IPC, were analysed by PXRF at the National Museum and Galleries of Wales, Cardiff, to test the PXRF method and to determine whether the PXRF analyses supported the IPC (thin section) assignments. Four of the implements had been classed by the IPC as "Group XIII", a dolerite group whose source has long been accepted as being a small area of the Preseli Mountains of South Wales, UK. Two axes had been assigned to IPC Group VIII, a rhyolite group with several suggested but mainly unconfirmed source localities in South Wales, and the final axe was reported to have some similarities to both Groups VIII and VII (microdiorite from North Wales) but had been assigned to neither. The implements were measured by PXRF following the guidelines summarized in the previous section, and the analytical data, corrected for surface relief variations, were compared directly with analogous geochemical data on the potential source areas in Wales. Our interpretation rested mainly on Zr, Y and Sr, elements which, as noted above, are amongst the more reliable indicators of bulk composition.

Figure 8.3, using concentrations of Sr, Y and Zr, illustrates the degree of similarity (or in some cases, dissimilarity) between axes and their proposed sources in South Wales. The data suggested that only two of the four previously assumed Group XIII (dolerite) implements actually are Group XIII. The Nevern and Llanfaethlu implements differ from the Group XIII source in Preseli. Netherwent is a better match for Preseli, though marginal for some elements. Arthog is more ambiguous, because the Sr is probably concentrated in certain parts of the rock and we did insufficient analyses to compensate for this effect; however, overall, it appeared likely that the Arthog and Netherwent implements come from a Preseli outcrop. One of the Group VIII (rhyolite) axes, from Barry, matched a known Group VIII source; the other rhyolite axe, Llangasty, did not match that source but could be likened to outcrops in the extreme west of Wales, an area that had previously been suggested, but not proven, to be a source of Group VIII axes. The Coygan axe, unprovenanced by thin section, could not be securely provenanced by PXRF either. Chemistry showed similarity to a Group VIII source, but the mineralogical features differ. However, Group VII could be excluded by the chemistry as a possible source for Coygan.

Several lessons were learnt from this first application of PXRF to axes. Identification of implement rock types was difficult, because of the possibility of surface alteration of diagnostic major elements (K, Fe, Ca, Ti). Rock identifications tended, therefore, to be fairly general, and were based on observations of mineralogy in hand specimen as well as on the chemistry. The selection of the most reliable elements for data interpretation was made more difficult in this application because Sr, one of the elements that appear to be relatively robust in slightly weathered surface layers of dolerites, proved to be concentrated, in this particular dolerite type, in some metamorphic mineralogical features present in the rock. As we had expected, mineralogical observations in hand

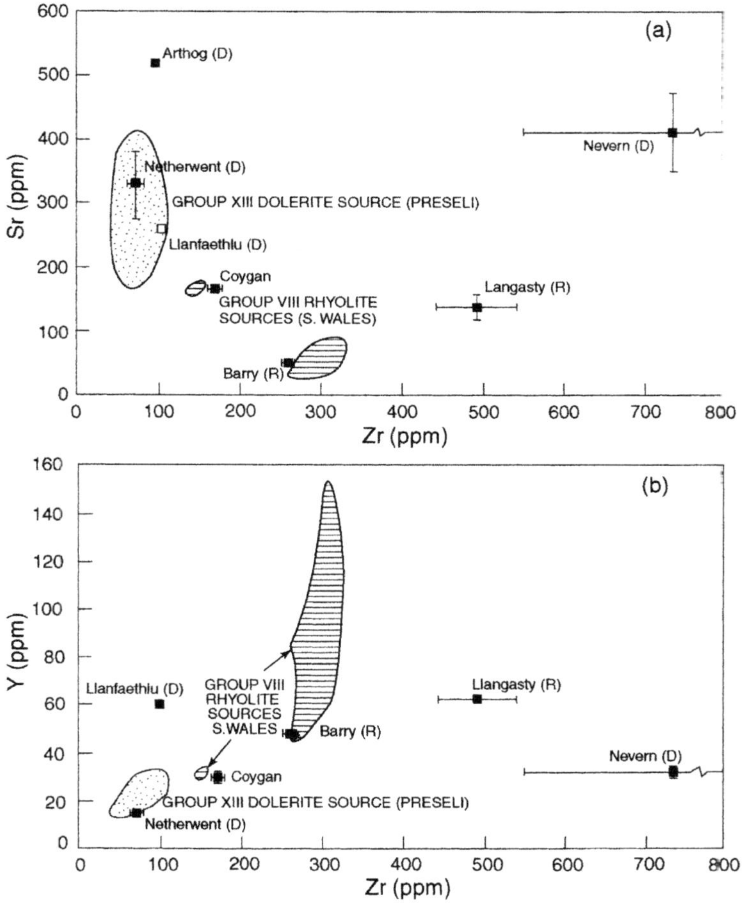

Figure 8.3 Graphs of (a) Sr *versus* Zr and (b) Y *versus* Zr, showing seven stone axes and similar implements compared with the source areas that had been previously proposed using petrographic criteria. The stone artefacts are named in lower case and the letters following (D, R) indicate dolerite or rhyolite artefact respectively (rock type of the Coygan implement is uncertain). The dotted ornament shows the field of samples from the Group XIII (Preseli) source area, and the lined ornament shows samples from Group VIII sources (confirmed sources only, near Carn Alw, South Wales; A. David and G. Williams, *Proc. Prehistoric Soc.*, 1995, **61**, 433–460). Error bars on implements are 1 standard deviation on repeat measurements where available (not shown where the error bars fall within the limits of the symbols). The Arthog implement is missing from (b) because Y data are below the detection limit. Source data are from R. Bevins *et al.*, *J. Geol. Soc. London*, 1989, **146**, 113–123; R.S. Thorpe *et al.*, *Proc. Prehistoric Soc.*, 1991, **57**, 103–157; O. Williams-Thorpe Appendix 1 in A. David and G. Williams, *Proc. Prehistoric Soc.*, 1995, **61**, 457–458; and O. Williams-Thorpe, 1992, unpublished report from the Open University. Artefact analyses are from PXRF, source analyses are from whole rock XRF. The diagram is adapted from O. Williams-Thorpe *et al.*, *J Archaeol. Sci.*, 1999, **26**, 234.

specimen, in addition to aiding rock identification, were also useful in either supporting or, in rare cases, excluding a source that was suggested by the geochemistry.

On the plus side, PXRF data were of sufficient quality for direct (graphical) comparison with the (laboratory) source data, and matches with previously proposed sources were, in some cases, very good. The PXRF added new information in several cases, *e.g.* the support (from the Llangasty axe) for the west Wales outcrops as a Group VIII source, and the exclusion of Group VII as a source for the Coygan axe.

Perhaps the most important result of this study, however, was that some reported Group XIII axes differ in chemistry from their supposed source in Preseli, suggesting a need for re-assessment of their petrographic assignment to Group XIII. Curiously, Group XIII is, supposedly, one of the easiest Groups to recognize by its mineralogy, because this dolerite is marked by characteristic, whitish, metamorphic spots. However, Thorpe *et al.*[24] have noted that these spots, frequently very clear in hand specimen, are much less apparent in thin section.

One of the problems in using PXRF data for provenancing stone implements is that the associated uncertainties, which result both from instrumental factors and from the analysis of unprepared artefact surfaces, are often greater than those associated with the analogous source data, which are nearly always determined from crushed and homogenized samples using well-tested laboratory-based methods. One way of compensating for this difference in uncertainty level is to use a statistical method (to compare implements and potential sources) that takes into account the relatively large errors on the PXRF data. We therefore extended our work on Welsh axes to include some more implements, and to employ a statistical index of atypicality that gives an indication of the probability that the chemical composition of a particular implement originated within a particular source distribution.[25] The atypicality index supported most of the source conclusions we had made earlier, on the basis of bi-variate graphs, for the seven implements shown on Figure 8.3. However, in one case (the Llanfaethlu dolerite axe-hammer) the index "allowed" the Preseli source that we had regarded as unlikely.

The conclusions on the provenances of eight implements that have been assigned to source groups by the IPC, analysed by PXRF, and for which atypicality calculations have been done, are given in Table 8.1. There is good agreement between the initial PXRF interpretation[15] and the subsequent atypicality study,[25] but some disagreement between these and the IPC provenances.[19] Notably, a further two IPC assigned Group XIII implements, Lamarsh and Thaxted, are added to the list of now doubtful members of that group. Also important is the fact that in two cases, Llanfaethlu and Thaxted, chemistry "allows" a source but mineralogical features make it most unlikely. This underlines the fact that some outcrops of relevance to our studies may not have readily distinguishable chemical compositions, and that both chemistry and mineralogy, in tandem, may sometimes be needed for provenancing.

It should also be borne in mind that atypicality (and other statistical) testing of source assignments does not always lead to unambiguous interpretation. Selection of a significance probability of 95% in hypothesis testing is common, yet this carries the implication that one in twenty predictions will be legitimately wrong. Within the context of the atypicality index, a value greater than 0.1 implies that an implement is not unusual relative to the source, while a value of <0.01 indicates an implement with measurements that are extremely unlikely to have arisen by chance from the source distribution. However, it is the values between 0.1 and 0.01 that, as with much hypothesis testing, give difficulty. As Jones and Williams-Thorpe[25] point out, " . . . it would be fairly surprising to find an implement with such extreme features if it did come from the source, but such measurements should occur occasionally".

8.3.1.2 Axes of Groups VI, VII and XVIII

The success of the initial work on Welsh axes, and also the fact that the results suggested that some previous (thin section) source assignments required re-assessment, prompted the extending of PXRF work to provenance implements of other groupings or rock types. We also took the opportunity, in collaboration with the University of Birmingham, to make a comparative test of PXRF, by provenancing a set of axes using the PXRF geochemical data on its own, and then comparing the conclusions with provenances carried out separately, based only on petrography.

Eleven axes from the English Midlands were analysed by PXRF at the OU.[26] No information was initially given to us concerning the archaeological provenance of the axes, or their petrographic characteristics, except that the axes were all of igneous rock types. We were able to examine the axes in hand specimen, and thus make some judgement about their likely rock type and degree of surface weathering prior to PXRF analyses. Meanwhile, Rob Ixer and colleagues at Birmingham and at the Museum of Wales, examined the thin sections of the axes using transmitted and reflected light microscopy. All the axes analysed had surfaces that were large, and smooth, enough for PXRF. Weathering of the surfaces appeared to be very slight.

Happily, the results from both methods, summarized in Table 8.2, identified the same two groupings that contained the majority of the axes, and which were identified independently by each method with IPC implement Groups VI (from the English Lake District) and VII (from North Wales) respectively. Similarly, each method picked out a further two axes as clearly different both from each other and from the Group VI and VII axes. These two axes had been recognized as different rock types during the initial examination of the axes, and both petrography and PXRF confirmed their distinction from Groups VI and VII. However, when it came to identifying the sources of these two axes, which could be matched in chemistry with no previously recognized IPC group, the petrography proved rather more successful than the PXRF. The limited

Table 8.1 Summary of PXRF and petrographic provenancing of stone implements previously assigned to IPC Welsh groups XIII and VIII.

Implement ref. (IPC) or name	*Object*	*IPC group and source assigned (Clough and Cummins 1988[a])*	*PXRF comments on group or provenance using bi-variate graphs (Williams-Thorpe et al. 1999[b])*	*PXRF comments on group or provenance using atypicality index (Jones and Williams-Thorpe 2001[c])*	*Comments*
Pembroke 13 Nevern	Mace	XIII (Preseli spotted dolerite)	Not XIII	Extremely unlkely to be XIII	Actual source not identified; mineralogy not consistent with XIII
Monmouth 8 Netherwent	Axe	XIII (Preseli spotted dolerite)	XIII	Could be XIII	Mineralogy consistent with XIII
Merioneth 8 Arthog	Axe-hammer	XIII (Preseli spotted dolerite)	Probably XIII	Could be XIII	Mineralogy consistent with XIII
Anglesey 10 Llanfaethlu	Axe-hammer	XIII (Preseli spotted dolerite)	Not XIII	Could be XIII	Mineralogy not consistent with XIII (no spots visible)
Essex 23 Lamarsh	Axe	XIII (Preseli spotted dolerite)	Not discussed	Very unlikely to be XIII	Mineralogy not consistent with XIII (no spots visible)
Essex 29 Thaxted	Axe-hammer	XIII (Preseli spotted dolerite)	Not discussed	Could be XIII	Mineralogy not consistent with XIII (no spots visible)

Glamorgan 14 Barry	Axe fragment	VIII (South Wales rhyolite, several possible sources, confirmed source near Carn Alw)	VIII Carn Alw type	VIII Carn Alw type	
Brecon 4 Llangasty	Axe	VIII (South Wales rhyolite, several possible sources, confirmed source near Carn Alw)	Not Carn Alw type; may be other west Wales Group VIII source	Not Carn Alw type	PXRF chemistry suggests origin in west Pembrokeshire coast (Ramsey Is or Abereiddi)

[a]T. H. McK. Clough and W. A. Cummins, eds, 1988, *Stone Axe Studies II*, Council for British Archaeology Report 67, London.
[b]O. Williams-Thorpe *et al.*, *Journal of Archaeological Science*, 1999, 26, 215–237.
[c]M. C. Jones and O. Williams-Thorpe, *Archaeometry*, 2001, 43, 1–18.
Table is abstracted from information in the references given.

Table 8.2 Comparison of provenancing results using total petrography and PXRF in a comparative test of eleven axes.

Implement reference	*Group and provenance from total petrography (including transmitted and reflected light microscopy)*	*Group and provenance from PXRF analysis*	*Comments*
WP1/187	VI[a]	VI	
WL90/123	VI	VI	
W91/80	VI	VI	
WF91/36	VI	VI	
WP1/25	VII[b]	VII	
WP2/18	VII	VII	
WP2/28	VII	VII	
WP1/44	VII	VII	
WP3/2	VII	Uncertain; HFS elements match Group VII, but other elements differ	Petrography shows altered surface layer, accounting for discrepancies in mobile elements (PXRF) from Group VII composition. PXRF provenance would remain uncertain without the petrography for this axe
WP3/3	Not a recognised IPC axe Group; source within the Borrowdale Volcanics of the English Lake District	Not a recognised IPC axe Group; high-Zr intermediate rock that could be from the English Lake District, or North Wales Ordovician volcanics	Petrography allows more precise provenancing than PXRF
WP1/236	Not a recognised IPC axe Group; source within English Lake District; superficial similarity with IPC Group XX dismissed on basis of calc-alkaline character of the axe material	Not a recognised IPC axe Group; superficial similarities with IPC Group XX dismissed by comparison with other (assigned) members of Group XX (unpublished PXRF analyses from the Open University); no source identified	Petrography allows more precise provenancing than PXRF

[a]Epidotised tuff from the English Lake District.
[b]Microdiorite from North Wales.
Table is after information in R. A. Ixer *et al.* in *Lithics in Action*, ed. E. A. Walker, F. Wenban-Smith, and F. Healy, Oxbow, Oxford, 2004, 105–115.

number of elements that we regarded as representative of bulk composition of the axes, and also the incomplete geochemical characterization of some potential source areas, made it impossible to be precise on provenance using chemistry alone. The petrographic information, in particular the identification of metamorphic character and mineral compositions, made it possible to suggest a Lake District origin.

Following this exercise, a further project[17] was initiated to characterize non-destructively members of one of the largest IPC implement groups, Group XVIII, axes and axe-hammers made from a dolerite identified with the Whin Sill intrusion in Northern England.[19] An important aim of this characterization was to establish guidelines for the future identification of members of this group without destructive sampling.

PXRF analysis of about 20 implements that had been previously assigned to Group XVIII by thin section petrography was used to establish a geochemical profile for the group. The data supported the Whin Sill as the source of most of the implements, and showed that only a small number (three implements) might have been previously mis-categorized as Group XVIII on the basis of petrography. As expected, those implements that appeared to be most weathered tended to show anomalous features both in chemistry and in petrography and were difficult to provenance with certainty by either method.

Two especially interesting aspects have emerged from the work on the Whin Sill/Group XVIII implements. Firstly, the limitations we had observed in the interpretation of PXRF data in previous projects prompted the use of a second non-destructive characterization technique to aid characterization and recognition of Group XVIII implements. This technique, magnetic susceptibility measurement, had already proved to be an effective method for provenancing certain other artefacts such as Roman architectural columns.[27] It proved also to be a useful adjunct to PXRF in the characterization of dolerite axes, though interpretation was limited by an incomplete comparative dataset for the Whin Sill source. Nevertheless, used in tandem, PXRF and magnetic susceptibility, both interpreted with the aid of the atypicality index (*cf.* above), provided a more complete characterization and more secure provenancing than either method used on it own.

Secondly, the PXRF provenancing enabled us to make some observations regarding the likely method of procurement of raw material for Group XVIII axes – whether by "trade" from primary source(s) along the Whin Sill or by opportunistic exploitation of locally-available glacial erratic boulders already dispersed over the UK. The use of erratics, already regarded as the most likely raw material of certain assemblages of Group XVIII axes,[28,29] gains support from our new work. Implements that we confirmed as Group XVIII were almost all found in areas where Whin Sill erratic boulders are common. In contrast, implements that proved not to be Group XVIII members were all found outside, or in marginal parts of, the distribution of Whin Sill erratics. Thus, the degree of correlation between the geographical distributions of Whin Sill erratics and Group XVIII implements is strengthened.

8.3.1.3 Cornish Axes

Cornwall has long been recognized as one of the most important sources of British stone axes, and ten petrographically defined groups of igneous axes have been provenanced to this region.[19] These groups, consisting of a range of altered doleritic and gabbroic rocks collectively known as "greenstones", show petrographic similarities to rock outcrops in Cornwall, and, in many cases, concentrations of finds of axes in and near Cornwall also support the hypothesis that this area was the source of their raw material. For two of the groups, specific outcrops have been suggested as sources, on the basis of close petrographic parallels. However, no axe manufacturing sites have been identified in Cornwall, and the precise locations of the sources have remained uncertain. Geochemical characterization of Cornish axes and potential source outcrops appeared likely to help in identifying source areas, and the use of PXRF would avoid destructive sampling of the implements.

Work on Cornish axes was, therefore, initiated at the OU[30] as a PhD project carried out by Mik Markham. The study encompassed nearly 300 greenstone axes and related implements, including previously thin-sectioned implements that had already been assigned to Cornish groups by the IPC and colleagues, and others that had not previously been studied. This represents the largest stone axe analysis project yet undertaken in England and Wales. PXRF was the main analytical technique employed, and magnetic susceptibility was also used to characterize implements and outcrops.[30,31]

Initial results on selected axes and a range of Cornish greenstone outcrops demonstrated the comparability of PXRF outcrop data with laboratory XRF data on the same rocks, and excluded certain outcrops as source of the major Cornish axe group (Group I).[32] More detailed accounts of the results of this project are in preparation.[33]

8.3.2 Roman Imperial Porphyry

Imperial Porphyry, the archaeological name given to a distinctive quartz andesite found only at Mons Porphyrites in the Eastern Desert of Egypt (located on Figure 8.4), was quarried by the Romans for sculptural and architectural stone. Most Imperial Porphyry artefacts are a striking purple colour, but a smaller number of artefacts are black (an unfortunate contradiction in terms, as the word "porphyry" means "purple"), and brownish, green and grey porphyry can also be found at the quarries. The four Roman quarrying areas of purple porphyry and the five that yielded black porphyry have recently been the subject of intensive archaeological investigations by the Universities of Southampton and Exeter,[34,35] and within the context of these investigations a re-assessment of the use and distribution of porphyry from the various quarries was undertaken at the OU.[36]

Imperial Porphyry artefacts include fine sculptures and reliefs, polished columns and decorated sarcophagi, objects for which destructive sampling is

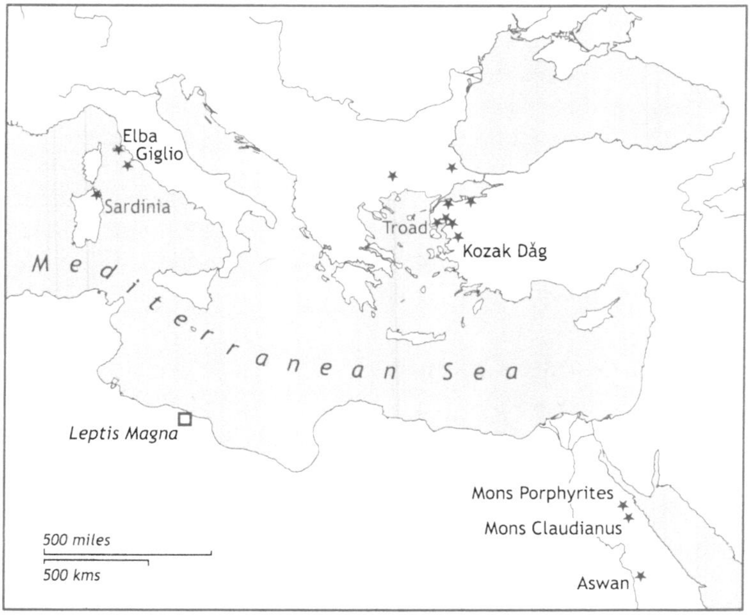

Figure 8.4 Map of part of the Mediterranean area, showing the locations of some important Roman granite quarries, the Imperial Porphyry quarries at Mons Porphyrites, and of Leptis Magna in North Africa.

particularly inappropriate. Characterization of Imperial Porphyry aimed at provenancing artefacts to individual quarries was, therefore, undertaken non-destructively using PXRF. Both quarry material and artefacts were analysed by PXRF, optimizing the analytical comparability between the two data sets. Thirty artefacts, including sculptures, wall veneers and relief carvings from sites in Britain and in the Mediterranean area, were analysed by PXRF.

The artefacts all appeared to have fresh, unaltered surfaces, no doubt partly due to the careful protection from the weather that many of these highly-prized artefacts have enjoyed, both in Roman and post-Roman times. In nearly all cases, the original surface polish was retained, giving a very smooth, and sometimes flat, surface for analysis. These factors, together with their relatively fine-grained mineralogy, makes Imperial Porphyry artefacts particularly suitable for PXRF analysis, and the data easier to interpret in terms of bulk composition than analogous data from coarser, more weathered artefacts.

The chemical compositions of all four purple porphyry quarries proved to be similar and/or overlapping, while compositions of the black porphyry quarries were only slightly less uniform. Even using linear discriminant analysis to optimize analytical distinctions between quarries, only one of the purple porphyry quarries could be completely distinguished from the others; results for the black porphyry were more promising, with two quarries distinguished from each other and from other black porphyry localities. Interestingly, the addition of magnetic susceptibility measurements improved discrimination slightly for the purple porphyry quarries, and significantly for black porphyry, in the latter case allowing three quarries to be unambiguously separated by discriminant analysis using both magnetic susceptibility and chemistry together.

Archaeological conclusions from the project were limited by the similarity of quarry compositions, but we were able to suggest that the majority of purple porphyry artefacts we had examined had originated in the "Lykabettus" quarries, and a small vase now in the Ashmolean Museum, UK, was assigned to the quarry area known as Rammius.

8.3.3 Roman Granite Columns

The final application of PXRF in archaeology described here concerns Roman granite columns, and illustrates the pitfalls of the method as well as its potential for reliable provenancing. Since only a brief account of this work has previously been presented,[37] analytical and interpretative details of the study will be given here.

Roman granite columns can be provenanced to their quarry sources in the Mediterranean area (Figure 8.4) using various methods, including visual recognition of mineral characteristics,[38] conventional laboratory analysis of samples,[39] and non-destructive measurement of magnetic and radioelement[40] characteristics. We chose to test the potential of PXRF for granite provenancing at the "Leptis Magna ruins" site next to Virginia Water in Windsor Great Park, London, UK, where large columns and other architectural stones brought from Roman Leptis Magna (North Africa) were set up in the early 19th century in a pseudo-Classical arrangement (Figure 8.5). The site was chosen because we already had a large amount of data on the granite columns: magnetic susceptibility measurements, portable gamma ray spectrometry analyses of radioelements, and laboratory WDXRF analyses together with some petrographic thin sections of samples that we had been permitted to remove from five columns. In this way, we had already identified the provenance of the columns at two quarries in western Turkey (Kozak Dağ and Troad; locations on Figure 8.4), and we had ample data with which to compare with PXRF analyses. The PXRF study was undertaken, by kind permission of the Crown Estate and English Heritage, between 1995 and 1996, at a fairly early stage of our work with the method. Five grey granite columns were measured, with the aim of comparing the PXRF element data with WDXRF analyses of samples taken from the columns, and, most importantly, assessing whether the PXRF data on its own enabled identification of the column sources.

The Leptis Magna columns represent an example of provenancing where not only the non-destructive capability of PXRF, but also the field portability, is essential, since clearly it would be impractical to move the columns. In addition, they presented several new problems specific to the rock type. The Leptis granite columns have a grain size ranging from medium to coarse, and, in common with many granitoid rocks, they contain mafic enclaves that differ

Figure 8.5 The Leptis Magna ruins in Windsor Great Park, London, UK. Phil Potts is using the portable XRF to measure one of the grey granite columns at the site. (Photograph reproduced with permission from The Crown Estate.)

from the bulk composition. The surface characteristics of the columns present several challenges for PXRF analysis. The columns are blackened due to pollution and lichen growth, and they show signs of weathering, including opaque-looking feldspars, orange (Fe?) staining, and exfoliation. Surface relief is variable, depending on the style and extent of weathering, and surfaces are slightly curved unless measurements are made on the flat column ends (these are sometimes accessible because some of the columns were arranged horizontally, or have fallen). The amount of weathering, and its position on a column, varies and is no doubt partly related to the location and orientation of each column within the site and its consequent exposure to wind and rain. Most of the columns do, nevertheless, retain substantial portions of their original polished surfaces.

The selection of surfaces for PXRF measurement was the result of careful compromise between the various problems presented by surface features. Wherever possible, we chose mineralogically representative surfaces that appeared to be smooth, fresh and unweathered, avoiding blackened areas, lichens, and enclaves. Polished surfaces typically offered the lowest surface relief (the column diameters are around 600 mm, so column curvature introduced only a very small gap (much less than 1 mm) between instrument measuring face and column), but broken ends of columns often appeared to be fresher albeit with greater surface relief. Our measurements included both cylindrical surfaces and column ends. At least five measurements per column were required to meet the guidelines established by Potts *et al.*[16] for medium-grained granites, and the limited areas of ideal surfaces on the columns meant that, in practice, some measured surfaces were slightly weathered, or had relief approaching the 3 mm limit suggested by Potts *et al.*[18] Total counting times were 170 s per measurement, and a particularly smooth and fresh-looking area on one of the columns (LM5) was selected as the reference surface for the relief correction.

Analytical data from PXRF and WDXRF are given in Table 8.3.

Comparison of the PXRF data, corrected and averaged, with WDXRF on samples taken from the same columns, revealed some unexpected, and unwelcome, discrepancies (Table 8.3; Figure 8.6). The PXRF data were mainly considerably lower than the WDXRF data; PXRF determinations for the major elements K, Ca and Fe were only about 70% of their WDXRF counterparts, and Ti showed an even greater discrepancy, being only 45%, on average, of the WDXRF concentrations. Zirconium, like Ti, showed large discrepancies, with PXRF concentrations being between 46 and 74% of WDXRF concentrations. However, Rb, Y, and Ba showed rather closer comparison with WDXRF data, at 75–91% of the latter values. Only Sr and Nb data proved to be within 5%, on average, of expected (WDXRF) values.

These discrepancies could not be explained by instrumental factors such as spectrum analysis uncertainties, the calibration of the PXRF (assessed by analysis of international reference materials), or lower limits of detection. Errors introduced by column surface relief, and by compositional (mineralogical) variation, had been compensated for by, respectively, applying the relief correction of Potts *et al.*[18] and averaging repeat (but non-overlapping) analyses of each column. The relative standard deviations of the means for each column were calculated at, typically, 8–12% (averaged for all elements; Table 8.3), values consistent with information (for rocks of similar grain sizes) in Potts *et al.*[16] Notably, the discrepant elements included some that had been assessed

Table 8.3 PXRF and WDXRF data for Roman granite columns from the Leptis Magna site, Windsor Great Park, UK.

	LM2 PXRF	*n = 7 rsd of mean*	*WDXRF*[a]	*LM5 PXRF*	*n = 5 rsd of mean*	*WDXRF*	*LM24 PXRF*	*n = 5 rsd of mean*	*WDXRF*	*LM25 PXRF*	*n = 5 rsd of mean*	*WDXRF*	*LM27 PXRF*	*n = 5 rsd of mean*	*WDXRF*
K_2O[b]	2.49	4.00	3.62	3.28	4.33	4.73	3.21	6.38	4.58	3.34	9.73	nd[c]	2.28	6.56	nd
CaO	2.53	8.26	4.12	2.73	4.42	3.82	3.11	13.39	3.97	2.78	7.63	nd	1.85	11.61	nd
TiO_2	0.25	23.21	0.54	0.22	15.09	0.58	0.25	16.34	0.54	0.26	10.89	0.48[d]	0.24	14.06	0.61
Fe_2O_3	2.85	8.19	4.07	2.71	4.02	3.95	3.18	3.93	4.11	3.08	10.94	3.42	2.21	9.60	4.34
Rb	85	7.17	120	177	4.02	215	165	3.75	205	203	5.49	211	79	9.33	121
Sr	583	5.40	629	728	8.08	688	732	5.54	728	824	7.90	804	561	8.35	660
Y	17	8.57	20.9	17	14.87	20.3	15	9.91	19.3	16	38.66	22.9	14	7.28	23.5
Zr	79	10.53	137	121	10.00	181	129	1.88	188	151	12.37	203	79	13.17	171
Nb	15	16.15	11.1	13	18.75	13.6	13	11.07	12.6	18	5.18	19.2	11	7.28	12.4
Ba[e]	1160	3.58	1309	1160	8.79	1216	995	7.90	1208	1100	15.71	1081	1170	11.92	1344
average rsd of the mean for each column, PXRF data		9.51			9.24			8.01			12.45			9.92	
Source of column	Kozak Dağ			Troad			Troad			Troad			Kozak Dağ		

[a]WDXRF is based on one sample per column; analyses from The Open University (analytical details in Potts & Webb, *J. Geochem. Expl.*, 1992, **44**, 251–296, and Ramsey *et al. Chem. Geol.*, 1995, **124**, 1–19).
[b]Major elements in weight %, trace elements in $\mu g\, g^{-1}$.
[c]nd is not determined.
[d]TiO_2 and Fe_2O_3 in LM25 and LM27 (WDXRF data) determined from pellet samples.
[e]Ba PXRF is given to 3 significant figures.

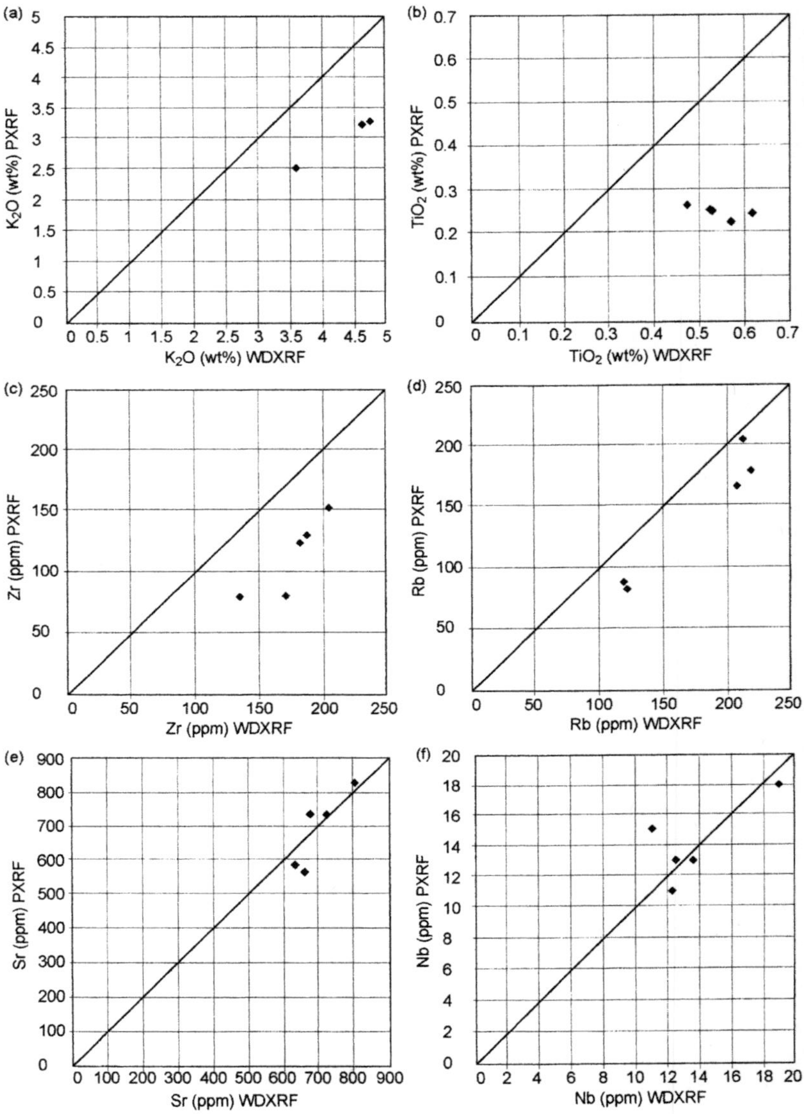

Figure 8.6 Graphs comparing PXRF (*in situ*, non-destructive) and WDXRF (bulk analysis of samples) analyses for five columns from the Leptis Magna ruins, for selected elements: (a) K_2O, (b) TiO_2, (c) Zr, (d) Rb, (e) Sr, and (f) Nb. WDXRF precision (1 sigma) is generally better than 1% relative, and 2% relative for Sr and Nb (K. Govindaraju *et al.*, *Geostand. Newsletter*, 1994, **18**, 211–300). PXRF precision (1 sigma) is mainly between 2 and 4% relative, and 27% for Nb (P. J. Potts *et al.*, *Analyst*, 1995, **120**, 1273–1278), and standard deviations of the PXRF means for the columns are given in Table 8.3. The discrepancies between the two sets of data for several elements are discussed in the text.

elsewhere as among the more reliable in PXRF estimations of bulk composition for dolerites and rhyolites, such as Ba, and Zr (*cf.* above).

The explanation for the discrepancies, therefore, appeared likely to lie either with the WDXRF samples, or with surface alteration of the granite columns that would affect preferentially the PXRF. WDXRF data were determined from relatively small (20–40 g) samples that were taken from the uppermost *ca.* 10–30 mm of each column, and therefore included some weathered material. Yet, despite these limitations, the WDXRF analyses of the samples matched well with the quarry sources that we had already identified using mineralogy, magnetic susceptibility and portable gamma ray spectrometry. It seems most likely, therefore, that the discrepant element concentrations observed from PXRF reflect depleted concentrations in the near surface layers of the columns, especially within the 1 mm layer nearest to the analyser, the layer that contributes 90% of the signal for many of the elements analysed.[16] The process by which this depletion might have taken place is uncertain, but may involve mechanical removal of mineral grains[41] such as titanite and zircon during the weathering process. This may offer an explanation for the depletion of elements that are often relatively immobile in granites, such as Ti, Zr and Y.[14]

The PXRF data from the Leptis columns were then compared with Roman quarry sources of granite columns in the Mediterranean area, including the quarries that we had already identified as the column sources, Kozak Dağ and Troad in Turkey. Major and trace element laboratory XRF analyses were available for the quarries from earlier published studies.[39] As expected, concentrations for elements that approach WDXRF levels compared well with the putative sources, while those severely depleted in comparison with WDXRF showed poor source matches. The comparison obtained for two of the less depleted elements, Rb and Sr, is illustrated in Figure 8.7. Distinction between the quarries is clear using these elements, and the columns lie close to Kozak Dağ and Troad, respectively, supporting our previous provenancing.

However, if we had not had the WDXRF analyses of the columns, would it have been possible to correctly provenance these columns? It is likely that, bearing in mind the weathered appearance of the columns, and also the unresolved problems with Ti PXRF data encountered in our other studies, we would not have relied on the major elements. We may well, however, have elected to use Zr and Y as discriminants, giving source indications that were ambiguous, and in contradiction to those suggested by the Rb and Sr data.

The study illustrates the importance of assessing the effects of weathering on PXRF data, for the particular rock type that is to be measured. Clearly, this is especially important when, as with the Leptis Magna columns, field examination of the mineralogy indicates that some measured surfaces have been subjected to some degree of weathering. Further study is now needed to determine whether the problems we have encountered with these columns are applicable to other granitoid columns in the UK and elsewhere.

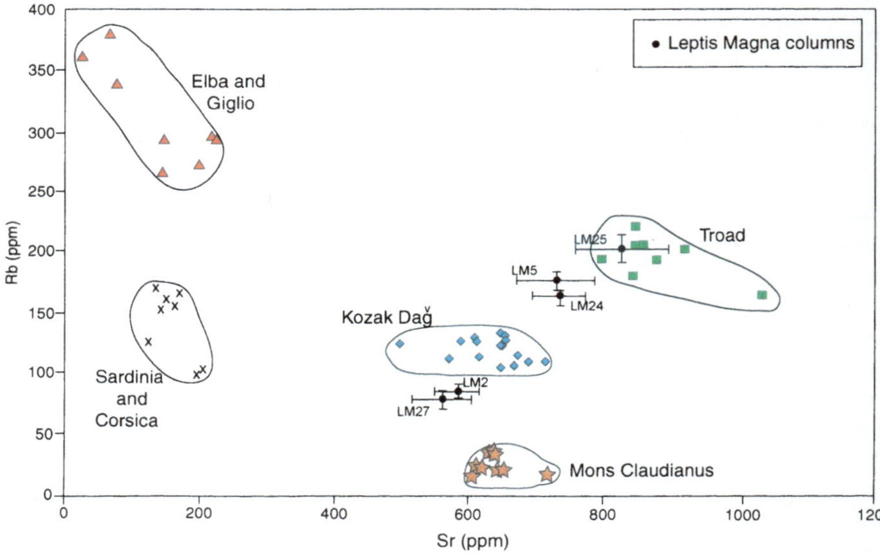

Figure 8.7 Graph of Rb *versus* Sr ppm, comparing five columns from the Leptis Magna ruins with the major Roman granite column quarry areas in the Mediterranean area. The quarry source data are by laboratory-based XRF (WDXRF and EDXRF; data from G. Poli, *J. Geol.*, 1992, **100**, 41–56; G. Galetti *et al.*, in Ancient Stones: quarrying, trade and provenance, *Acta Archaeologica Lovaniensa Monograph 4*, ed. M. Waelkens, N. Herz and L. Moens, Leuven, 1992, pp. 167–178; D. P. S. Peacock *et al.*, *Antiquity*, 1994, **68**, 209–230; and G. De Vecchi *et al.*, *J. Cultural Heritage*, 2000, **1**, 145–153. Typical uncertainties on XRF (quarry sources) data are precision of 2% relative or better (1 sigma), and accuracy of $<1\%$ to 2% relative. (After data in P. J. Potts *et al.*, *X-Ray Spectrom.*, 1984, **13**, 1–15, and K. Govindaraju e*t al., Geostand. Newsletter*, 1994, **18**, 211–300; accuracy based on EDXRF.) Leptis Magna column data are by PXRF (Table 8.3) and the error bars shown on the graph are 1 sd of the mean for each column. Ellipses surround samples from the same quarry area and have no statistical significance.

8.4 Assessment of the Contribution of PXRF to Lithic Provenancing, and Comment on its Future Potential

PXRF has been employed in lithic provenancing for little more than half a decade, but in that time it has made a significant contribution to the field.

An important area of contribution so far has been in the study of stone axes in England and Wales. Stone axe provenancing had, for over 60 years, been based on traditional methods of provenancing using petrographic thin sections. The desirability of bringing modern geochemical methods of analysis to the field has been recognized for some time,[42] but only with the advent of non-destructive PXRF has this begun to be achieved for a significant number of implements in England and Wales.

The extent of agreement between PXRF provenancing and that of the IPC and colleagues based on petrography is mainly good, but rather variable. This variation seems to be a function of the particular rock type, or IPC Group, of the implement. Thus, in the case of Group XVIII implements (made from a dolerite with rather characteristic mineralogy), PXRF supports the IPC work in the great majority of cases, suggesting that, for this Group, petrography is a generally robust method. Similarly, for two other well-recognized and numerous implement types, Groups VI (English Lake District tuffs), and VII (North Wales microdiorite), PXRF agrees well with the "total petrography" of Ixer (*cf.* Table 8.2 above). However, in the case of the numerically much smaller south Wales Group XIII, agreement between the two methods is less good (Table 8.1), while early indications from PXRF on a further Welsh type, Group VIII, suggest that the implements are of more than one chemical (and, probably source) type.[15,25] This last Group consists of implements manufactured from a very fine-grained rhyolite that often lacks distinctive features in transmitted light thin section, and for which several different source areas in south Wales had already been suggested on the basis of petrographic work (summarized in Williams-Thorpe *et al.*[15]).

These observations accord with Davis' comments some years ago on both the strengths and the limitations of petrography in stone axe provenancing.[42] He noted the usefulness of implement petrology in identifying distinctive rock types, while recognizing the problem presented by petrographically similar rocks outcropping in different geographical localities. This balance is summed up in his statement that " . . . in the general absence of more sophisticated geochemical investigations, the bulk of available implement petrology only provides indications and helpful suggestions . . . except, of course, where the data concern highly individualised rocks . . . Group 9, porcellanite, for example . . . ". Our more recent work supports the view that petrography remains a generally reliable method of provenancing for those implement groups that have well-known and distinctive mineralogical features, but is less diagnostic for rock types that either are not distinctive or are very variable, in thin section.

At present, therefore, PXRF has a role both in examining implements that have already been assigned to groups by the IPC and in examining previously unstudied artefacts. Examining IPC grouped artefacts allows problems with petrographic provenancing to be identified, and also provides a framework for establishing chemical characteristics for a particular Group that can be used in future, non-destructive, implement provenancing. As these two objectives are achieved, PXRF work can be aimed increasingly at the provenancing of previously unassigned implements.

We should, however, recognize here the limitations of PXRF in this aspect of assigning previously unstudied stone axes to sources. Given an implement that may be slightly weathered, and could come from any igneous outcrop in the British Isles, recognizing even its rock type from PXRF data and hand specimen alone can be very difficult, and may make the analyst long for a destructive thin section. The comparative outcrop database for UK rocks is certainly very extensive, but analyses found in the geological literature are rarely focussed on exactly those outcrops that the archaeologists believe to be important for axe studies. Nevertheless, it is often possible to use PXRF data to either allow or exclude the 35 or so source areas recognized by the IPC, and also to identify the tectonic type of the implement rock, and so constrain its possible source area.

The contribution of PXRF to other lithic artefacts is still in its infancy. The Imperial Porphyry PXRF work was limited by the close chemical similarity of many of the quarries, but nevertheless made a useful contribution to a research area where destructive sampling is particularly unacceptable. Granite columns proved to be perhaps the most difficult artefacts so far studied, in terms of interpreting the data, but even in that area useful results were drawn from the PXRF, and further work will enable clearer understanding of the problems encountered.

Some artefacts may be especially difficult to provenance using PXRF because of the weathered state of the artefact surfaces available for measurement; this is probably the most intractable problem of the method. However, progress has been made recently in a study by Potts *et al.*[43] at the OU, aimed at quantifying weathering effects for some dolerites and rhyolites from South Wales. Some of the "weathering factors" derived in this work are large – weathered layers in these particular rock samples are altered (mainly depleted) relative to fresh (unweathered) layers by 30% or more for many elements – but nevertheless point the way to a method for adjusting "weathered" PXRF data for comparison with bulk rock analyses of potential sources. Without such adjustment, the presence or absence of weathering has an important bearing on which elements are likely to reflect bulk composition. For a weathered axe, for example, only a few elements may be useful in assessing bulk composition, while for a fresh artefact all the elements analysed might be employed. Investigation of a weathered layer present on an artefact may, in itself, be informative of the processes that the rock has undergone, but is unlikely to enable specific geographical provenancing.

The potential of PXRF in lithic provenancing has only begun to be realized and there await extensive areas of research in which it might be usefully applied. Many stone axe groups await study, plus some thousands of unprovenanced implements manufactured from a large variety of rock types. Grindstones, whetstones, megaliths, building stones – all are potential subjects for PXRF. In addition to silicate rock artefacts, and given the appropriate assessment of analytical and other factors such as weathering, the method might be extended to non-silicate rocks such as marbles and limestones.

Non-destructive, precise, accurate and complete geochemical characterization of artefacts has often been described as one of the "holy grails" of archaeological science. Notwithstanding its limitations, PXRF is probably the nearest approach to this goal that has so far been achieved. As Professor Chris Stillman has suggested,[44] "This application may well point the way to the future of archaeo-geochemistry."

Acknowledgements

This chapter is largely based on work carried out at the Open University between 1994 and 2000 and was partly supported by grants F269/P from the Leverhulme Trust, and BR60 644 and BC15 1312 from the Open University Research Committee; this support is gratefully acknowledged. Archaeological PXRF work at the OU has been carried out in collaboration with Phil Potts, Peter Webb, Chris Jones and Mik Markham. For permissions and samples for the work at the Leptis Magna Ruins I thank English Heritage, and Richard Morrice, Andrew David, J. D. Bond, and The Crown Estate. The non-destructive characterization of Roman granite columns forms part of a project initiated by the University of Southampton (Professor D. P. S. Peacock), and the Imperial Porphyry study was carried out within the Universities of Southampton and Exeter Mons Porphyrites Project. Professor Peacock and R. Wilson kindly provided granite samples for analysis. Work on stone axes has benefited from collaboration with the Implement Petrology Group (formerly, Committee), and I am grateful to members of this group, especially Rob Ixer and Vin Davis, for their continuing support for the geochemical work. Individual museum permissions for stone axe work are in the papers referenced within this chapter. Thermo (FI) UK (formerly Thermo Electron and Thermo Unicam) gave valuable help and support during the early work with the PXRF. I am grateful to John Taylor for help with the diagrams.

References

1. E.T. Hall, F. Schweizer and P.A. Toller, *Archaeometry*, 1973, **15**, 53–78.
2. H. McKerrel, *Atti dei Convegni Lincei, Rome and Venice*, 1973, No. 11, 381–407 (Atti published in 1976, Accademia Nazionale dei Lincei, Rome).
3. R. Cesareo and M. Marabelli, *Atti dei Convegni Lincei, Rome and Venice*, 1973, No. 11, 409–420 (Atti published in 1976, Accademia Nazionale dei Lincei, Rome).
4. J. Henderson and S.E. Warren, *Archaeometry*, 1981, **23**, 83–94.
5. M.E. Hall and S. Shackley, *Al-Rafidan*, 1994, **15**, 25–32.
6. O.D. Hermes and D. Ritchie, *Geoarchaeology*, 1997, **12**, 31–40.
7. M.I. Weisler in *Prehistoric Long-Distance Interaction in Oceania: an Interdisciplinary Approach*, ed. M.I. Weisler, New Zealand Archaeological Association Monograph 21, New Zealand Archaeological Association, Auckland, 1997, pp. 149–172.
8. F.J. Hunter, J.G. McDonnell, A.M. Pollard, C.R. Morrice and C.C. Rowlands, *Archaeometry*, 1993, **35**, 69–89.
9. C.T. Yap, *Appl. Spectrosc.*, 1993, **47**, 330–333.
10. R. Cesareo, G.E. Gigante, J.S. Iwanczyk and A. Dabrowski, *Nucl. Instrum. Methods Phys. Res.*, 1992, **A322**, 583–590.
11. J. Lutz and E. Pernicka, *Archaeometry*, 1996, **38**, 313–323.
12. R. Cesareo, G.E. Gigante and A. Castellano, *Nucl. Instrum. Methods Phys. Res.*, 1999, **A428**, 171–181.

13. P.J. Potts, P.C. Webb, O. Williams-Thorpe and R. Kilworth, *Analyst*, 1995, **120**, 1273–1278.
14. H. Rollinson, *Using Geochemical Data*, Longman Scientific and Technical, Harlow, England, 1993.
15. O. Williams-Thorpe, P.J. Potts and P.C. Webb, *J. Archaeol. Sci.*, 1999, **26**, 215–237.
16. P.J. Potts, O. Williams-Thorpe and P.C. Webb, *Geostand. Newsletter* (now *Geostand. Geoanal. Res.*), 1997, **21**, 29–41.
17. O. Williams-Thorpe, P.C. Webb and M.C. Jones, *J. Archaeol. Sci.*, 2003, **30**, 1237–1267.
18. P.J. Potts, P.C. Webb and O. Williams-Thorpe, *J. Anal. Atomic Spectrosc.*, 1997, **12**, 769–776.
19. T.H. McK.Clough and W.A. Cummins, ed. *Stone Axe Studies II*, Council for British Archaeology Research Report 67, Council for British Archaeology, London, 1988.
20. C.I. Fell and R.V. Davis, in *Stone Axe Studies II*, Council for British Archaeology Research Report 67, ed. T. H. McK.Clough and W. A. Cummins, Council for British Archaeology, London, 1988, 73.
21. R.V. Davis, Stone implements of neolithic and Bronze Age date from the Northwest English highland zone and their origins: an archaeological assessment of recent petrological and experimental work, M. Phil. Thesis, University of Liverpool, UK, 1983.
22. G. Cooney and S. Mandal, *Antiquity*, 1995, **69**, 969–980.
23. S. Mandal, *Archaeometry*, 1997, **39**, 289–308.
24. R.S. Thorpe, O. Williams-Thorpe, D.G. Jenkins and J.S. Watson with contributions by R.A. Ixer and R.G. Thomas, *Proc. Prehistoric Soc.*, 1991, **57**, 103–157.
25. M.C. Jones and O. Williams-Thorpe, *Archaeometry*, 2001, **43**, 1–18.
26. R.A. Ixer, O. Williams-Thorpe, R.E. Bevins and A.D. Chambers in *Lithics in Action*, Lithic Studies Society Occasional Paper No. 8, ed. E. A. Walker, F. Wenban-Smith and F. Healy, Oxbow Books, Oxford, 2004, pp. 105–115.
27. O. Williams-Thorpe and R.S. Thorpe, *Archaeometry*, 1993, **35**, 185–195.
28. P. Phillips, W.A. Cummins and L. Keen, in *Stone Axe Studies II*, Council for British Archaeology Research Report 67, ed. T.H. McK.Clough and W.A. Cummins, Council for British Archaeology, London, 1988, p. 53.
29. O. Williams-Thorpe, D. Aldiss, I.J. Rigby and R.S. Thorpe, *Geoarchaeology*, 1999, **14**, 209–246.
30. M. Markham, Provenance studies of British prehistoric greenstone implements using non-destructive analytical methods, Ph.D. Thesis, The Open University, UK, 2000.
31. M. Markham, *J. Open Univ. Geol. Soc.*, 1997, **18**, 48–57.
32. M. Markham and P.A. Floyd, *Geosci. South-West England*, 1998, **9**, 218–223.
33. M. Markham *et al.*, papers in progress.

34. D.P.S. Peacock and V.A. Maxfield, *The Roman Imperial Porphyry Quarries: Gebel Dokhan, Egypt. Interim Report, 1996*, Universities of Southampton and Exeter, UK, 1996.
35. V.A. Maxfield and D.P.S. Peacock, ed., *The Roman Imperial Quarries, Survey and Excavation at Mons Porphyrites 1994–1998, Volume I: Topography and Quarries*, Egypt Exploration Society, London, 2001.
36. O. Williams-Thorpe, M.C. Jones, P.J. Potts and I.J. Rigby (based in part on field observations by D. Peacock, J. Phillips and N. Bradford), in *The Roman Imperial Quarries, Survey and Excavation at Mons Porphyrites 1994–1998, Volume I: Topography and Quarries*, ed. V.A. Maxfield and D.P.S. Peacock, Egypt Exploration Society, London, 2001, pp. 305–318.
37. O. Williams-Thorpe, P.J. Potts and P.C. Webb, Portable XRF analysis of Roman granite columns, poster presented at a Meeting of the Geochemistry Group of The Geological Society of London, 1996, London.
38. L. Lazzarini, *Boll. d'Arte Suppl.*, 1987, **41**, 157–172.
39. D.P.S. Peacock, O. Williams-Thorpe, R.S. Thorpe and A.G. Tindle, *Antiquity*, 1994, **68**, 209–230.
40. O. Williams-Thorpe, P.C. Webb and R.S. Thorpe, *Archaeometry*, 2000, **42**, 77–99.
41. M.A. Vicente, J. Delgado-Rodrigues and J. Acevedo, ed., *Degradation and Conservation of Granitic Rocks in Monuments*, European Commission Protection and Conservation of European Cultural Heritage Series Research Report No. 5, European Commission Directorate-General XII Science, Research and Development, Brussels, 1996.
42. R.V. Davis in *The Archaeologist and the Laboratory*, Council for British Archaeology Research Report No. 58, ed. P. Phillips, Council for British Archaeology, London, 1985, pp. 33–35.
43. P.J. Potts, F. Bernardini, M.C. Jones, O. Williams-Thorpe and P.C. Webb, Effects of weathering on *in situ* portable X-ray fluorescence analyses of geological outcrops: dolerite and rhyolite outcrops from the Preseli Mountains, South Wales, *X-Ray Spectrom.*, 2006, **35**, 8–18.
44. C. Stillman, *Geol. Today*, 1996, May–June issue, 89–90.

CHAPTER 9

Portable Systems for Energy-Dispersive X-Ray Fluorescence Analysis of Works of Art

ROBERTO CESAREO,[a] STEFANO RIDOLFI,[b] MAURIZIO MARABELLI,[c] ALFREDO CASTELLANO,[d] GIOVANNI BUCCOLIERI,[d] MARINA DONATIVI,[d] GIOVANNI E. GIGANTE,[e] ANTONIO BRUNETTI[a] AND MARCO A. ROSALES MEDINA[f]

[a] Dipartimento di Matematica e Fisica, Università di Sassari, via Vienna 2, 07100 Sassari, Italy; [b] Ars Mensurae, Rome, Italy; [c] Istituto Centrale del Restauro, piazza S. Francesco di Paola, 00184 Rome, Italy; [d] Dipartimento di Scienza dei Materiali, Università di Lecce, via per Arnesano, 73100 Lecce, Italy; [e] Dipartimento di Fisica, Università di Roma "La Sapienza", piazzale A. Moro 5, 00185 Rome, Italy; [f] Universidad de las Americas, Puebla 223, Col. Roma, 06700 Mexico D.F.

9.1 Introduction

Energy-dispersive X-ray fluorescence (EDXRF) analysis is a valuable technique for the study of works of art, because it is non-destructive, multi-elemental, simple and relatively inexpensive. Further, portable EDXRF equipment (PXRF) can be easily assembled and employed in museums, churches, excavation sites and so on. For this reason PXRF is a very popular analytical technique in the field of "archaeometry". Portability is, of course, extremely useful and almost mandatory in many cases, such as in the analysis of frescoes, large paintings, bronzes, brasses and gold alloys, *etc.*, especially when located in museums. In fact, in only a few cases is it possible to study a work of art outside

Portable X-ray Fluorescence Spectrometry: Capabilities for *In Situ* Analysis
Edited by Philip J Potts and Margaret West

its normal location (museum, church, excavation site, *etc.*) and, in any case, the bureaucratic problems and high costs of doing so are often prohibitive.

Various materials can be studied using PXRF instrumentation:

- paintings of all types, including frescoes
- alloys (bronzes, brasses, gold and silver)
- coins
- ceramic and porcelain artefacts, both for bulk composition and the analysis of decoration
- illuminated manuscripts and papers
- enamels
- stones, precious stones and marbles
- glasses
- stamps.

Furthermore, there are instances where a qualitative (or semiquantitative) analysis is sufficient (*e.g.*, a pigment or a surface treatment) and other cases where a quantitative approach is required (*e.g.* in the case of alloys).

PXRF analysis is generally undertaken on a sample area of a few mm^2, and a thickness of between μm and fractions of mm. The analysis is, therefore, superficial and dependent on the surface conditions. In special cases "capillary collimators" are employed to focus the radiation into smaller areas, of the order of 10^{-2}–10^{-4} mm^2.

Due to the reduced thickness of the analysed depth, XRF determinations are only representative of the bulk composition for homogeneous samples. This aspect is further explored in Chapter 8 on archaeological lithic provenancing.

Examples of the application of PXRF to works of art are given in this chapter: in particular the analysis of bronze statues (the statue of Perseo by Benvenuto Cellini and of Bartolomeo Colleoni by Andrea del Verrocchio), marble statues (the famous statue of David by Michelangelo), gold alloys (the golden altar of Saint Ambrose and precolombian artefacts) and the analysis of paintings (by De Chirico) and frescos (the chapel of the Scrovegni by Giotto).

9.2 Theoretical Background

9.2.1 Thick Samples

Artefacts such as statues, columns, alloys, *etc.* generally behave in PXRF analysis as "infinitely thick" samples, in the sense that the thickness of the object is infinitely thick with respect to the penetration depth of involved radiation (see Table 9.1 below).

When a "thick" sample is irradiated by photons of the appropriate energy,[1] it emits secondary photons that are characteristic X-rays from the elements that comprise the sample.

When a generic element a with concentration c_a, in an infinitely-thick and homogeneous sample is irradiated with incident photons of energy E_0 and

Table 9.1 Depth of an object irradiated with X-rays of the appropriate energy, giving rise to 90% of the fluorescent radiation.

Object	*Fluorescent radiation of element*	*Depth involving 90% of the fluorescent radiation*
Fresco[a]	Sulfur	25 μm
Fresco[a]	Iron	250 μm
Fresco[a]	Lead (L-lines)	1 mm
Fresco[a]	Tin (K-lines)	1 cm
Fresco[a]	Tin (L-lines)	70 μm
Bronze[b]	Copper	50 μm
Bronze[b]	Lead (L-lines)	15 μm
Bronze[b]	Lead (K-lines)	3 mm
Bronze[b]	Tin (K-lines)	120 μm
Gold[c]	Copper	5 μm
Gold[c]	Gold (M-lines)	1 μm
Gold[c]	Gold (L-lines)	10 μm
Gold[c]	Gold (K-lines)	500 μm
Ceramic[d]	Iron	130 μm
Ceramic[d]	Lead (L-lines)	0.5 mm
Stone or marble[e]	Iron	130 μm
Stone or marble[e]	Strontium	1 mm
Paper[f]	Zinc	3.5 mm
Paper[f]	Barium (L-lines)	0.6 mm
Paper[f]	Sulfur	200 μm
Gem or glass[d]	Iron	130 μm
Gem or glass[d]	Lead-Lα	0.5 mm
Wood (painting on)[g]	Copper	5 mm
Wood (painting on)[g]	Calcium	0.6 mm

[a]Assuming a composition similar to plaster, with a density of 1 g cm^{-3}; the real situation is, of course, much more complicated, and depending on the present pigment.
[b]Assuming a bronze with 100% Cu.
[c]Assuming a gold alloy with 100% Au; notably, measurement of M, L or K-X lines gives information from different penetration depths.
[d]Assuming a composition similar to SiO_2, with a density of 2 g cm^{-3}.
[e]Assuming stone or marble = $CaCO_4$ with a density of 2 g cm^{-3}.
[f]With a density of 0.7 g cm^{-3}.
[g]With a density of 0.5 g cm^{-3}.

intensity N_0, the secondary fluorescent X-ray intensity N_a is given by Equation (9.1):[1]

$$N_a = N_o k \omega_{\boldsymbol{a}} c_a [\mu_{\text{ph}.a}(E_0)/\mu_t(E_0) + \mu_t(E_a)] \tag{9.1}$$

where k is a constant representing the overall geometric and intrinsic efficiency; ω_a is the fluorescence yield of element a; $\mu_{\text{ph}.a}$ (E_0) represents the photoelectric attenuation coefficient of element a at incident energy E_0; $\mu_t(E_0) + \mu_t(E_a)$ represents the total attenuation coefficient of the sample at incident and fluorescent energies (E_0 and E_a) respectively; c_a is the concentration of element a, in percent by weight.

Besides fluorescent X-rays, given by Equation (9.1), the X-ray spectrum emitted by the irradiated "infinitely thick" sample also contains scattered photons, the intensity N_{sc} of which (mainly due to Compton scattering) is approximately given by:[1]

$$N_{sc} \approx N_0 k \mu_{sc}(E_0)/2\mu_t(E_0) \tag{9.2}$$

where $\mu_{sc}(E_0)$ and $\mu_t(E_0)$ are the scattering and the total attenuation coefficient of the sample at incident energy E_0, respectively. Scattered radiation is generally a disturbing effect that should be reduced as much as possible, but it can also be employed for normalization purposes.

Equations (9.1) and (9.2) are strictly valid for thick samples and for monoenergetic incident radiation. For bremsstrahlung radiation, Equations (9.1) and (9.2) should be integrated over the whole spectrum.

It may be calculated that, for low values of c_a ($<\approx 10$–15%)[i] or for limited c_a concentration ranges, Equation (9.1) yields approximately a linear relationship of N_a *versus* c_a.

Reference samples are required for an experimental test of Equation (9.1) and to establish the correlation between N_a and c_a.[1,2]

9.2.2 Thin Samples

When the penetration depth of the incident radiation emitted by the exciting source, and secondary X-rays emitted by the sample, is larger than (or of the same order of magnitude) as the sample thickness, then the sample is considered to be thin (infinitely thin when the sample thickness is negligible with respect to the penetration depth).

In this case, when a generic element *a* with concentration c_a is irradiated by incident photons of energy E_0 and intensity N_0, the secondary fluorescent X-ray intensity N_a is given by Equation (9.3):[1]

$$N_a = N_0 k \omega_a m_a \tag{9.3}$$

where m_a (in g cm^{-2}) is the mass of element *a* per unit area of the sample. In this case scattered radiation is negligible.

Works of art may often be considered as thin samples. Artefacts are not always "thick" samples. For example, in the case of frescos, there is a thick layer of plaster over which layers containing the pigments were painted (with a thickness from fractions of mm to mm). Furthermore, over the pictorial layer there is, in many cases, a layer caused by pollution effects (tenths of μm) containing sulfur, typically in the form of $CaSO_4$.[3]

When, for example, sulfur in frescos is to be analysed, the pathlength of the S K-line radiation is comparable to the thickness of the superficial sulfur layer. The same effect occurs when elements from iron to barium are analysed, when the penetration of the K-line radiation is of the same order of magnitude as the pigment layer. A quantitative approach is therefore difficult and generally not useful.

[i] Concentration is given in percent by weight, and in the following is simply indicated as %.

When radiation from an X-ray tube penetrates the pigments of a fresco or painting it is absorbed along its path. A fraction of the energy of the absorbed photons is converted into fluorescent photons of the various elements present and some of these photons, depending on the thickness of the intervening layers, reach the surface of the fresco and are detected.

For a fresco, the deepest layer corresponds to plaster. Superimposed is the preparation layer and above that one or more pigment layers, which are generally thin.

With a painting, the deepest layer corresponds to the canvas or wood, or any other support. Superimposed is, again, the preparation layer and above that one or more thin pigment layers.

As an example, in the case of Giotto's haloes in the chapel of the Scrovegni, the complexity of the X-ray spectra demonstrates the presence of various pigment layers below the gold leaf. Each layer behaves as a thin layer,[4] because elements such as strontium are also visible, originating from the deepest layer, which corresponds to the plaster. In this case of superimposed thin layers, fluorescent counts N_a from a generic element (a) may be written in the form:

$$N_a = N_o k \omega_a m_a A_{\mathrm{i}} \tag{9.4}$$

where A_{i} gives the attenuation of incident and output radiation if element a is in the internal layer (j), which is given by:[5]

$$A_j = \exp[-\Sigma_1^{j-1} \mu_i(E_0)x_j] \exp[-\Sigma_{j-1}^{1} \mu(E_{\mathrm{ph}\,.a})x_i] \tag{9.5}$$

where $\mu_i(E_0)$ and $\mu(E_{\mathrm{ph}.a})$ are the attenuation coefficient of the ith layer at incident energy E_0 and fluorescent energy $E_{\mathrm{ph}.a}$, respectively; x_i represents the thickness of the ith-layer.

A further consideration is the case of superimposed thin layers, when elements from various layers are visible. The attribution to the correct layer is possible, in many cases, when medium-heavy atomic number (Z) elements are present in a deeper layer, and the K-or L-lines of these elements (copper, gold, lead) are clearly visible in spectra.

9.2.3 Thickness Measurement in the Case of Thin Layers

In the case of superimposed thin layers (such as various pigment layers in a painting, or with gilding or silver plating) the approximate thickness of the pigments may also be calculated by the differential attenuation of K or L-lines.

Differential attenuation, and therefore the ratio $K\alpha/K\beta$ of a medium-Z element (b) present in a deeper (second) layer containing an element a depends on the thickness and composition of the more superficial (first) attenuating layer containing this element (typically the case of gilded copper in precolombian gold).

Differential attenuation, and therefore the ratio $L\alpha/L\beta$ of a heavy element (b') present in the second layer, depends on the thickness and composition of the first layer containing an element (a') (*e.g.* lead-lines attenuated by gold lines in the case of Giotto's gold halos).

Ratios Kα/Kβ or Lα/Lβ of elements b and b' *versus* thickness, density and attenuating properties of elements a and a' are given by:

$$[\mathrm{K}\alpha/\mathrm{K}\beta]_{b0} = [\mathrm{K}\alpha/\mathrm{K}\beta]_{b0}\exp - ([\mu_2 - \mu_1]\rho x)_a \quad (9.6)$$

$$[\mathrm{L}\alpha/\mathrm{L}\beta]_{b'0} = [\mathrm{L}\alpha/\mathrm{L}\beta]_{b'0}\exp - ([\mu_{2'} - \mu_{1'}]\rho x)_{a'} \quad (9.7)$$

respectively, where

- $[\mathrm{K}\alpha/\mathrm{K}\beta]_{b0}$ is the ratio Kα/Kβ in absence of any attenuating element a;
- $[\mathrm{L}\alpha/\mathrm{L}\beta]_{b'0}$ is the ratio Lα/Lβ in absence of any attenuating element a'; the terms $(\mu_2 - \mu_1)$ and
- $(\mu_{2'} - \mu_{1'})$ are the difference in attenuation coefficient difference at energies Kβ and Kα and Lβ and Lα, respectively.

Figure 9.1 shows the attenuation coefficients of copper, gold and lead as a function of energy,[6] and Figures 9.2 and 9.3 the ratios $R = \mathrm{K}\alpha/\mathrm{K}\beta$ and $R' = \mathrm{L}\alpha/\mathrm{L}\beta$ for copper K and lead L-lines, respectively, attenuated by gold leaf.

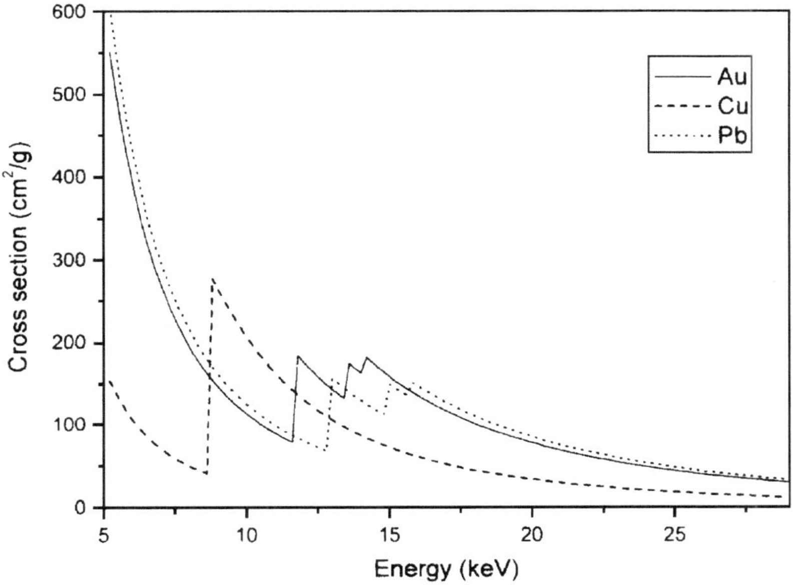

Figure 9.1 Mass attenuation coefficients of copper, lead and gold (in $\mathrm{cm^2\,g^{-1}}$) *versus* energy (keV). Pb-Lα and Pb-Lß lines bracket the Au-Lα-discontinuity, and Pb-Lß lines are therefore more absorbed by Au than Pb-Lα-lines; the opposite effect happens for Cu, and Kα-lines are more absorbed by Au. Au thickness can be determined in both cases.

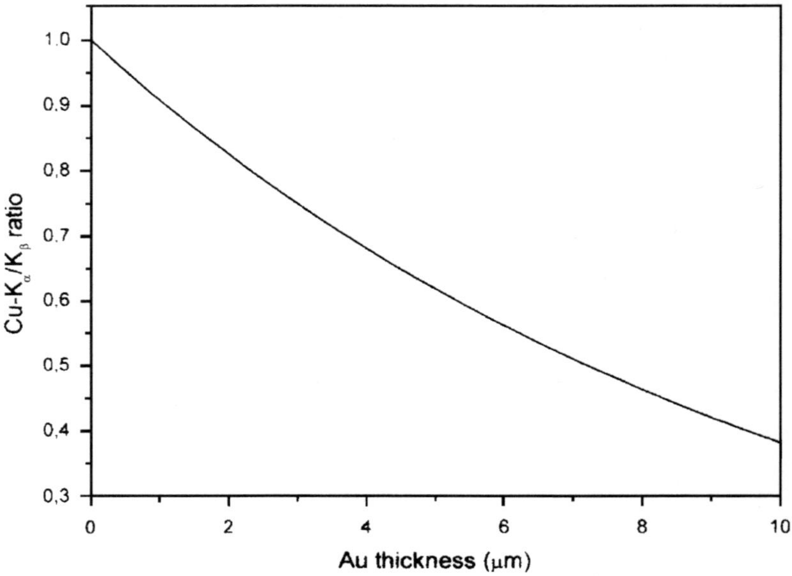

Figure 9.2 Ratio of Kα/Kβ Cu-lines attenuated by gold leaf of thickness *d* (μm).

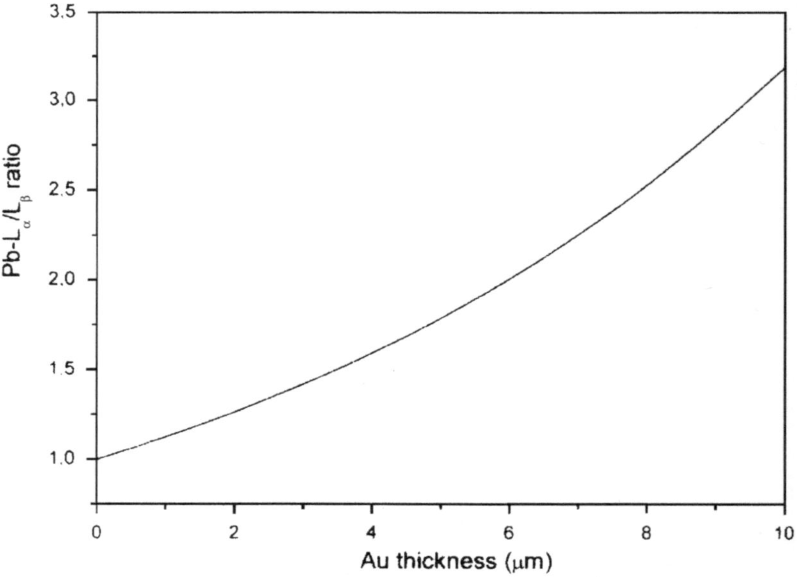

Figure 9.3 Ratio of Lα/Lβ Pb-lines attenuated by a gold leaf of thickness *d* (μm).

As observed above, PXRF is a "surface analysis" technique because of the limited penetration of both the primary and fluorescent radiation in the sample. As an example, Table 9.1 gives the sample depth from which 90% of the fluorescent radiation (FR) originates in the case of sulfur in frescos, copper, tin and lead in a bronze and gold, silver and copper in a gold alloy, and others. The values reported in Table 9.1 are approximate, because of the assumption of homogeneity of the samples and that incident and output radiation are normal to the sample surface.

9.3 Objects, their Preparation and Elements that can be Analysed

First of all it is important to classify the many and different artefacts in terms of *thin* or *thick* sample, according to Sections 9.2.1 and 9.2.2.

All alloys (bronzes, brasses, gold except gildings, silver and others), ceramics and porcelains, stones including gems, and glasses are thick samples; paintings of all types, illuminated manuscripts, papers, inks and stamps generally behave as thin samples (inks and stamps are infinitely thin samples).

Ideal sample conditions for PXRF analysis are given when the sample needs no preparation, and only qualitative or semiquantitative measurements are required. This is the general case for all types of paintings. In such situations, PXRF-analysis really has no competitors.

In some cases quantitative analysis is required, but with no sample preparation. This is, for example, often the case with the analysis of gold alloys, when surface and bulk composition coincide, and also with ceramics, glasses and others.

In other cases a quantitative analysis is required and the sample requires preparation. For example, in the analysis of bronzes and silver alloys, where surface patina, enrichment and corrosion processes affect the surface and, therefore, inhomogeneities are very common. For a possible correct analysis, the patina must be completely removed from the analyzed area (about 1 cm^2 or a little less), in the hope that corrosion is not too deep.

However, comparison between PXRF- and chemical analysis carried out on micro-withdrawals show that, despite the surface cleaning, the PXRF-analysis is sometimes still affected by uncertainties. PXRF-analysis is, therefore, not ideal for such samples.

It is important, in the field of works of art, to delimit the elements that are useful to analyse by means of PXRF. In order of atomic number Z, and considering that elements with $Z < 15$ can hardly be analysed by PXRF, because the energy of their X-rays is low, and are absorbed by the air and by the detector Be-window. Taking this limitation into account Table 9.2 shows the elements that can be analysed, the best conditions for their analysis and typical material (work of art) in which they can be found.

Table 9.2 Elements in works of art that can be determined by PXRF, X-rays emitted, suitable components of the equipment and typical materials in which they are present.

Z	*Element*	*Energy of emitted X-ray (keV)*	*X-ray tube*	*Detector*	*Material (see also ref. 7)*
14	Silicon	1.74	5 kV, Ca-anode[a]	TBW[b] Cooled[c]	Pottery, ceramics, glass
15	Phosphor	2.0	5 kV, Ca-anode[a]	TBW[b] Cooled[c]	Pottery, caramics, glass
16	Sulfur	2.31	5 kV, Ca-anode[a]	TBW[b] Cooled[c]	Index of pollution in frescos and monuments, pigments, glasses
17	Chlorine	2.62	5 kV, Ca-anode[a]	TBW[b] Cooled[c]	Index of pollution, enamels
19	Potassium	3.3	5 kV, Ca-anode[a]	TBW[b] Cooled[c]	Pottery, glass, pigments
20	Calcium	3.7	10–30 kV[d] Mo,W,Ag-anode	TBW[b] Cooled[c]	Pottery, glass, plaster, pigments
22	Titanium	4.5	10–30 kV[d] Mo,W,Ag-anode	TBW[b] Cooled[c]	Ti – white pigment, potteries, gems
24	Chromium	5.4	10–30 kV[d] Mo,W,Ag-anode	Cooled[c]	Pigments, pottery, glass, gems
25	Manganese	5.9	10–30 kV[d] Mo,W,Ag-anode	Cooled[c]	Pigments, pottery, glass, gems, ink, oxydian
26	Iron	6.4	10–30 kV[d] Mo,W,Ag-anode	Cooled[c]	Pigments, pottery, alloys, glass
27	Cobalt	6.9	10–30 kV[d] Mo,W,Ag-anode	Cooled[c]	Pigments, glass, pottery, gems, oxydian
28	Nickel	7.5	10–30 kV[d] Mo,W,Ag-anode	Cooled[c]	Rarely, with Co, in paintings, ink, modern alloys
29	Copper	8.04	15–30 kV[d] Mo,W,Ag-anode	Cooled[c]	Pigments, alloys, glass, gems, ink, pottery
30	Zinc	8.6	15–30 kV[d] Mo,W,Ag-anode	Cooled[c]	Zn – white pigment, alloys, glass, pottery
33	Arsenic	10.5	15–30 kV[d] Mo,W,Ag-anode	Cooled[c]	Pigments, bronzes
34	Selenium	11.2	15–30 kV[d] Mo,W,Ag-anode	Cooled[c]	Cd-red pigment
35	Bromine	11.9	15–30 kV[d] Mo,W,Ag-anode	Cooled[c]	Organic pigments

Table 9.2 (*Continued*).

Z	*Element*	*Energy of emitted X-ray (keV)*	*X-ray tube*	*Detector*	*Material (see also ref. 7)*
37	Rubidium	13.4	20–30 kV[d] Mo,W,Ag-anode	Cooled[c]	Pottery, glass
38	Strontium	14.1	20–30 kV[d] Mo,W,Ag-anode	Cooled[c]	Sr – yellow pigment, pottery, glasses, oxydian
39	Yttrium	14.9	20–30 kV[d] Mo,W,Ag-anode	Cooled[c]	Pottery, glass, oxydian
40	Zirconium	15.7	20–30 kV[d] W,Ag,Pd-anode	Cooled[c]	Pottery, glass, oxydian
41	Niobium	16.6	20–30 kV[d] W, Ag, Pd-anode	Cooled[c]	Pottery, glass, oxydian
47	Silver	2.98 (L)	5 kV, Ca-anode[a]	TBW[b] cooled[c]	Ag-alloys, Au-alloys
47	Silver	22.1 (K)	30–35 kV[d] W-anode	Cooled[c]	Ag-alloys, Au-alloys
48	Cadmium	3.13 (L)	5 kV, Ca-anode[a]	TBW[b] cooled[c]	Cd – yellow (or red) pigment
48	Cadmium	23.1 (K)	30–35 kV[d] W-anode	Cooled[c]	Cd – yellow (or red) pigment
50	Tin	3.44 (L)	5–7 kV, Ca-anode[a]	TBW[b] cooled[c]	Pigments, alloys, glass, pottery, ceramics
50	Tin	25.2 (K)	35–40 kV[d] W-anode	Cooled and efficient[e]	Pigments, alloys, glass, pottery, ceramics
51	Antimony (L)	3.6 (L)	5–7 kV, Ca-anode[a]	TBW[b] cooled[c]	Pigments, bronzes, glass, pottery, ceramics
51	Antimony	26.3 (K)	35–40 kV[d] W-anode	Cooled and efficient[e]	Pigments, bronzes, glass, pottery, ceramics
56	Barium	4.46 (L)	6–8 kV, Ca-anode[a]	TBW[b] cooled[c]	Pigments, pottery, ceramics
56	Barium	32.1	≈45 kV W-anode	Cooled and efficient[e]	Pigments, potteries
74	Tungsten	8.4–9.7 (L)	15–30 kV[d] Mo,Ag-anode	Cooled[c]	X-lines from X-ray tubes
78	Platinum	9.4–11.1 (L)	15–30 kV[d] Mo,Ag-anode	Cooled[c]	Impurity in gold
79	Gold	2.12 (M)	5 kV, Ca-anode[a]	TBW[b] cooled[c]	Au-alloys, halos in Middle Age frescos
79	Gold	9.7–11.4 (L)	15–30 kV[d] Mo,Ag-anode	Cooled[c]	Au-alloys, halos in Middle Age frescos

Table 9.2 (*Continued*).

Z	*Element*	*Energy of emitted X-ray (keV)*	*X-ray tube*	*Detector*	*Material (see also ref. 7)*
79	Gold	68 (K)	W-anode, 100 kV	Cooled[c]	Au-alloys, halos in Middle Age frescos
80	Mercury	2.2 (M)	5 kV, Ca-anode[a]	TBW[b] cooled[c]	Cinnabar (HgS), gold
80	Mercury	10–11.8 (L)	17–30 kV[d] Mo,Ag-anode	Cooled[c]	Cinnabar (HgS), gold
80	Mercury	70 (K)	W-anode, 100 kV tube	TBW[b] cooled[c]	Cinnabar (HgS), gold
82	Lead	2.35 (M)	5 kV, Ca-anode[a]	TBW[b] cooled[c]	Pb-white pigment, other pigments, alloys, glasses, pottery
82	Lead	10.5–12.6(L)	17–30 kV[d] Mo,Ag-anode	Cooled[c]	Pb-white pigment, other pigments, alloys, glasses, pottery
82	Lead	74 (K)	W-anode, 100 kV tube	Cooled[c]	Pb-white pigment, other pigments, alloys, glasses, potteries
92	Uranium	3.2 (M)	5 kV, Ca-anode[a]	TBW[b] cooled[c]	Glasses
92	Uranium	13.5–16.8	17–30 kV[d] Mo,Ag-anode	Cooled[c]	Glasses

[a]Ca-anode X-ray tube or other X-ray tube working at 5–10 kV, according to Section 9.4.1.1.
[b]TBW = thin Be-window ≤50 μm.
[c]N_2-cooled Si(Li) or HpGe or thermoelectrically cooled Si-PIN or Si-drift.
[d]Every X-ray tube working in this energy interval, drawing attention to the anode-material.
[e]The "normal" Si-PIN or Si-drift detectors with 300 μm thickness have not enough efficiency; at least 500 μm thickness Si is required, or CdTe or HgI_2 thermoelectrically cooled detectors or CdTe or HgI_2 thermoelectrically cooled detectors.

9.4 Instrumentation for PXRF Analysis

The portable XRF system is composed of an X or γ-ray source (*i.e.* an X-ray tube or a sealed radioactive source), an X-ray detector (generally a semiconductor detector, thermoelectrically or nitrogen cooled) and a multichannel analyzer.

9.4.1 Radiation Sources

9.4.1.1 Sealed Radioactive Sources

Despite several limitations and difficulties related to their use, radioactive sources are still employed for PXRF-analysis. A selected number of radioactive sources can be used, the most important of which are listed in Table 9.3.[8]

Table 9.3 Radioactive sources for portable EDXRF instrumentation.

Source	*Principal photon energies (keV)*	*Half-life*	*Elements that can be analysed*
^{55}Fe	5.9	2.7 years	$Z < 23$ (K-lines)
^{109}Cd	22, 88	453 days	$Z < 42$ (K-lines) $Z = 50-92$ (L-lines)
^{241}Am	59.5	433 years	$Z < 69$ (K-lines) $Z = 70-92$ (L-lines)
^{57}Co	122	272 days	Heavy elements (K-lines)

Radioactive sources are small and stable, and have a fixed energy with intensity output that decays with respect to half-life (Table 9.3). However, the photon flux is too low for many applications, including the field of archaeometry.

9.4.1.2 X-ray Tubes

Due to the relatively low cost of small size, low-power X-ray tubes, appropriate X-ray tubes are available for each application, taking into account the chosen detector characteristics.[9,10]

For low atomic number elements (11–19, including sulfur and chlorine), a low-power Ca-anode X-ray tube may be usefully employed,[11] working at 6–8 kV maximum voltage and 0.1–0.3 mA current. Alternatively, a Pd or Ag -anode X-ray tube may be used, working at about 5–6 kV and hundreds of μA.[9] In this case, the L-lines of Pd or Ag are used for excitation, with energies of 2.8–3.2 keV, which are close to the excitation energy of these elements.

Both Pd or Ag-anode X-ray tubes, working at 30–40 kV, can also be employed for excitation of medium (K-lines) and high (L-lines) atomic number elements. A W-anode X-ray tube working at 40 kV, 0.1 mA may also be used, especially when elements from Ag to Sn are to be determined.

As observed above, the photon output from an X-ray tube is generally collimated to irradiate an area of about 10–100 mm^2. The material and dimensions of the collimator must be carefully matched, to the detector, and the dead time of the system depending on the specific application.

There are cases when only a very small area or extremely small amounts of material must be irradiated and analysed. In these cases the incident radiation can be collimated using a "capillary collimator", and areas as small as 10^{-2}–10^{-4} mm^2 may then be irradiated.[12,13] This technique is called μ-XRF analysis and is currently applied to the analysis of works of art.[14]

9.4.2 X-ray Detectors

As traditional Si or Ge liquid nitrogen-cooled semiconductor detectors are not very compatible with portable systems, several small, thermoelectrically cooled

X-ray semiconductor detectors have increasingly been employed in recent years, including:

1. Si-PIN detector, 300 or 500 μm thick, with an area of 6 or 10 mm^2 and a typical energy resolution of about 180–200 eV at 5.9 keV.[15] This detector type has an efficiency that rapidly decreases above about 20–25 keV due to the relatively thin detector material. Good energy resolution of the detector is obtained with a relatively long amplifier shaping time of 12–24 μs and performance, therefore, deteriorates rapidly at high count rates.
2. Si-drift detector, 300–500 μm thick, with an area of about 4 mm^2, and an energy resolution of approximately 130–160 eV at 5.9 keV.[16–18] This detector also has an efficiency that rapidly decreases above 20–25 keV, but it responds much better than Si-PIN's at high count rates, due to the low shaping time constants of the amplifier (1–2 μs). Si-drift detectors are, therefore, better than Si-PIN, but are also much more expensive.
3. CdTe and CdZnTe (CZT) detectors,[15] with typical dimensions of $5 \times 5 \times 1$ mm^3, and an energy resolution of about 300 eV at 5.9 keV, can be used over the whole X-ray energy range with good efficiency.
4. HgI_2 detector,[19] with typical dimensions of $7 \times 7 \times 1$ mm^3, and an energy resolution of about 180 eV at 5.9 keV, can also be employed to cover the whole X-ray energy range.

The entrance of the detector generally requires a collimator, to avoid spurious and background photons. Also in this case, the material and size of the collimator must be carefully examined.

9.4.3 Multi-channel Analyser

Due to the small size and flexibility of modern multi-channel analysers, these devices can easily be coupled to sources and detectors to form a portable system.[15]

Multichannel analysers are also generally equipped with software for element identification, background subtraction, peak evaluation, *etc.*

9.4.4 Capillary Collimators

A polycapillary X-ray optics[20] consists of an array of numerous small, hollow glass tubes formed into a certain shape. The optics collects a large solid angle of X-rays that emerge from an X-ray source and redirects them by multiple external total reflection to form either a focused beam or a parallel beam. Small spots of 20–50 μm diameter can be obtained.

The use of polycapillary optics has become widespread in various X-ray analysis applications and also in archaeometry, where it is often used to collimate the incident beam to perform "micro-analysis". The rapid development of capillary optics has also triggered the development of related X-ray equipment such as micro-focus X-ray sources.

Monocapillary optics (single tapered channel optics) also offer further improvement in resolution, *e.g.* spots of 5–25 μm diameter can be achieved, beyond that currently achieved with polycapillary optics. An example of using capillary collimators is given later (see Figure 9.12 below).[21]

9.5 Experimental Set-Up

After the above brief review of the theory and instrumentation used for archaeometric applications, the following sections illustrate the use of PXRF in several projects undertaken by the authors' laboratories.

For the analysis of low atomic number elements (sulfur, chlorine, potassium, calcium) a portable XRF system was assembled, consisting of a Ca anode X-ray tube working at 5–8 kV^2 (Figure 9.4). In this case, the incident radiation consisted of the Ca-K lines at 3.7 keV and the bremsstrahlung radiation. As an alternative, a Pd anode X-ray tube was also employed, working at 4–6 kV^2. In the latter case the incident radiation was the Pd-L lines at 2.9 keV and the bremsstrahlung radiation. In both systems an AMPTEK thermoelectrically cooled Si-PIN detector was employed, having an energy resolution of 200 eV at 5.9 keV and coupled to a pocket AMPTEK multi-channel analyser.

To analyse the pigments of both the Scrovegni and De Chirico paintings at Giotto's Chapel, a portable XRF was assembled using a small, low-weight W anode Oxford X-ray tube, working at 30 kV and 10–50 μA, a small, thermoelectrically cooled AMPTEK Si-PIN detector with an energy resolution of 200 eV at 5.9 keV and a pocket AMPTEK multi-channel analyser.

For the analysis of the 14 authenticated De Chirico paintings, a thermoelectrically cooled Si-drift detector was employed, having an energy resolution of about 140 eV at 5.9 keV (Figure 9.5). The same instrumental configuration was used for the analysis of the equestrian statue of Bartolomeo Colleoni.

Precolombian gold was analysed with equipment consisting of an Eclipse II X-ray tube working at 30 kV and 0.1 mA maximum voltage and current, and a Si-PIN detector (Figure 9.6).

In all cases, both the X-ray tube and detector were placed about 20–30 mm from the object to be analysed, with tube and detector at angles of about 55° and 90°, respectively, from the object surface. An area of approximately 3–5 mm^2 was irradiated and analysed with a typical measuring time of about 100–200 s. When a capillary collimator was employed, an area of about 0.03 mm^2 was irradiated.

9.6 Results

This section reports results obtained in the field of archaeometry for the analysis of bronzes, frescos, gold objects and paintings.

9.6.1 Bronze Statues

9.6.1.1 The Perseo by Benvenuto Cellini

The huge statue of Perseo by Benvenuto Cellini, completed in 1554 and located in the Piazza della Signoria in Florence, consists of 1800 kg of bronze. It was

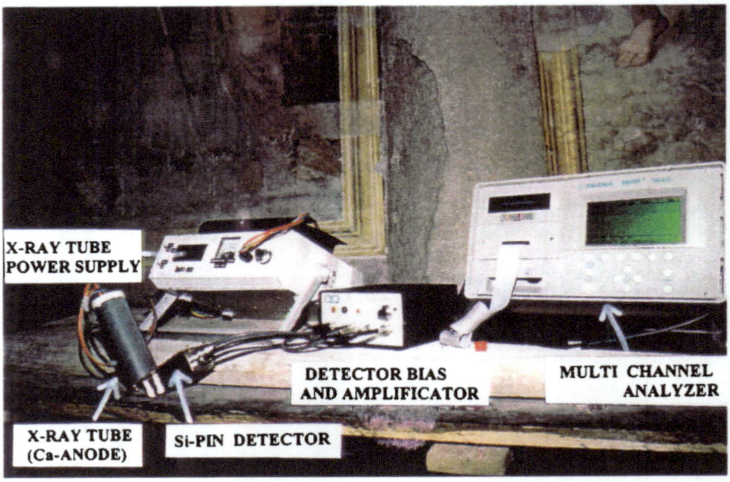

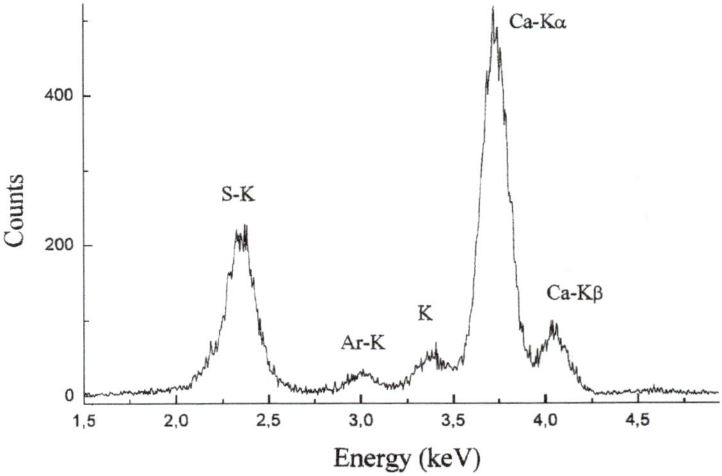

Figure 9.4 Experimental set-up for the analysis of sulfur and chlorine in the frescos of S. Stefano Rotondo in Rome, Fano's cathedral and the chapel of the Scrovegni in Padua. The equipment comprised of a Ca-anode X-ray tube working at 5–10 kV, 300 μA and a thin window Si-PIN detector. The X-ray spectrum of a typical area containing sulfur is also shown.

analysed during the restoration process in the Uffizi Museum in Florence.[22] The Perseo (Figure 9.7) is composed of two parts: the Perseo itself and the Medusa; in addition, some accessories such as the sword were analysed.

These bronzes were covered by a thick patina and, therefore, ten areas of about $5 \times 5\,mm^2$ were cleaned before analysis. In addition, four micro-samples and two metallographic samples were collected.

All these samples were analysed by PXRF and metallographic and electric conductibility studies were performed as well.[23] The results were

Figure 9.5 Experimental set-up of the equipment employed for the analysis of the bronze statue of Bartolomeo Colleoni. It is composed of a W-anode (40 kV), 1 mA X-ray tube and a high-resolution Si-drift detector.

Figure 9.6 EDXRF equipment employed for the analysis of the bronze globe of Peter's church in Rome (in this figure) and for precolombian gold objects in Peru. It consists of an Eclipse II X-ray tube and a Si-PIN detector.

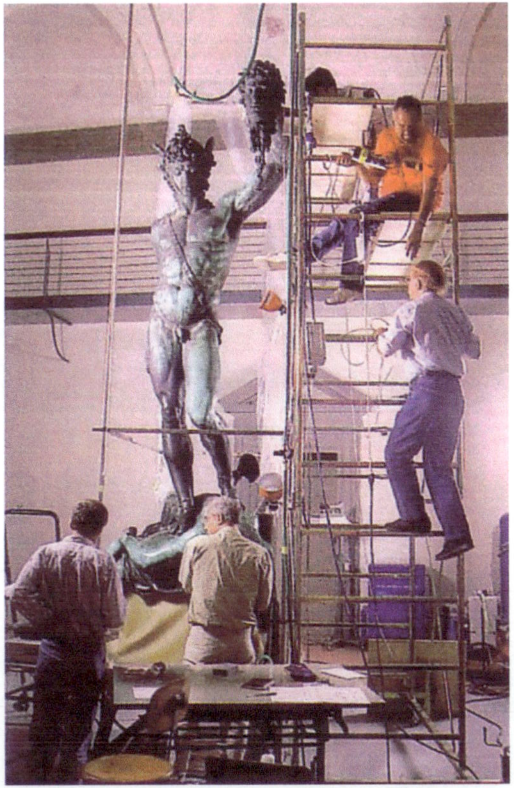

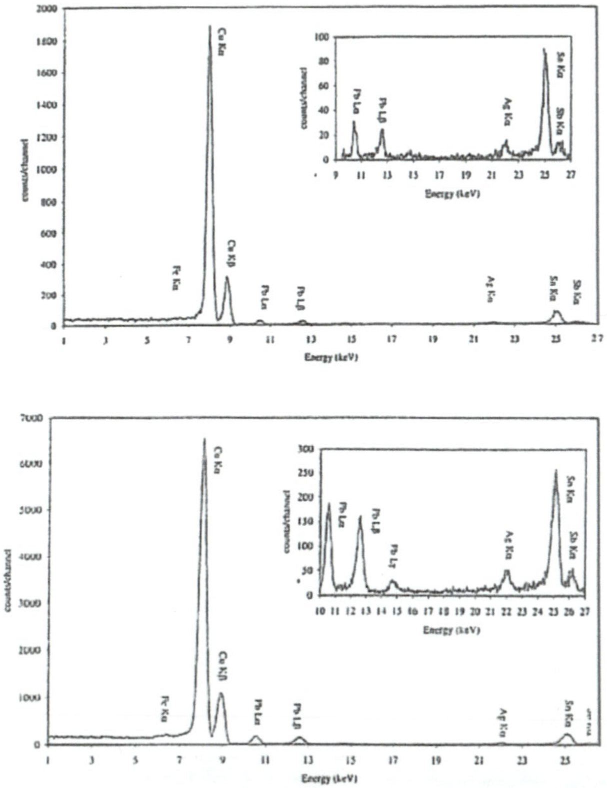

Figure 9.7 The huge statue "Perseo" by Benvenuto Cellini, Piazza degli Uffizi, Florence during the measurements, and two typical X-ray spectra of Perseo (top right) itself and of the Medusa head (bottom right), showing the different composition of the two parts of the huge statue (mainly concerning Pb- and Sn-concentration).

Table 9.4 Mean concentration values (% m/m[a]) of elements determined in Perseo and Medusa bronzes.

Object	*Cu (%)*	*Fe (%)*	*Pb (%)*	*Ag (%)*	*Sn (%)*	*Sb (%)*
Perseo	94 ± 1.5	0.5 ± 0.3	2.2 ± 0.3	0.05 ± 0.1	2.5 ± 0.3	0.6 ± 0.3
Medusa	90 ± 2	0.4 ± 0.3	1.2 ± 0.3	0.05 ± 0.1	7.0 ± 0.3	0.4 ± 0.2

[a]In the subsequent text, concentrations are also given in percent by m/m.

generally consistent. Figure 9.7 shows typical X-ray spectra of Perseo and Medusa. Table 9.4 gives the mean values of Perseo and Medusa bronzes.

Clearly, from the results in Table 9.4, Perseo and Medusa have different compositions, in terms of the Cu, Pb and Sn concentrations.

According to Benvenuto Cellini,[24] due to problems during the casting of the Perseo, 60 pounds of Sn and 22 English dishes (containing about 10% Sb) were added to the fusion melt. In some areas of the left side of the Perseo there was evidence of abnormal values of IACS conductivity, which are possible signs of the interruption of the casting process, as mentioned by Cellini.

9.6.1.2 The Equestrian Statue of Bartolomeo Colleoni by Andrea del Verrocchio

The huge equestrian statue of Barrtolomeo Colleoni (Figure 9.8) was created, but not completed, by Andrea del Verrocchio around 1480.[25] It is a gilded bronze, some 4 m high (without base), weighing about 4000 kg and located in Campo S. Giovanni e Paolo in Venice. During recent restoration work, the statue was systematically analysed at 21 points for the following purposes (see Table 9.5):

1. To determine the presence of sulfur and chlorine caused by pollution and the influence of sea water.
2. To analyse the patina composition and thickness.
3. To determine the composition of the soldering areas (3, 6, 7, 9, 12, 13 and 15 in Table 9.5).

To achieve this aim, 21 points were selected and first analysed without any cleaning of the surface, to clarify points 1 and 2. Additionally, areas of a few mm^2 around these 21 points were completely cleaned to remove the patina and were then analysed.

The following results were obtained (Table 9.5):

1. Sulfur and chlorine were present almost everywhere on the surface of the statue. It was not easy to determine exactly their concentration but it ranged from percent to tens of percent levels; Figure 9.9 gives a typical X-ray spectrum, showing the presence of sulfur and chlorine.
2. The behaviour of the patina composition *versus* "bulk" composition strongly depends on the degree of exposure of the analysed area, *i.e.*, if it has been exposed to rain, sun, humidity and so on. The patina contained much

Figure 9.8 The huge equestrian statue of Bartolomeo Colleoni by Andrea del Verrocchio (about 1480), located in Campo S Giovanni and S Paolo in Venice.

Table 9.5 Analysis of the equestrian bronze statue of Bartolomeo Colleoni (concentrations are in % by m/m).[26]

Area[a]	*Cu* (% *m/m*)	*Sn* (% *m/m*)	*Pb* (% *m/m*)	*Sb* (% *m/m*)	*Ag* (% *m/m*)	*Fe* (% *m/m*)	*Zn* (% *m/m*)
1-cleaned with patina	73.3	20.2	4.0	2.1	0.08	0.35	0
	22.8	64.7	6.0	5.9	0.1	0.5	0
2-cleaned with patina	78.1	17.1	2.6	1.8	0.08	0.3	0
	41.5	42.8	9.3	5.3	0.5	0.6	0
3-cleaned with patina	84.0	2.3	2.7	0.25	<0.01	0.3	10.4
	84.1	5.3	5.9	0.6	0.1	0.8	3.2
4-cleaned with patina	67.6	24.6	4.9	2.5	0.1	0.3	0
	71.9	22.7	1.6	2.6	0.1	1.2	0
5-cleaned with patina	84.6	11.9	2.0	1.2	0.02	0.3	0
	78.6	16.2	2.3	1.9	0.1	1.0	0
6-cleaned with patina	88.4	4.1	1.4	0.6	0.02	0.4	5.1
	84.4	7.7	2.1	1.3	0.1	1.5	3.0
7-cleaned with patina	88.0	0.5	2.5	<0.05	<0.01	0.45	8.5
	89.7	0.5	2.5	0.4	0.1	1.0	5.9
8-cleaned with patina	84.0	12.0	1.9	1.5	0.05	0.3	0
	69.6	22.5	4.2	2.5	0.1	1.1	0
9-cleaned with patina	81.4	12.0	1.9	1.0	0.03	0.3	3.4
	41.4	43.6	8.8	3.9	0.5	0.9	0.9
10-cleaned with patina	84.9	11.3	1.9	1.6	0.03	0.25	0
	80.1	14.8	1.8	2.2	0.1	1.1	0
11-cleaned with patina	86.5	10.2	1.8	1.3	0.03	0.2	0
	80.4	14.6	1.5	2.1	0.1	1.3	0
12-cleaned with patina	83.9	0.9	1.6	<0.05	0.02	0.35	13.2
	90.0	1.4	4.1	0?	0?	1.5	3.0
13-cleaned with patina	86.9	0.9	2.9	0.15	0.03	1.7	7.4
	84.9	3.2	5.5	0.5	0?	3.2	2.7
14-cleaned with patina	87.1	9.1	1.8	1.8	0.03	0.2	0
	81.1	12.4	2.5	2.6	0.1	1.4	0
15-cleaned with patina	89.8	4.4	1.6	0.8	0.05	0.35	3.0
	79.3	11.5	5.0	2.3	0.1	0.9	1.0
16-cleaned with patina	87.8	8.9	1.6	1.4	0.03	0.3	0
	82.1	13.1	2.1	1.4	0.1	1.1	0
17-cleaned with patina	73.8	21.2	1.7	3.0	0.04	0.25	0
	35.7	49.8	5.7	7.8	0.5	0.6	0
18-cleaned with patina	80.4	15.0	2.6	1.8	0.04	0.2	0
	38.1	45.8	9.7	5.6	0.6	0.3	0
19-cleaned with patina	81.1	14.8	2.2	1.6	0.02	0.25	0
	41.2	44.2	8.4	4.6	0.5	1.1	0
20-cleaned with patina	79.9	12.7	1.6	5.5	0.07	0.25	0
	57.0	26.7	5.3	9.6	0.6	0.8	0
21-cleaned with patina	<57.0	36.6	3.8	2.6	LS	LS	0
	35.1	52.2	7.3	4.4	0.5	0.5	0

[a]Key to the areas in the statue: (1) Horse, right posterior leg, exposed to rain. (2) Horse, right posterior haunch, exposed to rain. (3) Horse, right posterior haunch, soldering. (4) Horse, body, exposed to rain. (5) Horse, left posterior leg, exposed to rain. (6) Horse, left posterior leg, under welding, exposed to rain. (7) Horse, left side of belly, soldering area, not exposed to rain. (8) Horse, left anterior leg, exposed to rain. (9) Horse, left anterior leg in soldering area, exposed to rain. (10) Horse, left anterior leg, not exposed to much rain. (11) Horse, right anterior leg, internal area under soldering, not exposed to rain. (12) Horse, right anterior leg, soldering area. (13) Horse, right anterior leg, other soldering area. (14) Horse, neck, over soldering, exposed to rain. (15) Horse, soldering area between neck and body, exposed to rain. (16) Knight, left calf, exposed to rain. (17) Knight, dress in area of left haunch, exposed to rain. (18) Knight, left hand, exposed to rain. (19) Knight, left forearm. (20) Knight, chest area close to hand. (21) Knight, helmet, exposed to rain (measurement not reliable as PXRF hand-held in precarious position).

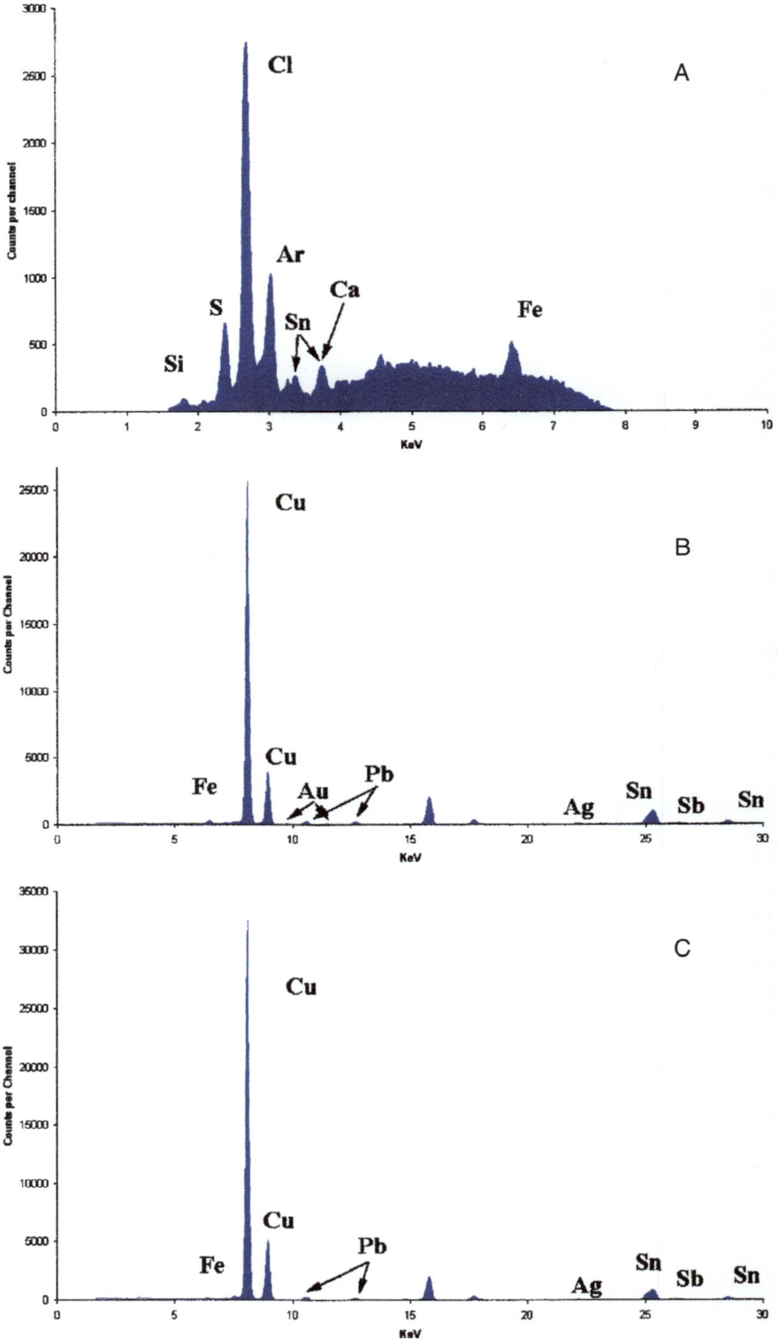

Figure 9.9 Typical X-ray spectra of area 10 of the equestrian statue of Bartolomeo Colleoni: (A) at 7.5 kV before cleaning, for analysis of S and Cl; (B) at 35 kV before removing patina; and (C) at 35 kV after removing patina.

more tin, lead and antimony, and less copper and zinc in areas that suffer extensive exposure (see, for example, areas 1, 2, 9, 15 and 17–20 in Table 9.5) whilst the composition of the patina was similar to that of the cleaned surface when the area was protected (see, for example, areas 7, 10 and 11).

3. Soldering areas, which are junctions between various parts of the horse (areas 3, 6, 7, 9, 12, 13 and 15), are basically brasses, and in some cases they have a composition similar to that of the neighbouring areas, with zinc substituting tin (areas 5 and 6 and areas 11–13). Area 9 is an exception, in which a small quantity of zinc (3.4%) seems to have been added to the bronze of the neighbouring areas 8 and 10.
4. Areas 1, 2 and 5 (posterior legs of the horse) show – with some doubt for area 5 – a similar composition; the same for areas 8, 10, 11 and 14 (the two anterior legs of the horse and neck), and for areas 18–20 (arm and chest of the knight). The welding areas 3, 6, 7, 12, 13 and 15 showed a similar content of tin and lead (but are different for zinc). However, these latter areas were not completely reliable, because the PXRF instrumentation was not fixed in position during the analysis but held by hand. It was concluded that the anterior part of the horse, and possibly the posterior legs, have the same composition. The left leg of the knight seems to be a different composition whilst the helmet was found to be quite different in composition. However, in area 17, the tin content was strongly dependent on the extent of cleaning, and it may be that a small part of the irradiated area had not been sufficiently cleaned. The same comment could be made for areas 1 and 2 in Table 9.5.
5. Iron composition was almost constant in all areas of the statue, with a mean value of $0.29 \pm 0.07\%$; the only exception being area 13, where the concentration was 6 times higher.
6. The Sb/Sn ratio was quite constant in all areas except 20 and 21.
7. The copper content did not vary too much ($83.3 \pm 4.7\%$ for the two statues; $84.3 \pm 4\%$ for the horse and $80 \pm 5\%$ for the knight).

9.6.1.3 The Statue of David by Michelangelo

The statue of David by Michelangelo is probably the most famous statue in the world. The artist began to work on the colossal figure in 1501; once the statue was completed, in 1505, it was decided to place it in the main square of Florence, opposite "Palazzo Vecchio". The statue remained there until 1873, when it was decided to protect the statue by moving it to the Accademia Gallery. After 500 years the condition of the statue was investigated for about 200 days using different diagnostic techniques, including PXRF. In particular, PXRF analysis is suitable, as shown above, for determining the sulfur content, which is one of the most important indicators of possible deterioration of stone monuments and frescos. Large S-concentrations become dangerous for stone in the presence of humidity and can irreversibly damage a work of art.

Figure 9.10 shows the PXRF-equipment during measurement of S-content, and a low-energy XRF-spectrum, showing the presence of sulfur, and its removal after a cleaning treatment with Solvigel.

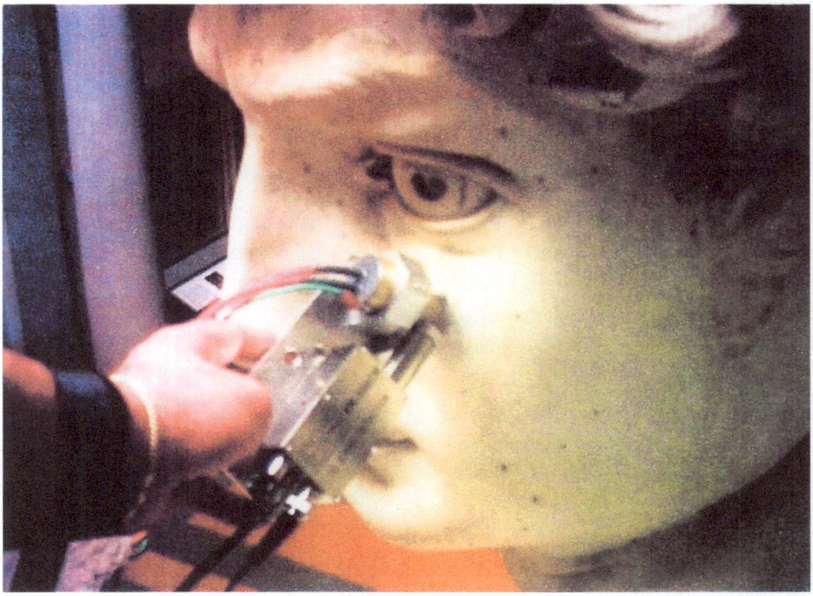

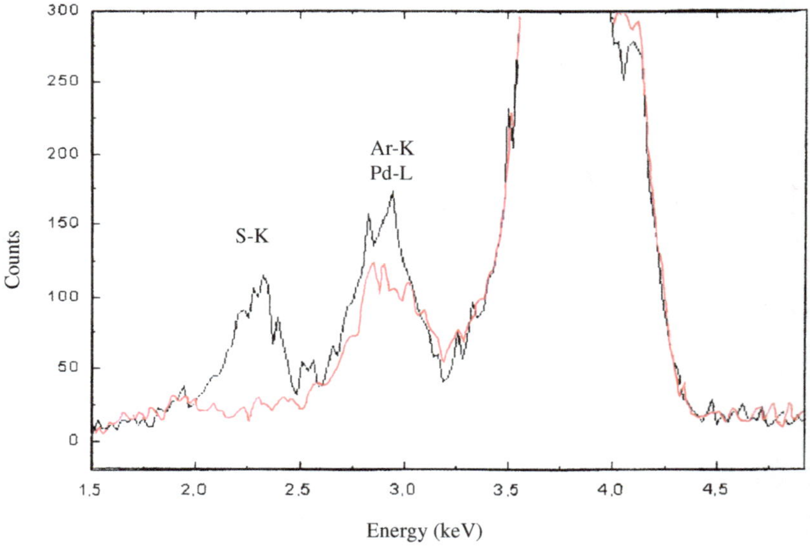

Figure 9.10 Analysis of sulfur in the statue of David by Michelangelo. The equipment is composed of a Pd-anode X-ray tube working at 4.5 kV and 60 μA and a Si-PIN detector. The eye of David looks "diffidently" at the equipment. X-ray spectra are shown, carried out at the right thigh under the gluteus, before and after cleaning with Solvigel.

9.6.2 Mural Paintings

9.6.2.1 Pollution Effects in Mural Paintings due to the Presence of S or Cl

Superficial sulfur, in the form of $CaSO_4$ (gypsum), is an index of pollution. It is often present on the surface of frescos and monuments, producing a black colouring and damage.[3] Sulfur and chlorine were analysed using the equipment described in Section 9.4.

By using a Ca-anode X-ray tube, the detection limit for sulfur was 0.1% m/m with 100 s measuring time, at 3 SD from the background.[2] By using a Pd-anode X-ray tube, the detection limit was approximately the same.[2] When both S and Cl are present, the Ca-anode configuration is much better for Cl, because the separation between Cl-lines and the exciting peak is greater.

The following frescos were analysed:

- Frescos attributed to Pomarancio in the church of S. Stefano Rotondo in Rome, which was under restoration by the "Istituto Centrale del Restauro" of Rome (Figure 9.4). Many areas were analysed[27] with the result that (1) in unrestored areas, sulfur was found everywhere at concentrations up to about 12% m/m; (2) in areas that were simply sponged with an appropriate solution, sulfur was found at concentrations between 2% and 4% m/m; and (3) in areas correctly treated to remove pollution layers, no sulfur was found.
- Frescos of Piero della Francesca, Church of S. Francesco, Arezzo; sulfur was found practically everywhere.[28]
- Frescos of Domenichino, Nolfi Chapel, Cathedral of Fano; again sulfur was found everywhere, except in the restored areas.[29]
- Ancient Roman frescos, Church of S. Clemente, Rome. In this church, both the ancient fresco in the mithraic school and the fresco in the lower basilica were analysed.[3,30] Large quantities of sulfur were detected in the lower basilica, at the upper level, close to the outside air. No sulfur was detected in the mithraic school, which is underground, in an isolated location.
- The famous frescos by Giotto in the "chapel of the Scrovegni" in Padua were systematically analysed in July and September 2001, at about 300 points, before and during restoration, to detect the possible presence of sulfur and to characterize the pigment compositions employed by Giotto.

With reference to the last campaign of analysis, the chapel of the Scrovegni was begun in 1303 and consecrated on March 25, 1305. The chapel, dedicated to Our Lady of the Annunciation, was commissioned by Enrico Scrovegni in suffrage for the soul of his father, Reginaldo, accused of usury. It was E. Scrovegni who commissioned Giotto to execute the frescos in the interior of the chapel, where the Master attained the height of his artistry, whereby this cycle of paintings is described as "a point of no return in the entire history of western painting" (Figure 9.11).[31]

Sulfur was analysed with two different instrumental configurations: first, using a Ca-anode X-ray tube (Figure 9.4) and, second, using a Pd-anode

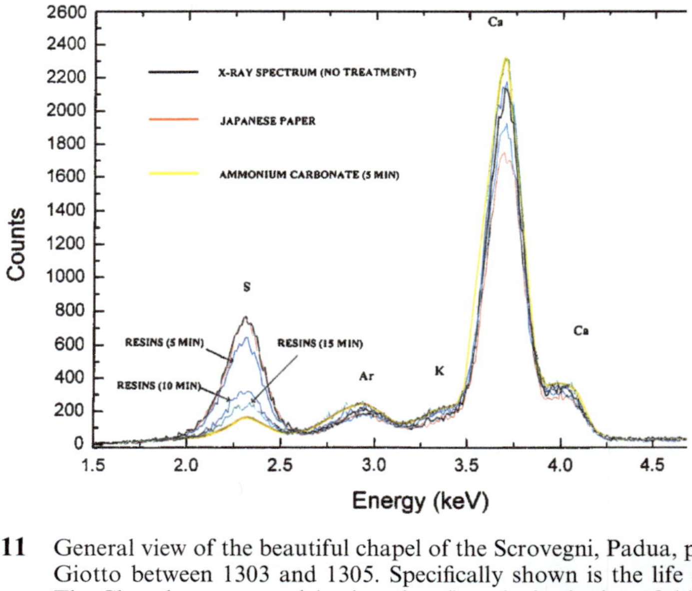

Figure 9.11 General view of the beautiful chapel of the Scrovegni, Padua, painted by Giotto between 1303 and 1305. Specifically shown is the life of Christ. The Chapel was restored (and analyzed) at the beginning of this century. X-ray spectra of an area of the fresco is also shown, and the results of various types of treatment for removing sulfur.

operating at low voltages to excite selectively Pd-L lines, with an energy of about 2.8 keV (suited to the excitation of sulfur). The fresco-pigments were analysed with the same Pd X-ray tube working at about 10 kV, and with a W-X ray tube working at 30 keV.

The following results were obtained:

- Sulfur was detected everywhere at a concentration level from about 1% m/m to about 10%, depending on the exposure and on the underlying pigment; the sulfur content was, for example, lower in azurite pigments, higher in the white and green pigments; the sulfur content strongly decreased after using a cleaning process based on ion-exchange resins;[30] the use of the Ca-anode X-ray tube gave rise to a "cleaner" spectrum compared with the Pd-L X-ray tube, but the count rates were much lower, due to the large output window of the first tube (the X-ray tube output was strongly collimated to irradiate an area of about 1 cm^2).
- The S-cleaning process, of great importance for the restoration of the fresco, was continuously monitored with the PXRF equipment. Various cleaning procedures were carried out and it was found that cleaning based on ammonium carbonate gave the best results.
- Chlorine was detected only once, in an area that had possibly been recently cleaned.
- Titanium was detected in many white areas, also indicating recent restoration; in fact titanium started to be used as a white pigment after the World War I.
- The gold of the halos shows a complicated X-ray spectrum, including, Au-L lines, Ag-K lines and Cu-K lines, which are all typical of gold, S (due to pollution), Ca, Fe, Ni, and Pb, with the latter at high levels. These elements may be present due to an under-layer of pigment.[32]

Figure 9.12 shows, as an example, the analysis of the halo of Good, both carried out *in situ* and, with the help of a capillary collimator, on a very small fragment.[33]

9.6.3 Gold Artefacts

9.6.3.1 The Golden Altar of Saint Ambrose

The golden altar of S. Ambrose in Milan is considered to be one of the most important examples of goldsmith's work. It was constructed in the period 824–859 AD by the artisan Volvinius, of whom almost nothing is known.

Ancient silver normally contains copper, lead, gold and iron (gold and iron at concentrations below 1% m/m) and in some cases tin, nickel and zinc at trace levels. In silver alloys, there is a frequent occurrence of surface enrichment, especially for base silver or for copper-rich alloys.

The altar of Volvinius was analysed in about 200 areas (Figures 9.13 and 9.14). Table 9.6 shows mean values of the results.[34]

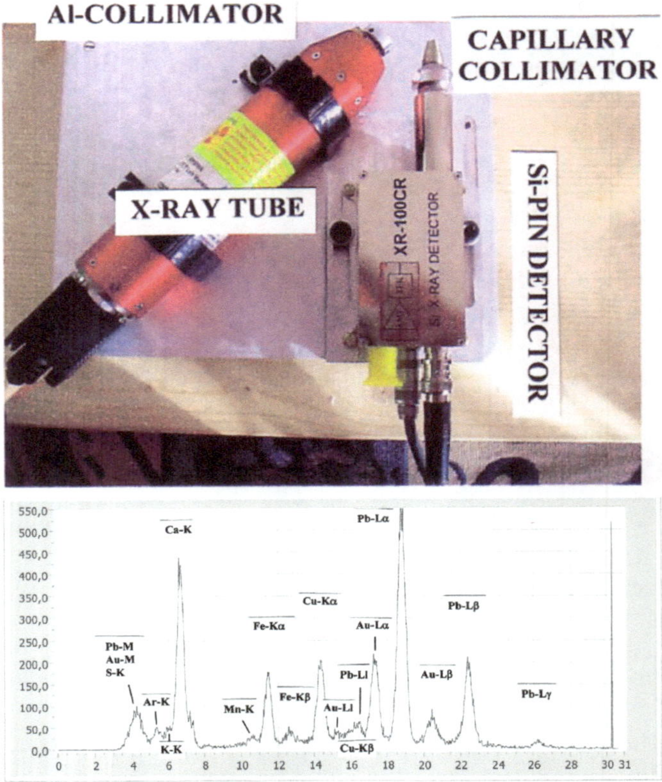

Figure 9.12 PXRF-spectrum of a fragment from the halo of Good in the chapel of the Scrovegni painted by Giotto in Padua, analysed with equipment (top figure) consisting of an Eclipse II X-ray tube and a Si-PIN detector with a capillary collimator on the detector window.

From the above results the following conclusions can be made:

In the three sides made of silver, the silver has the same composition, *i.e.*, (Ag ≈ 94.6% m/m, Au ≈ 1.6% m/m, Cu ≈ 3.4% m/m and Fe ≈ 0.4% m/m); the presence of gold in the silver alloys is remarkable.

Gilding on the silver sides appears to have a similar mean composition, *i.e.*, Fe ≈ 0.2% m/m, Cu = 0% m/m, Ag ≈ 12.7% m/m, Au ≈ 87.1% m/m, despite complications due to its reduced thickness.

In the gold panel, the gold has a mean composition of Au ≈ 93.5% m/m, Ag ≈ 2.1% m/m, Cu ≈ 3.8% m/m and Fe ≈ 0.6% m/m, except in the three panels of the 20th century.

9.6.3.2 Precolombian Gold from Northern Peru

Gilded Copper from the "Museo de las Tumbas Reales de Sipán", Lambayeque. One of the most remarkable cultures that flourished in ancient Peru was that of the Moche, which evolved on the north coast between the first

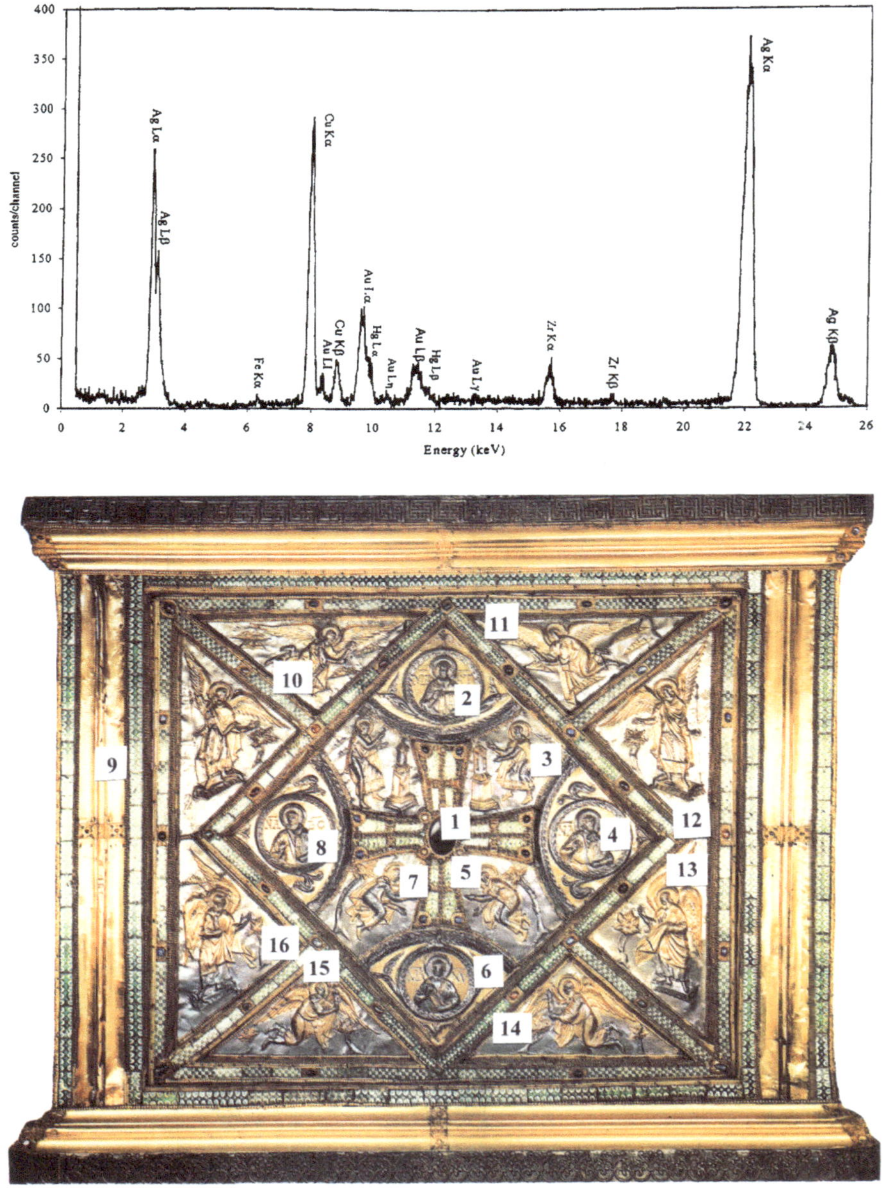

Figure 9.13 Silver panels with gildings on the east side of the altar of Volvinius, and a typical X-ray spectrum of this panel, obtained with a W-anode X-ray tube operating at 30 kV and 0.2 mA, with a Si-PIN detector.

and eighth centuries AD, close to present-day city of Chiclayo.[35] Moche metal-workers knew how to plate copper with silver or gold, and how to treat a low grade gold alloy (known as tumbaga) so that the surface appears to consist of high carat gold.[36–38] About 20 years ago, the excellent ability of Moche's

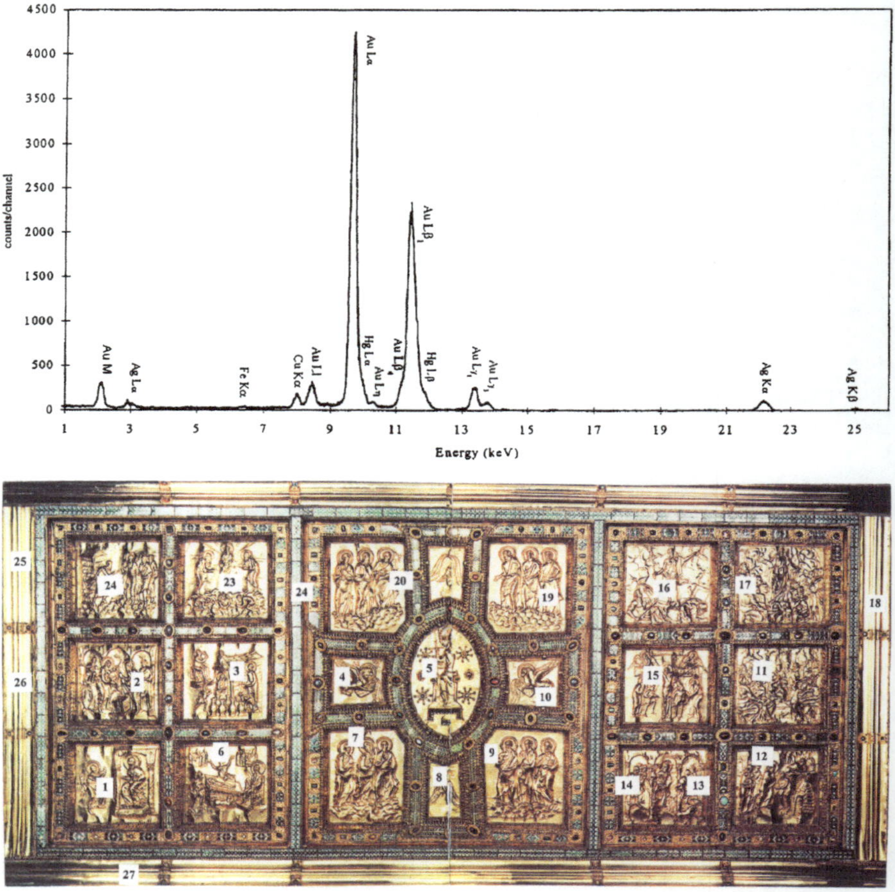

Figure 9.14 Gold panels on the west side of the altar of Volvinius, and a typical X-ray spectrum of this panel, obtained with a W-anode X-ray tube working at 30 kV and 0.2 mA, with a Si-PIN detector.

jewellers was impressively demonstrated by the gold and silver ornaments and ceremonial artefacts found by W. Alva in the Royal Tumbs of Sipán[39] and now exhibited in the "Museo Tumbas Reales de Sipán", in Lambayeque.

About 40 objects have been analysed, each of them at several different points (Figure 9.15).[40] From PXRF spectra it could be deduced that many of the objects are on gilded copper. In this last case, the gold thickness could be determined, using the methods explained in Section 9.2.3, *i.e.* using the (Au-Lα/Au-Lβ) and the Cu(Kα/Kβ) ratios. The thickness-distribution of the gilded copper object from Sipán Museum is shown in Figure 9.16.

Gilded Copper and Silver from the "Museo de Sicán", Ferrañafe, Peru. The name Sicán has recently been adopted to refer to the culture that flourished in the Lambayeque region between approximately 750 and 1350 AD.[41]

Table 9.6 Summary of the PXRF-measurements on the altar of Volvinius (concentrations in % m/m).[34]

	Fe (% m/m)	*Cu (% m/m)*	*Au (% m/m)*	*Ag (% m/m)*
NORTH SIDE				
Silver	0.4 ± 0.3	3.1 ± 1.8	1.8 ± 0.4	94.7 ± 1.7
Gildings[a]	0.1	0	85	15
SOUTH SIDE				
Silver	0.38 ± 0.3	2.9 ± 1.8	1.8 ± 0.4	94.8 ± 2.2
Gildings[a]	0.3	0.1	87.6	12
WEST SIDE				
Gold	0.6 ± 0.65	3.8 ± 2.5	93.5 ± 2.5	2.1 ± 1.7
20th century panels	0	0	98.5	1.5
EAST SIDE				
Silver	0.32 ± 0.25	4.1 ± 0.9	1.3 ± 0.5	94.3 ± 1.4
Gildings[a]	0.23	0	88.8	11

[a]Only mean values without errors are given because the gilding concentration values are subject to large fluctuations, due to its thin and variable depth (10–15 μm) in all areas.

This culture traces its roots in the Moche culture (see Section 9.6.3.2) and in other contemporary cultures.

The most outstanding technology from Sicán can be admired in the pottery metals, produced with arsenical copper and tumbaga (low carat gold or silver). Figure 9.17 shows a beautiful gold mask displayed in the Museum of Sicán and two typical PXRF spectra, showing the presence of cinnabar at the surface. The mask is on gilded copper, with a mean Au-thickness of about 3 μm.

Other objects from Sicán are on gilded silver, and both from the Ag(Kα/Kβ) ratio = 0.83 and (Au-Lα/Ag-Kα) ratio = 3.1, a gilding thickness of about 6–7 μm could be deduced.[40]

9.6.4 Paintings

Portable X-ray fluorescence analysis is particularly suited to the analysis of paintings because it is non-destructive and multi-elemental. Furthermore, it can be carried out easily with portable equipment.

PXRF analysis of paintings generally gives the following information:

- possible presence of sulfur and chlorine at the surface, due to pollution effects (Section 9.6.2.1);
- identification of the elements, and often chemical compounds, identifying the pigments employed by the artist (Section 9.6.4.1);

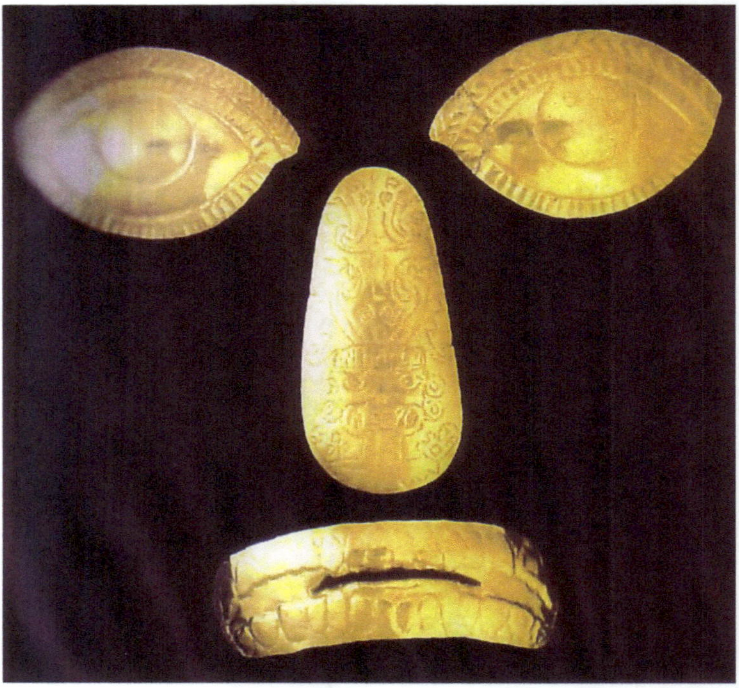

Figure 9.15 Face decoration of the "Senhor de Sipàn", Museum "Tumbas reales de Sipàn", Lambayeque, Peru. These objects are on gilded copper.

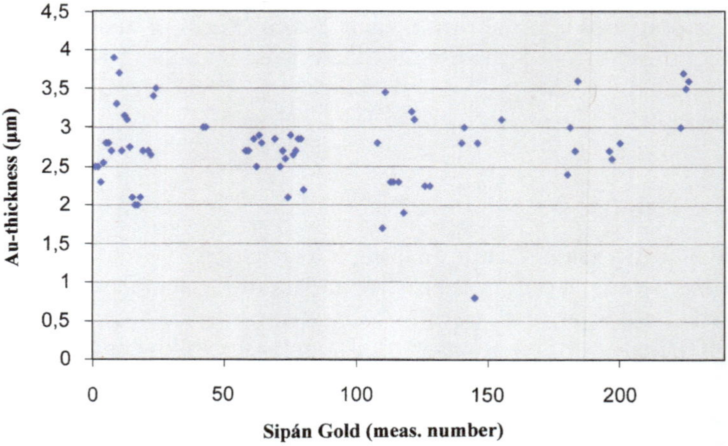

Figure 9.16 Distribution of gold-leaf thickness in gilded copper objects from the Royal Tumbs of Sipán.

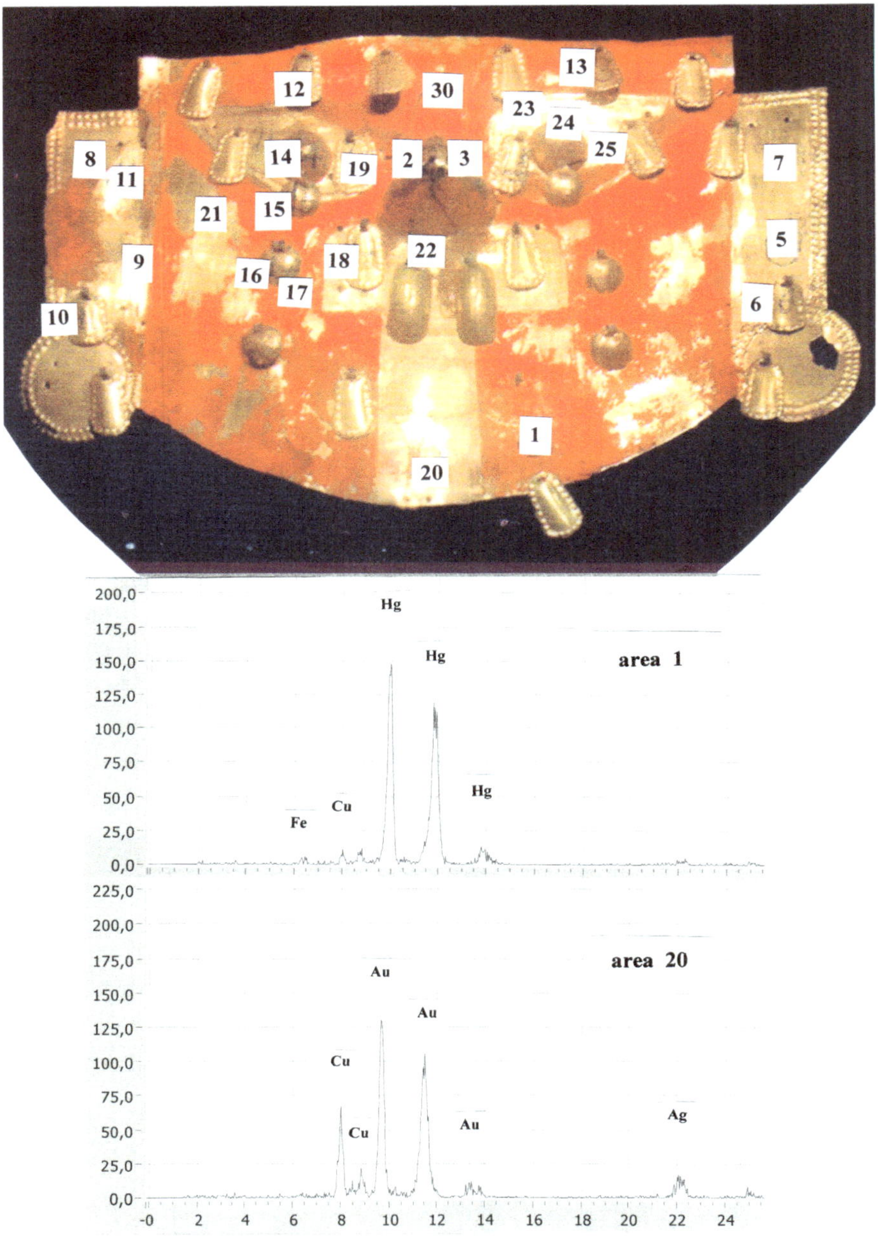

Figure 9.17 Mask from the Museum of Sicàn, Ferrañafe, Peru, made of gilded copper, and the XRF-spectra of areas 1 (top) and 20 (bottom).

- identification of previous restoration areas, through the presence of "modern" elements, such as titanium, zinc, cadmium and so on;
- identification of fakes (see Section 9.6.4.2).

9.6.4.1 Pigments and Chemical Compounds

The PXRF technique shows the chemical elements present in the sample and not the chemical compounds. In many cases this is not a big drawback in the determination of the pigments analyzed, since many of those are easily recognized by the visual colour and the presence of some strong characterizing single inorganic chemical element.

It is easy to identify the white pigment, white lead "biacca" ($PbCO_3$), because of the presence of high quantities of Pb in the X-ray spectrum. The same can be said for a red pigment of which the XRF spectrum shows the presence of element mercury (cinnabar: HgS) or for a yellow pigment where arsenic is found (orpiment: As_2S_3) or a brown pigment with manganese (MnO_2).

Slightly different is the case where more than one inorganic element characterizes the pigment. As an example, we can talk about the lead-tin yellow (one form of which is the compound Pb_2SnO_4) recognized in an XRF spectrum, as the name suggests, by the simultaneous presence of lead and tin. Other examples are Naples yellow ($Pb_2Sb_2O_7$), recognized in an XRF spectrum by the presence of lead and antimony, and emerald green ($Cu[C_2H_3O_2]3Cu[AsO_2]_2$), recognized in an XRF spectrum by the simultaneous presence of copper and arsenic, and even smalt blue (a potash glass containing cobalt) that can be recognized by the simultaneous presence of cobalt, and often by potassium, arsenic, bismuth, nickel and other impurities. Figure 9.18 shows, as an example,

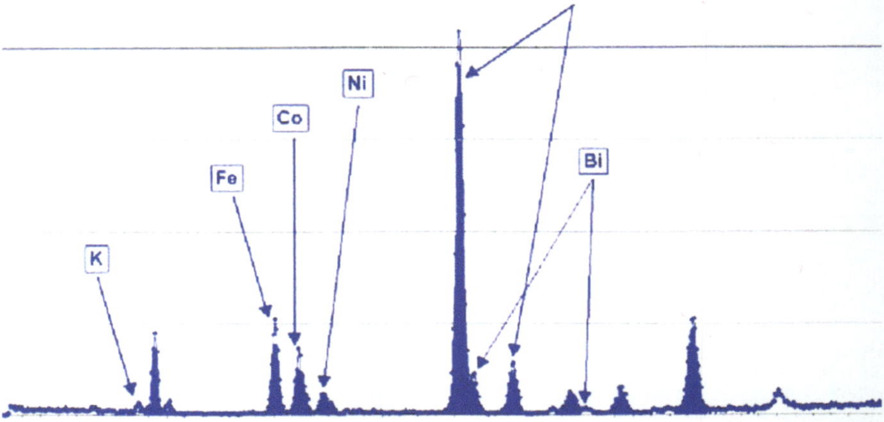

Figure 9.18 XRF-spectrum of a smalt blue pigment belonging to the frescos in the Villa "La Tesoriera" created in 1713 in Turin.

the spectrum of a smalt blue pigment belonging to the frescos in the Villa "La Tesoriera" created in 1713 in Turin.

Although from the spectrum in Figure 9.18 it could be suggested that the pigment is smalt blue, this is not quite correct. More significantly, this spectrum could easily be reproduced by a mixture of modern cobalt blue ($CoOAl_2O_3$) and orpiment (As_2S_3). The presence in a single XRF spectrum of the different elements that belong to a chemical compound is not enough to assert that the compound is actually present. The elements found in the spectrum could belong to different pigments mechanically mixed or even belonging to different layers. In this case it is necessary to make more measurements on the same "campitura" with the goal of establishing the precise relationship among elements. Only by making more measurements of the same "campitura" and plotting the counts related to the elements we want to connect is it possible to assert the presence or not of a chemical relationship between the elements considered.

Figure 9.19 shows scatter-plots of cobalt *vs.* arsenic, nickel, bismuth and potassium belonging to these frescos. The linear relationship between the elements attests to the searched for relationship that demonstrated the presence of smalt blue pigment in the frescos.

9.6.4.2 *Identification of Fakes*

Fifteen paintings from the last period 1960–1970 of Giorgio De Chirico, from the "Fondazione De Chirico, (piazza di Spagna, Roma) were analysed to identify the pigments typically employed by the artist during this period.[42] All these paintings appeared to have a similar composition, with the following characteristics:[43]

- a preparation layer made with a mixture of lead and zinc white,
- the systematic use of lead, not only for the preparation layer,
- red colours based on the use of cinnabar (HgS),
- moderate use of organic paints.

Figure 9.20 shows a typical painting by De Chirico with corresponding X-ray spectrum.

Eleven other paintings attributed to De Chirico were also successively analysed to establish their authenticity. The analysis confirmed that all the paintings were by one artist with characteristics as follows:

- a preparation layer made with zinc oxide,
- the almost complete absence of lead,
- the red colour made from cadmium red,
- frequent use of organic pigment.

Figure 9.21 shows a typical painting and X-ray spectra from this group. The eleven paintings have very different characteristics when compared with the group of 15 authenticated paintings and are thought probably to be fakes.[44]

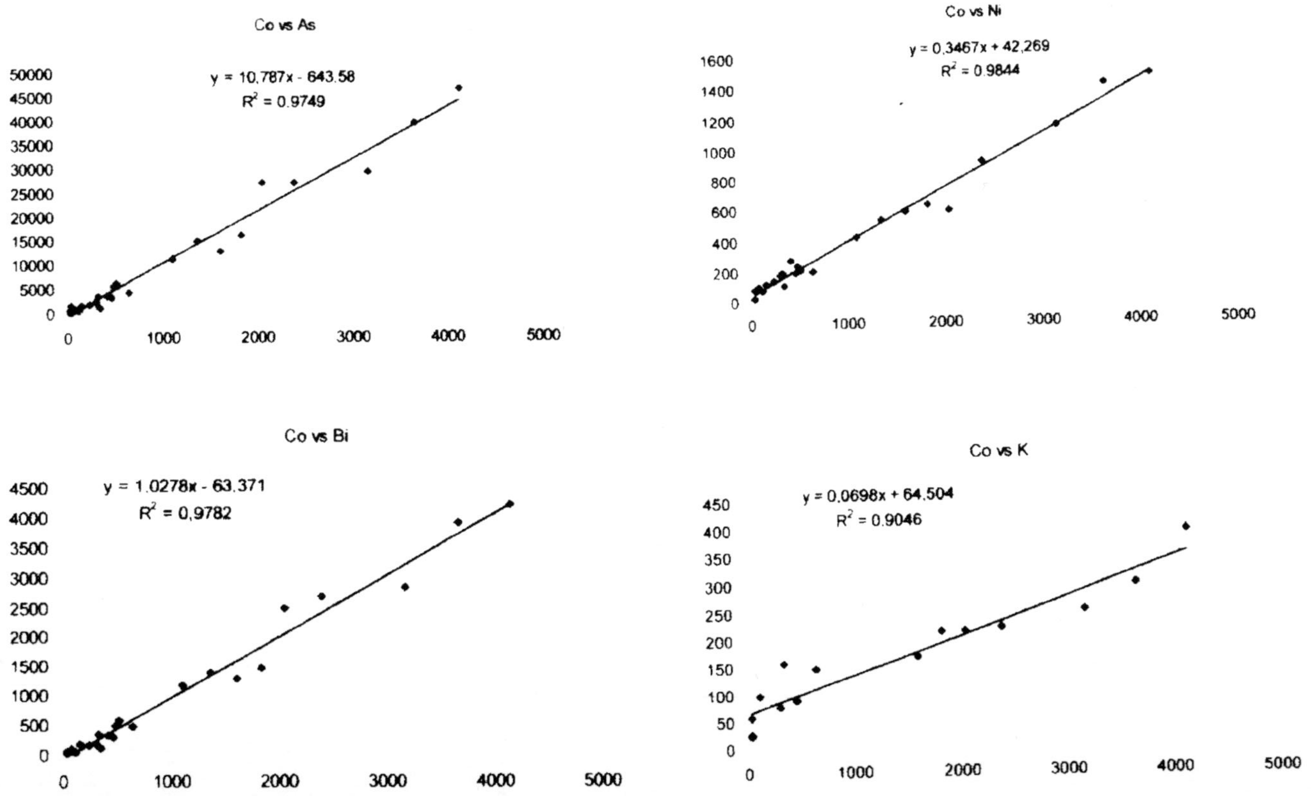

Figure 9.19 Scatterplots of cobalt *versus* arsenic, nickel, bismuth and potassium from the frescos in the Villa "La Tesoriera", showing a linear relationship between these elements.

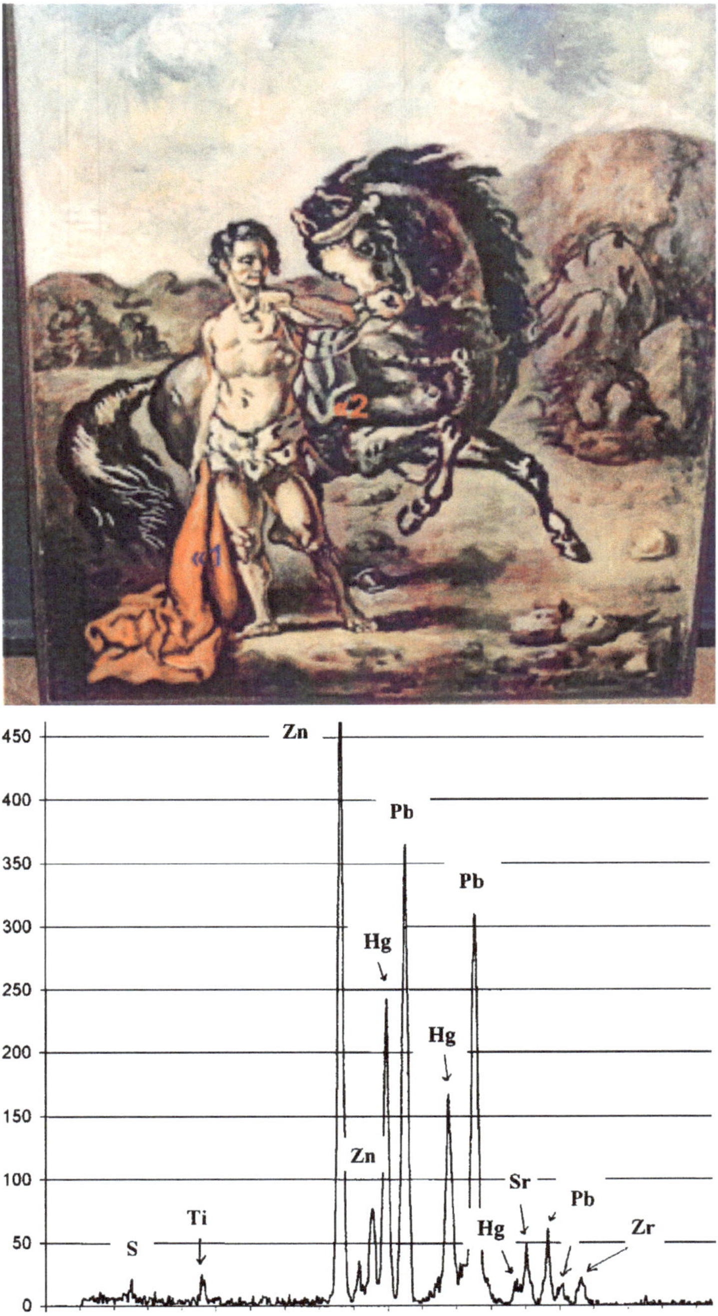

Figure 9.20 Child with horse by Giorgio de Chirico, Fondazione de Chirico, Piazza di Spagna, Rome, and the X-ray spectrum of a red pigment containing cinnabar.

Figure 9.21 Child with horse supposedly by Giorgio de Chirico and the X-ray spectrum of a red pigment composed of cadmium red.

9.7 Conclusions

Works of art such as paintings, frescos, alloys, coins, illuminated manuscripts, enamels, *etc.* should generally be analysed in a non-destructive manner and on site. PXRF analysis is particularly suited for the analysis of artefacts because it is non-destructive and multi-elemental.

PXRF equipment of reduced size and weight can be currently assembled, which includes a miniaturized, dedicated X-ray tube (or, in some cases, a radioactive source emitting X-or γ rays), a thermoelectrically cooled X-ray detector and a small pulse height analyser.

The properties of such equipment in terms of elements that can be analysed (from sulfur to lead) and minimum detection limits (often better than 100 $\mu g\, g^{-1}$) make PXRF suitable for the analysis of almost all types of artefacts.

In the last few years PXRF equipment has been successfully used for the qualitative analysis of pigments in paintings, frescoes, enamels, illuminated manuscripts and so on, and for quantitative analysis of bronzes, brasses, gold alloys, *etc.*

It is expected that future equipment for PXRF analysis will be completely battery equipped and further reduced in size and weight to increasingly extend areas of applications.

Acknowledgements

This work was partially supported by the Consiglio Nazionale delle Ricerche, Programma Finalizzato "Beni Culturali".

This work was partially performed within the framework of IAEA-CRP (G4.20.02/1371) "Unification of nuclear spectrometries: integrated techniques as a new tool for material research".

One of the authors (R.C.) acknowledges A. Bustamante, S.J. Fabian, M. dos Anjos, C. Calza, R.T. Lopes and M. Delgado for participating in the measurements on precolombian golds in Peru, W. Alva, L. Chero, F. Gutiérrez, M. Seclén, R. Dominguez from the "tumbas reales de Sipán", C. Elera, V. Curay, L. Cajo from the Museum of Sicán, and I. Veloachaga, E. Gamarra, S. Chávez, J. Berrios, F. Campos, A. Manrique from the "Museo Nacional" of Lima for their kindness and helpful cooperation during the measurements.

References

1. R. Cesareo, Photon induced X-ray emission, in *Nuclear Analytical Techniques in Medicine*, ed. R. Cesareo, Elsevier, Amsterdam, New York, Tokyo 1988, pp. 19–119.
2. R. Cesareo, C. Cappio Borlino, G. Stara, A. Brunetti, A. Castellano, G. Buccolieri, M. Marabelli, A.M. Giovagnoli, A. Gorghinian and G.E. Igante, A portable energy dispersive X-ray fluorescence apparatus for the analysis of sulfur and chlorine in frescos, *J. Trace Microprobe Techniques*, 2000, **18**, 23–33.

3. M. Laurenzi Tabasso and M. Marabelli, Il degrado dei monumenti in Roma in rapporto all'inquinamento atmosferico, BetaGamma Ed., Viterbo (1992).
4. M. Marabelli *et al.*, Metal leaves utilized for decoration of Giotto's mural paintings, *Bollettino d'arte del Ministero per i Beni e le Attività Culturali*, Poligrafico dello Stato, Roma 2005, pp. 121–144.
5. R. Cesareo, A. Brunetti and S. Ridolfi, Pigment layers and precious metal sheets by EDXRF-analysis, submitted to *X-ray Spectrometry*.
6. M. Berger and J.H. Hubbell, *XCOM: Photon Cross Sections on a Personal Computer*, US Dept. of Commerce, NBSIR 87-3597.
7. C. Seccaroni and P. Moioli, *Fluorescenza X: Prontuario per l'analisi XRF portatile Applicata a Superfici Policrome*, Nardini Ed., Firenze 2002.
8. See, for example, *Industrial Gauging and Analytical Instrumentation Sources*, The Radiochemical Centre, Amersham, UK 1989.
9. Oxford Analytical Systems Division, 275 Technology Circle, Scotts Valley, CA 95066, USA.
10. Moxtek Inc., 452 West 1260 North, Oren, UT 84057, USA, www.moxtek.com.
11. Hamamatsu Photonics, 1126-1 Ichino-cho, Hamamatsu City, 435 Japan, www.hamamatsu.com.
12. X-ray Optical Systems Inc., Corporate Circle, Albany, NY, 12203.
13. IfG Institut für Gerätebau GmbH, Rudower Chaussee 29/31, 12489 Berlin, Germany, www.ifg-adlershof.de.
14. K. Janssens, Use of microscopic XRF for non-destructive analysis in art and archaeometry, in *16th International Conference on X-ray Optics and Microanalysis, Vienna, (July 2–6, 2001)*, p. 44; H. Bronk, S. Röhrs, A. Bjeoumikhov, N. Langhoff, J. Schmalz, R. Wedell, H.E. Gorny, A. Herold, U. Waldschläger, ArTAX: a new mobile spectrometer for micro-X-ray fluorescence spectrometry on art and archaeological objects, *16th International ICXOM*, Vienna, July 2–6, 2001, p. 41.
15. AMPTEK Inc., 6 De Angelo Drive, Bedford, MA 01730-2204, USA.
16. Röntec GmbH, Rudower Chaussee 6, D-12489, Berlin, www.roentec.de; KETEK GmbH, G.Heinemann Ring 125, D-81739 München, Germany; www.ketek.net.
17. Eis, via Silvani, Roma, eissrlrm@tin.it.
18. C. Fiorini and A. Longoni, Application of a new non-cryogenic X-ray detector in portable instruments for archaeometric analysis, *Rev. Sci. Instrum.*, 1998, **69**, 1523.
19. J.S. Iwanczyk, B.E. Patt, Y.J. Wang and A. Khusainov, Comparison of HgI_2, CdTe and Si-PIN X-ray detectors; *Nucl. Instrum. Methods*, 1996, **A380**, 186. Constallation Technology Co, 7887 Bryan Dairy Rd., Largo, FL 33777, USA; R. Cesareo, A. Castellano, C. Fiorini, G.E. Gigante, J.S. Iwanczyk, A. Longoni, J.A. Pantazis, J.L. Pena Chapas and M.A. Rosales, Thermoelectrically cooled semiconductor detectors for portable EDXRF equipments, SPIE, San Diego (31 July–1 August 1997), *Proc. SPIE-Int. Soc. Opt. Eng.*, **3115**, 274–283; R. Cesareo, G.E. Gigante, A. Castellano, C.

Fiorini, A. Longoni and A. Brunetti, Comparison between semiconductor detectors for portable EDXRF systems; SPIE, Denver, CO, July 19, 1999, *Proc. SPIE-Int. Soc. Opt. Eng.*, 112–116.

20. N. Gao and K. Janssens, Polycapillary X-ray optics, in *X-Ray Spectrometry: Recent Technological Advances*, ed. K. Tsuji, I. Injuk and R. Van Grieken, J. Wiley & Sons, 2004; K. Janssens, Use of microscopic XRF for non-destructive analysis in art and archaeometry, in *16th International Conference on X-ray Optics and Microanalysis, Vienna*, (July 2–6, 2001), p. 44.
21. R. Cesareo, S. Ridolfi, A. Bjeoumikhov, N. Langhoff and A. Brunetti, X-ray fluorescence microanalysis of works of art with capillary collimators, in preparation, 2008.
22. J. Pope-Hennessy, *Benvenuto Cellini*, Hazan Publ., Paris, 1985.
23. Cesareo, M. Cordaro, G.E. Gigante, G. Guida and M. Marabelli, Indagini non distruttive sulla statua di bronzo del Perseo mediante EDXRF, metallografia non distruttiva e conducibilità IACS; unpublished work.
24. B. Cellini, La Vita, Sonzogno Ed., Milano, 1907.
25. L. Butterfield, *The Sculptures of Andrea del Verrocchio*, Yale University Press, 1998.
26. S. Ridolfi, R. Cesareo and M. Marabelli, The equestrian statue of Bartolomeo Colleoni: diagnostic analysis by means of a portable EDXRF system, *Proceedings European Conference on X-ray Spectrometry*, EXRS 2006, Paris.
27. A. Castellano, G. Buccolieri and M. Donativi, Portable EDXRF surface mapping of sulphate concentration on Michelangelo's David, *X-ray Spectrom.*, 2006, **35**, 276.
28. R. Cesareo, A. Castellano, G. Buccolieri and M. Marabelli, A portable apparatus for EDXRF-analysis of S and Cl in frescoes and stone monuments, *Nucl. Instrum. Methods B*, 1999, **155**, 326–330.
29. A. Castellano and R. Cesareo, A portable instrument for energy dispersive X-ray fluorescence of S in frescoes, *Nucl. Instrum. Methods B*, 1997, **129**, 281.
30. G. Buccolieri, A. Castellano, R. Cesareo and G.E. Gigante, Un'apparecchiatura portatile per l'analisi dell'inquinamento e per l'individuazione di pigmenti degli affreschi, *Atti I Congresso AIAR*, Verona, Dicembre 2–4, 1999, Bologna, pp. 99–106.
31. R. Cesareo, C. Cappio Borlino, G. Stara, A. Brunetti, A. Castellano, G. Buccolieri, M. Marabelli, J. Teixeira de Assis and G.E. Gigante, Analysis of S and Cl in frescoes and lapideous monuments by using a portable EDXRF apparatus, *Proceedings 2nd International Congress on Science and Technology for the Safeguard of Cultural Heritage in the Mediterranean Basin*, Parigi, 2000, ed. A. Guarino, Elsevier, Amsterdam, pp. 673–678.
32. Giotto in the chapel of the Scrovegni, Medoacus 99, 30030 Oriago (VE, Italy).
33. R. Cesareo, Le aureole d'oro di Giotto nella Cappella degli Scrovegni, *Il Nuovo Saggiatore*, 19 (2003) 74–77; R. Cesareo, A. Castellano, G. Buccolieri, S. Quarta and M. Marabelli, Giotto in the Chapel of the Scrovegni:

EDXRF-analysis of the golden haloes with a portable equipment, Berlin EXRS-2002, *X-Ray Spectrom.*, 2004, **33**, 289.
34. L'altare d'oro di Sant-Ambrogio a cura di C. Capponi, Banca Agricola Milanese, 1996.
35. R. Cesareo, A. Castellano, M. Marabelli, S. Bandera, C. Fiorini, A. Longoni and G.E. Gigante, The golden altar of S. Ambrogio in Milan: non destructive XRF-analysis with a portable apparatus, *Proc. 2nd International Congress on Science and Safeguard of Cultural Heritage in the Mediterranean Basin, Paris, July 5–9, 1999*, ed. A. Guarino, Elsevier, Amsterdam, p. 678–687.
36. H. Lechtman, *Sci. Am.*, 1984, **250**, 38–46.
37. G. Hörz, M. Kallfass, The treasure of gold and silver artefacts from the Royal Tombs of Sipán, Peru, *Materials Characterization*, 2000, **45**, 391–420.
38. E.A.O. Saettone *et al.*, Plasma clearing and analysis of archaeological artifacts from Sipán, *J. Phys. D Appl. Phys.*, 2003, **36**, 842–851.
39. J. Jennings, Treasures of Sicán, Ancient Peru at the Royal Ontario Museum, Minerva, March/April 2007.
40. W. Alva and C.B. Donnan, The Royal Tumb of Sipán, Los Angeles Fowler Museum of Cultural history, University of California 1993.
41. R. Cesareo *et al.*, EDXRF-analysis of precolombian golds from the "Tumbas Reales de Sipán", Museo de Sicán, Museo de la Nacion, to be presented at *ART 2008, Intern. Conference, Jerusalem, May 25–30, 2008*.
42. Sicán National Museum; http://www.go2peru.com/chiclayo_sican.htm
43. G. De Chirico, *Piccolo trattato di tecnica pittorica*, V. Scheiwiller Publ., Milano, 1983.
44. R. Cesareo, A. Brunetti, S. Ridolfi and M. Marabelli, *L'analisi di Fluorescenza X nell'identificazione di falsi*, Giornata di Studio "Vero e falso nelle opere d'arte", Nov. 8, 2006, to be published.

CHAPTER 10

Extraterrestrial Analysis: Planetary X-Ray Fluorescence from Orbiting Spacecraft and Landers

G.W. FRASER

Space Research Centre, Michael Atiyah Building, Department of Physics and Astronomy, University of Leicester, Leicester LE1 7RH, UK

10.1 Introduction

The three essential components of any X-ray fluorescence (XRF) experiment are: (1) a sample of unknown elemental composition, (2) a primary source of exciting radiation and (3) an X-ray detector. These essentials can be found not only in the usual familiar terrestrial contexts but also in planetary geology. Landing on the surface of another planet to make *in situ* geochemical measurements takes XRF to the limits of portability, but operating the spectrometer from an orbiting spacecraft opens up the possibility of XRF as a remote sensing technique on global scales. Here, solar coronal X-rays provide the primary excitation and the airless bodies of the inner solar system – the Moon, Mercury and the Near Earth Objects (NEOs) – provide samples whose surface composition and differentiation may contain clues to the solar system's origin and evolution. Farther out in the solar system, the primary excitation is provided, not by the Sun, but by the intense proton flux in the environment of Jupiter. Future targets for planetary XRF could well be the exotic surfaces of the Galilean moons – Io, Europa, Ganymede and Callisto.

Portable X-ray Fluorescence Spectrometry: Capabilities for *In Situ* Analysis
Edited by Philip J Potts and Margaret West

The present chapter describes progress in the XRF of planetary surfaces (and, to a much lesser extent, atmospheres) using instruments carried on both landers and orbiting spacecraft. According to Blake,[1] the first lander experiments were developed for the lunar programme of the Jet Propulsion Laboratory (JPL) as long ago as 1960. Rieder *et al.*[2] describe the pioneering work of Turkevich at the University of Chicago, beginning in 1961 and culminating in the *Surveyor* V, VI and VII robotic lunar landings in 1967–1968. Subsequent development of *in situ* planetary XRF, where the primary excitation is provided by radioactive sources, has been extensively reviewed (with emphasis on the Soviet programme) by Surkov[3] and in the monograph by Adler and Trombka,[4] and is updated below with specific reference to Mars (Section 10.2) in a modern exploration era that began with *Pathfinder*. One is struck by the complexity and ambition of the earliest instrument designs,[5] whose science goals are only now being met, 40 years later.

The parallel field of "X-ray remote sensing" first opened up in the Apollo era, suffered a fallow period of almost 25 years duration and is currently enjoying a strong resurgence. In Section 10.3, we summarize the achievements of past orbiting missions using collimated, non-imaging detectors working in the energy band 1–10 keV and look forward to the deployment of true imaging instruments, based on novel, low mass grazing incidence optics developed originally for X-ray astronomy (Section 10.3.2).

In the pages available, we are unable to describe in any detail the field of planetary gamma ray spectroscopy – which also has its beginnings in the lunar programmes of the 1960s[4] – from either orbiting spacecraft or landers. Gamma rays provide information on both local and planet-wide distributions of elements such as K, U and Th[3,4] and at rather greater sub-surface depths ($\sim$1–10 mm compared to $\sim$1–10 μm) than X-ray fluorescence. Thermal and epithermal neutron spectroscopies, searching for water ice, and alpha particle detection, searching for radon emission from the surfaces of airless bodies like the Moon, also have their place in the suite of planetary analytical techniques. The emphasis throughout this chapter is on the enabling X-ray technologies and their implementation in space missions, rather than on the (sometimes far-reaching) results recently achieved in planetary geology.

10.2 *In Situ* XRF Analysis of Planetary Surfaces

10.2.1 Instrumentation Principles

The figures-of-merit that characterize laboratory XRF spectrometers (major and trace element detection limits, signal-to-noise and so on) compete in the planetary context with the space science constraints of:

- available volume (1–2 L), available mass ($\sim$1 kg) and the need for low power consumption ($\sim$10 W),
- survival through the vibration loads associated with launch and landing impacts ($>$10 g),

- tolerance of high levels of ionizing radiation (>10 Gy),
- pre-launch biological sterilization to comply with planetary protection regulations.

The instrument must operate semi-autonomously after landing, preferably without the use of mechanisms or external cooling, in a harsh environment. Possibly the most heroic measurement in the history of XRF was the spectrum of the surface regolith of Venus obtained by the USSR's *Venera 14*[1,3] at an ambient temperature of 500 °C in a 65 bar atmosphere loaded with corrosive sulfuric acid. Even on the relatively benign surface of Mars, the diurnal temperature swing may exceed 60 °C and instruments are subjected to atmospheric dust, cosmic rays unimpeded by any significant planetary magnetic field and an intense flux of solar ultraviolet radiation, unhindered by any significant ozone layer.

The basic design choices for an *in situ* planetary X-ray spectrometer are:

1. The type, number and layout of the radioisotope source(s) that provide the primary X-ray excitation
2. The detector, its active area and entrance window
3. The materials of the detector body, which must be selected to minimize spectral artefacts due to self-fluorescence.

Conventional X-ray tubes are not presently favoured because of their demands on electrical power, although new generations of long-lived, cold X-ray source – based on the piezoelectric effect[6] or on field emission from carbon nanotubes or using focused electron beams, with the source vacuum isolated from the (Martian) atmosphere by a thin membrane[7] – will be flight tested in the near future. Given that β-emitters such as ^{3}H (end-point energy 18.6 keV) produce unwanted bremsstrahlung background from a solid target in addition to the desired characteristic line emission, the current choice of source, in practical terms, lies between (a) an annular array of α-emitters – now usually ^{244}Cm (half-life 18.1 years) (as used on *Pathfinder/Sojourner* (Section 10.2.2.2) and the Mars Exploration Rovers (MERs) *Spirit* and *Opportunity* (Section 10.2.2.4) as well as on the European Space Agency (ESA) mission to land on the nucleus of a comet – *Rosetta*[8] – and (b) paired electron capture sources (*e.g.* ^{55}Fe and ^{109}Cd as flown on both *Viking* (Section 10.2.2.1) and *Beagle 2* (Section 10.2.2.3). The latter solution is the more generally-available of the two.

The preparation of high specific activity, sealed, stable, dual-mode ^{244}Cm sources based on curium silicide is described in ref. 9. Such sources provide complementary excitation from both the 5.8 MeV alpha impact (X-ray production cross-section varies as the α-particle energy squared and inversely as the sixth power of Z, the atomic number of the target atom[4]) and the 12.5–21.4 keV L series X-rays of plutonium, the daughter product of curium. The cross-over between excitation of low-Z elements by the α-particles and of high-Z elements by Pu X-rays occurs at about 6 keV, corresponding to $Z \sim 26$ (iron). Omand *et al.*[10] describe in detail the ^{244}Cm sources used in the *Athena* Alpha Particle

X-ray Spectrometers (APXSs) that have operated successfully on the surface of Mars since January 2004 (see Section 10.2.2.4).

Adopting the terrestrial goals of high sensitivity for all the major rock forming elements (Na to Zn) and for trace elements such as Sr and Rb, one can use, as in conventional portable XRF, the Mn K-series X-rays at 5.9 and 6.4 keV from sealed ^{55}Fe sources, together with the 22.16 and 24.95 keV Ag K X-rays and 87.7 keV nuclear gamma-ray from ^{109}Cd. The major difference between the planetary and terrestrial contexts is that there is unlikely to be the available mass or power in the space instrument to implement a source interchange mechanism or shutter, so that X-ray spectra are acquired (and must be analysed[11]) with both source types active at once. Using ^{55}Fe excitation, importantly, compromises the instrument's sensitivity to the rock iron content, because of the overlap between elastically backscattered Mn Kβ with fluorescent Fe Kα. The Ag L-shell emission at ~2.3 keV from ^{109}Cd must also be absorbed by a source "cap" [carbon fibre in the case of the *Beagle 2* XRS (Section 10.2.2.3)] to avoid interference with the K-series of chlorine and neighbouring elements.

With the source activities scoped for radioactive decay during the (~0.5–6-year long) interplanetary cruise from Earth, with the instrument's working distance (typically a few cm) selected and the materials [*e.g.* Cu-Be or carbon fibre reinforced plastic (CFRP)] of the instrument body chosen for strength, shock resistance (particularly for a Mars landing based on parachute and airbag braking only) and minimal self-fluorescence, the X-ray detector itself must be identified. As in terrestrial XRF, there has been a steady transition from the sealed, thin-windowed gas proportional counters of the *Viking* era – with an energy resolution $\Delta E/E \sim 0.4E^{-1/2}$ (where the X-ray energy E is expressed in keV) – to thermo-electrically-cooled silicon PIN diodes[12] or semiconductor drift detectors (SDDs)[13] with intrinsic full-width-at-half-maximum (FWHM) resolution $\Delta E/E \sim 0.05E^{-1/2}$ (practically, 100–200 eV at 1–2 keV) and a high energy cut-off determined by the depletion depth d_{Si} of the silicon – which is usually found to be somewhat less than the 300–500 μm wafer thickness.[14] An energy resolution of ~150 eV is sufficient to clearly separate the Kα lines of the major rock-forming elements Mg, Al and Si, whereas one requires multiple proportional counters equipped with different absorption filters to unambiguously measure the important geochemical ratios Mg/Si and Al/Si. Irrespective of manufacturer, the solid-state detector is packaged, together with its preamplifier, on a thermoelectric cooler chip based on a semiconductor such as bismuth telluride (Bi_2Te_3) and sealed in a nickel can filled with an inert atmosphere (1 bar) of nitrogen. The entrance window is typically beryllium of thickness 5–8 μm, protected from water vapour by a passivation layer of boron or parylene. X-Ray absorption in the window limits the low energy cut-off to about ~1 keV, *i.e.* to elements above sodium in the periodic table. The detector active area is typically 5–10 mm^2, with the best energy resolution achieved using only the central area of the detector to minimize charge loss due to edge effects.

The elements used in detector manufacture may include gold (in wirebonds), lead (in solder), nickel (in the detector housing) and zirconium (in the detector collimator). Taken together, these elements give rise to a complex self-fluorescence spectrum that can limit the instrument's sensitivity to certain trace and major elements. For example, Zr L shell emission from the detector collimator complicates the response of the *Spirit* and *Opportunity* APXS to phosphorus.[15] An instrument's ultimate sensitivity can, however, be limited by other, external factors. The co-mounting of both the *Beagle 2* XRS and the MER APXSs with Mössbauer spectrometers containing strong (*i.e.* > 100 MBq) ^{57}Co sources, half-life 271 days, significantly increased the internal background of the X-ray instruments, despite the introduction of 1–2 mm tantalum shielding around the cobalt sources to absorb their penetrating 14 and 122 keV γ-ray emission. Other possible local sources of enhanced background are Radioisotope Thermal Generators (RTGs)[16] powering the lander and/or smaller (< 1 W as opposed to ~100 W) Radioisotope Heater Units (RHUs)[17] used, particularly in a Martian context, to keep the night-time temperatures of electronic circuitry above survival limits.

10.2.2 Missions

10.2.2.1 Viking 1 and 2

The landing of *Viking 1* on Chryse Planitia (22.8° N, 49.97° W) on the 20th July 1976 began a detailed exploration of the surface of Mars that lasted for nearly 1900 Sols (1 Martian day or Sol corresponds to $24^{h}\ 40^{m}$). *Viking 2* landed on the opposite side of Mars in Utopia Planitia (49.97° N, 225.74° W) in September 1976 and operated for 1281 Sols. The scientific emphasis of the *Viking* mission was the detection of life by means of labelled release, pyrolytic release and gas exchange experiments – the first ever experiments in exobiology, whose ambiguous results are still being reinterpreted.[18] The X-ray Fluorescence Spectrometers (XRFSs)[19–21] carried by the large (~600 kg), complex *Viking* landers, each powered through the dark and cold of the Martian winters by two 30-W ^{238}Pu RTGs, found that the composition of the fine-grained, wind-blown Martian soil was identical at sites ~6500 km apart. The XRFSs quantified about a dozen major and minor elements, and measured both the soil density and the atmospheric argon mixing fraction.[22] The XRFS operated in the thermally-controlled enclosure of the lander, rather than in the extreme cold of Martian night, and was fed ~25 cm^3 soil samples by means of a funnel loaded by a surface sampler scoop. The XRFS instrument mass was 1.5 kg and the dissipated power was 4 W. *Viking 1* obtained nine soil spectra and *Viking 2*, eight. The four proportional counters were filled with distinct mixtures of Xe : Ne : He with carbon dioxide and sealed with Al or Be windows. Calibration targets were built into the sample chamber walls opposite paired electron capture X-ray sources, so that when the chamber was empty of soil the ^{55}Fe sources fluoresced an aluminium target and the ^{109}Cd sources a silver–zinc plaque. The limiting sensitivity for major elements was 1–2% while the sensitivity for trace elements lay in the 1–10 $\mu g\, g^{-1}$ range. The limiting factors in these XRFS sensitivity levels included not only background γ-ray fluxes from the RTGs but, remarkably, from the decay of thorium in the magnesium-based alloy used in the detector housings. Retrospective re-analysis of the XRFS data sets alongside more modern data is described in refs. 23 and 24.

10.2.2.2 *Pathfinder/Sojourner*

Nearly 15 years elapsed from the end of *Viking* surface operations until our understanding of Martian geology could be extended – by a very different X-ray spectrometer[25–27] carried by a very different spacecraft, one built in a new era of economic stringency, the era whose mantra was "faster, better, cheaper". While the landings of *Viking 1* and *2* were moderated in their final stages by retro-rockets, the arrival of *Pathfinder* at Ares Vallis (19.28° N, 33.52° W) was a much more violent affair. The tiny spacecraft, built for one-sixth the cost of *Viking*, bounced in its airbag cushion several times before coming to rest and deploying the engineering testbed micro-rover, *Sojourner*. On July 6th 1996, the Alpha Proton X-ray Spectrometer (APXS) carried by *Sojourner* acquired the first ever *in situ* XRF spectrum of an extraterrestrial rock – nicknamed *Barnacle Bill* – in a nocturnal spectrum obtained over some 13 780 s, with a peak count rate for the Si Kα line of only 0.6 s^{-1}. The instrument, based on an Amptek[28] silicon PIN diode, weighing only 0.57 kg and with a power consumption of 0.34 W, accumulated 16 X-ray spectra in total, in a mission lasting 83 Sols in which the rover covered 104 m around the landing site. The agreement between *Viking* and *Pathfinder* derived soil compositions is generally good.[23,24]

The primary source for the *Pathfinder* APXS was an annulus of nine individual ^{244}Cm capsules, with a total activity of 1.85 GBq at beginning-of-mission. The development of the APXS originated, as did that of several payload elements of the later *Mars Express/Beagle 2* mission, with an instrument designed for Russia's failed *Mars 96*.[29] The principles governing the use of curium α-particle sources for *in situ* planetary XRF have already been described in Section 10.2.1. Using separate detector channels, however, one can gain further information on the light element content of a geological target by recording the Rutherford backscattered α-particle flux (quantifying carbon, oxygen) and protons arising from (α,p) nuclear reactions (quantifying the elements sodium to sulfur). Thus, in principle, a three-channel APXS instrument as deployed on *Pathfinder* should have high sensitivity throughout the entire periodic table. In practice, the presence of the ($\sim$7 mbar carbon dioxide) Martian atmosphere is a severe complication to the interpretation of the α-particle spectra, as is the strong dependence of the backscattered intensity on sample–detector distance.[30,31] Only a weak upper limit (0.8 wt%) could be placed on the carbon content of the rocks and soils at the *Pathfinder* landing site.[27] Nevertheless, developments of the *Pathfinder* APXS design continue to dominate the field of *in situ* planetary XRF, as described in Section 10.2.2.4.

10.2.2.3 *Beagle 2*

The *Beagle 2* X-ray Spectrometer (XRS) was originally conceived as a single channel counter for the registration of potassium K X-rays. In conjunction with the ^{40}A measurements made by the Open University's Gas Analysis Package (GAP),[32] estimating the potassium content of a Martian rock would permit the first estimate to be made of the age of the planet's surface.

The headline science goal of the XRS, therefore, was to measure, in a single overnight (~8 hour) acquisition, the potassium content to 5% relative accuracy of prepared rocks containing 1% by weight of K_2O. By the standards of lunar surface dating – based on *Apollo* and *Luna* returned samples – this was a modest goal, but any age estimate would have been a first for the surface of Mars. The laboratory performance of the XRS, built at the University of Leicester with the present author as Principal Investigator, is described by Talboys *et al.*[33] A large number of pure element samples and planetary analogues in the form of hand samples, pressed powder pellets and other forms were used to fully characterize the XRS response.

As launched onboard the UK-led lander *Beagle 2*, part of the payload of ESA's *Mars Express*, on July 1st 2003, the XRS consisted of two sub-units, in addition to the multi-element "spot painting" calibration target created by the noted UK artist Damien Hirst and mounted separately on the actual body of the lander. These sub-units were:

(a) the Detector Head Assembly (DHA – see Figure 10.1), containing four radioactive sources, two ^{55}Fe and two ^{109}Cd, arranged in a carbon fibre structure around a single central Amptek silicon PIN diode with an area of 7 mm^2 and a beryllium window 7.5 μm thick. An aluminium contact ring with an inner diameter 2 cm defined the measurement area and prevented the abrasion of the carbon-containing DHA structure. The outer surface of the assembly was flashed with a thin layer of gold, also to

Figure 10.1 *Beagle 2* X-ray Spectrometer (XRS) Detector Head Assembly (DHA), showing the gold-coated CFRP source/detector holder at left and the front-end electronics cards on the right. The stand-off distance between the contacted rock surface and detector is 12 mm and the sampled area is 20 mm in diameter. In this view, the ^{55}Fe sources and the Si pin diode have been removed.

prevent contamination of the *Beagle 2* landing site with terrestrial carbon. The DHA was co-mounted on the lander's robotic arm – the Payload Adjustable Workbench or PAW – close to the University of Mainz Mössbauer spectrometer,[34] whose large primary source and secondary calibration source produced a significant leakage flux of 122 keV ^{57}Co gamma rays which, despite the introduction of 2 mm thick Ta shielding between the spectrometers, led to a significant increase in the irreducible XRS background. The handling of large radioactive sources during the laboratory phase and final integration of a lander payload is discussed by Butcher *et al.*[35]

(b) the Back End Electronics (BEE), containing the main amplifier and 512 channel analog-to-digital conversion stages. The XRS's operational energy range was chosen to be 1–24 keV, the upper energy limit being sufficient to retain the information[36] contained in the Thomson- and Compton-scattered Ag K line complex. The temperatures of both the DHA and BEE were expected to vary significantly during each night-time observation and so the estimate of each photon's energy was to be corrected according to pre-launch look-up tables of the temperature coefficients of electronic gain, before storage in memory and later transmission to the orbiting *Mars Express*.

The total mass of the XRS was 340 g and the total power requirement, 5 W. A typical pre-launch XRS calibration spectrum (Figure 10.2) shows relatively

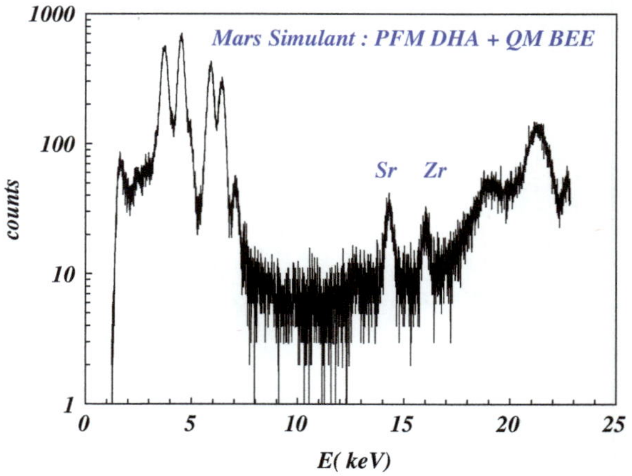

Figure 10.2 Logarithmic XRS calibration spectrum obtained from a pressed powder sample of "Mars Simulant". PFM denotes "proto-flight model" and QM is "qualification model". The principal peaks at low energies are due to Ca, Ti, Mn (backscattered), Fe and Ni (from the detector "can"). L-shell artefacts due to gold detector components are visible in the region of low background as are the trace element peaks of Sr (740 $\mu g\, g^{-1}$) and Zr (330 $\mu g\, g^{-1}$). Backscattered Ag K X-rays dominate the high energy part of the spectrum.

moderate energy resolution (~250 eV at Si K) but rather good sensitivity for the trace elements from yttrium–strontium. The L-shell emission from the gold wire-bonds to the detector chip is clearly visible, as are the Ni K lines from the detector enclosure. A broad peak at about 2 keV is at least partially due to "Compton escape" events, where the electron from the Compton interaction of an Ag K photon leaves the depletion layer of the silicon diode without being detected. The relatively weak signals observed from Al (present as 14 wt% oxide in the sample in question) and Si (52 wt% oxide) are intrinsic to the use of X-rays to stimulate fluorescence from targets of low atomic number.

Beagle 2 was released from *Mars Express* on December 19th 2003 and entered the Martian atmosphere early on Christmas morning. No signal was ever received from the Isidis Planitia landing site and the mission was declared lost in February, 2004. The potassium-argon dating of the Martian surface awaits a future lander, still unbuilt. Realizing many of the other scientific goals of the XRS was, when *Beagle 2* met its fate, only weeks in the future, as NASA's Mars Exploration Rovers (MERs) closed in on the Red Planet.

10.2.2.4 *Spirit and Opportunity*

The remarkable APXS[37–41] carried by MERs *Spirit* and *Opportunity* is a development of earlier *Mars 96* and *Pathfinder* instruments, but without the original proton channel, made redundant by continuing advances in silicon detector technology. The APXS sensor head contains, as the source of primary excitation, six (as opposed to nine for *Pathfinder*) ^{244}Cm α-sources of initial total activity of 1.11 GBq. These sources excite fluorescent X-ray emission from rock, soil, atmospheric or calibration targets by means of both Pu L characteristic X-rays and 5.8 MeV α-particles, slowed in the new design to 5.2 MeV by passage through a thin (2.5 μm) titanium foil, windowing the source capsules. The X-ray conversion efficiency (summed over all the lines of the Pu L series) is 0.092 X-rays per decay.[10] The half-life of ^{244}Cm is 18.1 years, so that the APXS sources cannot quite be treated as sources of constant activity, given that the time from manufacture through delivery of the APXS instruments to their respective spacecraft through launch, interplanetary cruise, landing and extended surface operations – the latest release of the MER database[42] now extends to over 1000 Sols – is now much in excess of 4 years.

Each 1–16 keV, 512-channel X-ray spectrum is detected by a single semiconductor drift detector (SDD) produced by the Ketek[13] company. The SDD is centrally located within the ring of curium sources and preceded by a zirconium collimator, made up of two annuli. The effective half-angle of the collimator is $\theta_0 = 26.4°$.[37] The sensor head is protected by a pair of reclosable doors, whose gold/kapton-coated detector-facing surfaces are intended to act as a calibration target. The doors have in fact been left permanently open since the early part of the MER mission because of concerns over cumulative mechanism unreliability.[38] APXS integrations are mostly made overnight, at ambient temperatures of *ca.* –40 °C, with an energy resolution of ~140 eV at an X-ray energy of 6 keV. A beryllium window thickness of 5 μm translates to a lower energy threshold of

~0.8–1.0 keV. The X-ray SDD is surrounded by an annular array of silicon-based α-particle detectors, each preceded by a multilayered stainless steel collimator that has radial symmetry, the channels joining the centre of the backscatter (*i.e.* sample) plane with the centre line of the detector annulus.[37]

In nearly four years on the Martian surface, the *Spirit* and *Opportunity* APXSs have helped to generate an enormously detailed geochemical picture[43–45] of their landing sites – Gusev Crater and Meridiani Planum, respectively. Although the former landing site was selected for its similarity to a terrestrial dried-up lake bed, it was the latter terrain that is now interpreted as having been once inundated with liquid water. Both rovers have far exceeded their design lifetimes and travelled much further than the 600 m that would have been acceptable to mission planners before landing. New observation strategies have given unexpected insights into the dynamics of the Martian atmosphere *via* measurement of its argon content.[40] The next generation of APXS is already under construction for flight on the 2009 *Mars Science Laboratory* (MSL).[46]

10.2.3 Future Developments

One important development for the NASA, ESA and JAXA (Japan) programmes of planetary exploration is the deployment of XRF/XRD spectrometers – the combination, in a single instrument, of X-ray fluorescence (providing information on elemental abundance) and X-ray diffraction, providing crystal structure and, thus, mineralogy. Dual mode X-ray instruments covering lattice spacings of 1–7 Å and atomic numbers $11 < Z < 29$ date from the 1960s,[1,4,5] but their excessive mass (~20 kg), and their dependence on complex mechanisms to scan a non-imaging Xe side window proportional counter detector in a circle in relation to the sample, militated against deployment on the Moon. Without the XRD channel, planetary geologists must rely on infrared or optical reflectance spectra to help determine mineral class.

The first XRF/XRD instrument to fly will be the CHEMIN (*CHEmistry-MINeralogy*) experiment developed by the NASA Ames Research Centre for the MSL mission.[47,48] An Italy/UK/Netherlands team is developing a similar instrument for the ESA EXOMARS lander (launch *ca.* 2013).[49] The addition of the diffractometer introduces several new technical challenges alongside the now familiar problems of mass, power and volume:

(a) The need for a two-dimensional energy-resolving *imaging* sensor, such as a cooled silicon Charge Coupled Device (CCD) not only to register the forward-scattered Laue diffraction pattern with high sensitivity and sufficient spatial resolution to facilitate order separation, but also to detect the isotropic fluorescent X-ray flux. The CCD must operate in the cooled (~230 K) slow-scan spectroscopic mode (*i.e.* with an equivalent noise of ~4 electrons rms) developed originally for X-ray astronomy.[50]
(b) The need to robotically prepare and deliver an appropriate powdered sample for diffraction measurement in either reflection or transmission.

(c) The generation of a collimated, ideally parallel, X-ray beam of constant spectral profile to fire at that sample. CHEMIN employs a novel "cold" X-ray source design based on a field emitter array while the EXOMARS instrument uses conventional radioisotope sources.

In the longer term, the development of large band-gap semiconductor materials such as silicon carbide,[51,52] capable of ~100 eV spectroscopy even at elevated (+100 °C) temperatures, may find particular application in inner solar system missions and, more generally, in so-called Highly Integrated Payload Suites (HIPS)[53] for future (1–10 kg) *nanolanders*, where extreme mass and power constraints will make cooling using passive radiators or thermoelectric elements difficult to achieve.

10.3 X-Ray Remote Sensing of Planetary Surfaces

Whereas *in situ* spectrometers address absolute elemental abundances on local (~cm) scales, differentiating structures within individual rock specimens, orbiting instruments have – to date – mostly provided element ratios (*e.g.* Mg/Si and Al/Si) on the scales of major planetary terrains (~100–1000 km). Figure 10.3 illustrates the basic principles of planetary XRF from an orbiting spacecraft.

Viewed from orbit, the X-ray fluorescence from planetary bodies is always very faint, even under extreme solar flare conditions, when the primary excitation source – the Sun – is at its brightest. It is therefore remarkable that, with

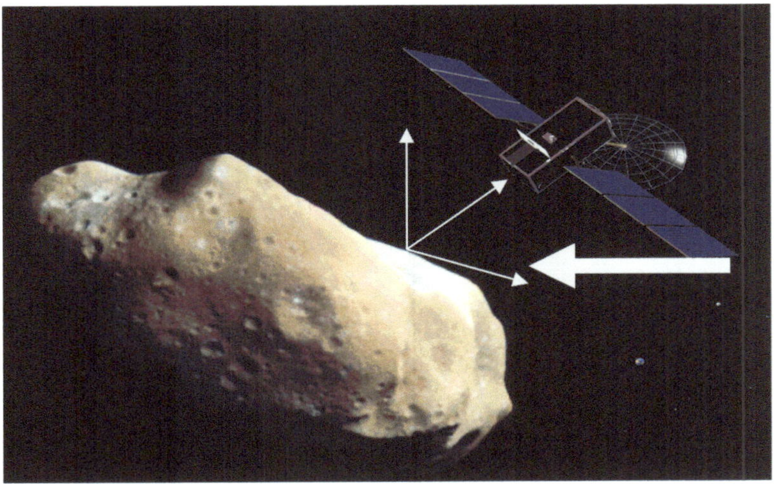

Figure 10.3 Principles of planetary XRF. The primary solar X-ray flux (thick arrow) excites isotropic fluorescent emission from the surface of a planetary body. This fluorescent line flux, together with a continuum of both elastically and inelastically backscattered solar X-rays (thin arrows), is received by a detector on an orbiting spacecraft.

the recent detection of oxygen K-shell fluorescence from the atmosphere of Mars[54] at a count rate of $9.3 \times 10^{-3}\,s^{-1}$ and a distance of 70 million km by the *Chandra* X-ray observatory, all the planets from Venus to Uranus, along with several moons and comets, have now been observed in X-rays by instruments in *Earth* orbit.[55] The first such detection was of the Moon by *Rosat* in 1990,[56,57] nearly 30 years after the famous sounding rocket flight[58] of the American Science and Engineering/MIT Geiger counter array, which failed to measure the composition of the lunar maria but, by serendipitously scanning across the bright cosmic source Sco X-1, gave rise instead to the new field of X-ray astronomy. After *Rosat*, the Moon was next detected by the Japanese *ASCA* X-ray astronomy satellite in 1993.[59] The ASCA observations are remarkable in that, for the first time, the geochemically important Al/Si and Mg/Si ratios, which relate to optical albedo, were measurable over the lunar disk from low Earth orbit. *ASCA* also provided evidence for particle-induced X-ray emission (PIXE) – possibly bremsstrahlung emission from low energy (>2 keV) electrons – from the lunar night side. This interpretation has been challenged by Wargelin *et al.*[60] who attribute the signal observed by *Chandra's* ACIS CCD detector to a foreground – *i.e.* local to the spacecraft – charge transfer interaction of the solar wind ions with neutral hydrogen in the Earth's geocorona. The relative importance of particle excitation and solar X-ray excitation is discussed in the next section.

10.3.1 Sources of Primary Excitation

10.3.1.1 The Solar Corona

The tenuous, million degree outer atmosphere of the Sun – the corona – is a highly variable X-ray source. The Solar X-ray spectrum is rich in lines superimposed on an underlying continuum intensity that falls off very rapidly with energy – by some five orders of magnitude between 0.5 and 3.0 keV in conditions of solar quiet, the spectral index hardening with increased solar activity.[61] These few fundamental facts introduce several important practical considerations into the conduct of XRF mapping from orbit around an airless planet:

1. Even in an elevated flare state, the K-shell fluorescence from elements with atomic numbers above $Z = 20$ is very inefficiently excited by the solar flux. For orbiting detectors flown to date, mostly sealed gas proportional counters sensitive only to energies above 1 keV, the important (*i.e.* readily measurable) elements are Mg (Kα emission at 1.25 keV), Al (1.49 keV) and Si (1.74 keV), with Ca (3.69 keV), Ti (4.51 keV) and Fe (6.4 keV) requiring very long integration times. The state of the Sun is usually described in terms of the intensity of the 1–8 Å X-ray flux monitored by instruments on the US GOES satellite, with decades of intensity denoted, in descending order, by the letters X (*i.e.* $10^{-4}\,W\,m^{-2}$), M, C, B and A – the last letter refers to the "quiet" Sun (*i.e.* $10^{-8}\,W\,m^{-2}$). Equivalently, the coronal state

Table 10.1 Ca, Ti and Fe sensitivity [$\sim I\omega$] using L-shell compared to K-shell fluorescence, before instrument factors are taken into account. $I_{\rm sun}$ in units of photons $\rm cm^{-2}\,s^{-1}\,keV^{-1}$ appropriate to normal mid-solar-cycle-activity. L-shell fluorescent energies: 0.34, 0.45 and 0.705 keV, respectively. Corresponding K-shell fluorescent energies: 3.69, 4.51 and 6.40 keV, respectively.

Element	$I_{sun}(E_K)$	ω_K	$I_{sun}(E_L)$	ω_L	$[I\omega]_K$	$[I\omega]_L$
Ca	10^2	0.17	2×10^8	1.8×10^{-4}	17	3.6×10^4
Ti	20	0.22	2×10^8	1.0×10^{-3}	4.4	2×10^5
Fe	1	0.35	2×10^8	5×10^{-3}	0.35	1×10^6

may be described by the electron temperature ($\sim$3–5 $\times$ 10^6 K) of a model isothermal plasma. The fraction of time spent in elevated states varies significantly within the eleven-year solar cycle.

2. There is potentially a great advantage in designing solid-state spectrometers to have good sub-keV response in order that the L-shell fluorescence of the higher-Z metals can be detected independently of the prominent oxygen K line at 0.525 keV. Table 10.1 records, for Ca, Ti and Fe, the product of the quiescent solar flux times the fluorescent yield ω_L evaluated at the threshold energy for L-shell emission.[62] These calculations show that this product is much higher than the same figure-of-merit for the equivalent K-shell energy. For iron, the apparent sensitivity advantage is very large indeed. Measuring iron from its L-shell emission is a design goal of the MIXS instrument on ESA's *BepiColombo* mission to Mercury (Section 10.3.3.5).
3. Solar X-ray variability on all timescales, from minutes to days to the eleven years of the nominal solar cycle, demands that the principal energy resolving detector must be supplemented by a wide-field-of-view or sun-pointed solar monitor to properly normalize fluorescent X-ray data obtained for different planetary fields at different times. Solar monitor designs are discussed in Sections 10.3.3.2 and 10.3.3.5.
4. The inverse square law dependence of the solar intensity on distance essentially confines "classical" planetary XRF to the inner solar system. The solar flux at Mercury, for example, is about one order of magnitude greater than that at one astronomical unit (AU), the mean Earth-Sun distance. The elliptical nature of Mercury's orbit (0.31–0.48 AU) adds a further cyclic variability to the X-ray intensity falling on that planet's surface.
5. The complicated energy- and time-dependence of the primary solar X-ray spectrum demands that instrument development should always be accompanied by a detailed numerical model of the X-ray emission from the planetary surface. The first computations of such detector input spectra, following X-ray interactions in the topmost few micrometres of the planetary surface and producing estimates of both line and underlying

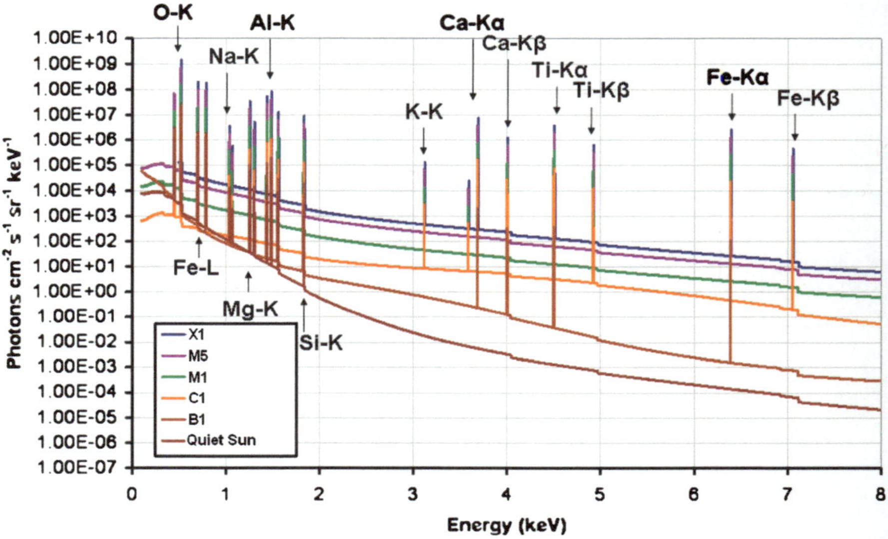

Figure 10.4 Simulated solar-induced X-ray spectra from Hermean basalt on Mercury as function of solar state Bottom-to-top: solar quiet, B-flare, C-flare, M-flare. The major flare states differ by an order-of-magnitude in X-ray power, with B corresponding to $10^{-7}\,\mathrm{W\,m^{-2}}$ in the 2–10 keV band.[63]

(Thompson or Compton scattered) continuum, were reported by Adler and Trombka.[4] Figure 10.4 shows the X-ray spectra expected from a hypothetical basalt mare (*i.e.* lava sea) on Mercury. These curves were produced at Leicester by a computer code embodying the "fundamental parameters" formulae of Clark and Tromka.[64] Quantitative Monte Carlo models of X-ray production at this planetary surface have also been produced by Truscott *et al.* under contract to the European Space Agency (ESA).[65] The Truscott analysis extends to the case of X-ray production by charged particles.

10.3.1.2 Charged Particle Fluxes

With the evidence already discussed for particle-induced X-ray emission (PIXE) from the dark side of the Moon and the expectation that PIXE will be important both in the auroral zones of Mercury[66] and at the Galilean Moons of Jupiter,[67] it is important to be able to estimate the likely magnitudes of solar energetic proton (SEP), galactic cosmic ray (GCR) and electron-induced X-ray line fluxes, in comparison with the expected fluorescent fluxes induced by solar X-rays.

Such calculations depend on $\sigma_{K,Z}$, the K-shell ionization cross-sections for the postulated elements in the planetary surface and a knowledge of the relevant incident particle flux $\varphi(E)$. For the limiting case of a material where the optical depth is much less than the proton range, the number of X-rays emanating from element of atomic number Z per unit time in direction θ_o is given by:[65,68]

$$Y(\theta_o, Z) = \frac{N_{Av} P_Z \omega_{K,Z} f_Z}{2 A_Z \mu_Z} \int_{E_K}^{\infty} \varphi(E) \sigma_{K,Z} \, dE \qquad (10.1)$$

where N_{Av} is Avogadro's number; P_z is the fraction of element Z by weight; A_z is the atomic weight of that element and $\omega_{K,Z}$ is its K-shell fluorescence yield; f_z is the fractional intensity for the particular line in the K series and μ_z is the mass absorption coefficient of the surface material for the fluorescent X-rays in question. E_K is the energy of the absorption edge. The effective upper limit of integration is set by the fall-off with energy of the ionization cross-section and of the particle flux. For most purposes, an upper limit to the proton energy of 200 MeV is appropriate.

In fact, the K-shell ionization cross-section for 4 MeV protons in aluminium is about $2 \times 10^{-20}\,\text{cm}^2$ (ref. 68), the same as for 3.5 keV X-rays. One can quickly show from Equation (10.1) that the efficiency (in terms of fluorescent X-rays per incident particle) of the PIXE process is rather high – of order $\sim$2%. The SEP-induced X-ray flux cannot, therefore, always be neglected in the estimation of instrument performance or in the reduction of instrument count rates to surface composition. The simultaneous presence of solar X-ray and solar proton stimuli certainly complicates the design of the solar monitor for missions to Mercury (Section 10.3.3.5).

10.3.2 Non-imaging and Imaging Instruments

According to Surkov,[3] the first orbital X-ray measurement was made by the Russian *Luna 10* spacecraft, which entered a (350 × 1017 km) orbit around the Moon with initial inclination of 71.9° in March 1966. Such an orbit – near-polar, so that the spacecraft ground track moves in longitude over the planet's surface from one orbit to the next, and somewhat elliptical – is typical of the planetary mapping missions that have followed *Luna 10*. Little information is available on the Luna 10 X-ray instrument (and indeed it may be that the first soft X-ray – rather than hard X-ray/γ-ray – instrument actually flew on *Luna 11*, in August 1966). Whatever the details of the design, it is safe to assume that the *Luna* X-ray spectrometers were collimated devices.

10.3.2.1 Collimator Principles[69]

A planar collimator consists of a multiplicity of parallel channels of length L and diameter D. Let T_o be the X-ray transmission of the collimator at normal incidence, $\theta = 0°$. Then, in the absence of reflection from the channel interior walls, over a range of off-axis angles:

$$|\theta| < \theta_w \tag{10.2}$$

The overall transmission is given by the triangular function:

$$T(\theta) = T_o[1 - |\theta|/\theta_w] \tag{10.3}$$

where:

$$\theta_w = \arctan(D/L) \tag{10.4}$$

The parameter θ_w is the full-width-at-half-maximum (FWHM) transmission of the collimator, the best measure both of the instrument field of view and of its angular resolution. The collimator gives no information on the angular location of a source, except that it is within the field of view. Typically, $\theta_w \sim 5°$. The average transmission for an ideal planar collimator with this triangular response is simply $T_o/2$. Equation (10.3) holds for X-ray energies E less than some upper limit E_{max}, beyond which the collimator structure becomes generally transmissive. This high energy leakage may be significant if the collimator structure is produced by microfabrication techniques (see Section 10.3.3.2) but, in general, the collimator is useful for all the fluorescent lines – up to and including Fe K – expected in planetary spectra.

It is simple to estimate, for a generalized collimated instrument of area A_c, the detector count rate N_c corresponding to a fluorescent line flux F (photons $cm^{-2}\,s^{-1}\,sr^{-1}$) incident at an orbiting spacecraft. We find:

$$N_c = 1/2 F A_c T_o T_f Q \theta_w^2 = F A_{eff} \theta_w^2 = F G_c \tag{10.5}$$

where Q is the detector quantum efficiency and T_f is the transmission of its associated UV/optical bandpass filter. The product of the effective area A_{eff} and the aperture $\theta_w{}^2$ is the collimator grasp, G_c. The fluorescent X-ray signal is accompanied by a continuum flux scattered from the planet's surface and by a charged-particle induced background whose event rate in the detector is given by:

$$B_c = b[1 - r_c] A_c \Delta E \tag{10.6}$$

where b (counts $cm^{-2}\,s^{-1}\,keV^{-1}$) is the raw background rate, r_c is the particle rejection efficiency for the X-ray detector plane and ΔE is the detector energy resolution. Equations (10.5) and (10.6) together permit the estimation of instrument sensitivity. The signal-to-noise ratio for a planetary observation of duration t seconds from a nadir-pointing platform is:

$$S = (N_c t)/(B_c t)^{1/2} \propto A_c^{1/2} \theta_w^{5/2} \Delta E^{-1/2} \tag{10.7}$$

The five-halves power of θ_w in Equation (10.7) arises from the fact that any object on the planetary surface stays in the field of view for a time:

$$t = h\theta_{\mathrm{w}}/v \tag{10.8}$$

where h is the orbital altitude and v is the velocity of the sub-spacecraft point.

If the fluorescent emission from a planetary surface is perfectly isotropic, it follows immediately that the count rate in a collimated instrument of effective (*i.e.* X-ray collecting) area A_{eff} and angular aperture θ_w by θ_w is *independent* of orbital altitude, provided that a uniformly-illuminated planetary surface element fills the aperture at all times. Defining φ_x to be the fluorescent line flux emitted from the planet's surface – per unit area per unit time per unit solid angle – and recognizing, as in Equation (10.8), that the instrument's "footprint" on the surface is a square of side length $h\theta_w$, the fluorescent emission from the surface element of interest is $\varphi_x[h\theta_w]^2$ photons $\mathrm{s}^{-1}\,\mathrm{sr}^{-1}$. The solid angle subtended by the instrument at the surface, however, is A_{eff}/h^2, giving a count rate:

$$N_{\mathrm{c}} = \varphi_{\mathrm{x}}[h\theta_{\mathrm{w}}]^2[A_{\mathrm{eff}}/h^2] = \varphi_{\mathrm{x}}\theta_{\mathrm{w}}^2 A_{\mathrm{eff}} \tag{10.9}$$

which is independent of altitude and which is once again expressed as the product of an emitted fluorescent flux and the instrument grasp. The identity of F and φ_x follows from the conservation of flux.

The parameter that *does* vary with altitude along with the exposure time is the surface resolution of the spectrometer – both depend on the aperture of the instrument. An understanding of how the coverage of the planet builds up with mission elapsed time is a complicated problem in orbital dynamics, especially for Mercury, where the planet's day and year are in a 3 : 2 resonance (see Section 10.3.3.5). The departure from isotropy of the X-ray emission from real, topographically-irregular planetary surfaces composed of porous, granular regolith at arbitrary solar illumination and detector viewing angles is the subject of much current laboratory research.[71]

The design rules implicit in Equation (10.5) demand, for maximum sensitivity, the largest possible collimator area, the best possible energy resolution and the widest collimator aperture. For a given orbit, however, large aperture leads to poor resolution of surface terrains while the one-to-one registration of a silicon-based detector plane with a large-area, planar collimator leads to problems with regard to detector cost, complexity, radiation shielding, mass and cooling. Using a *radial* collimator geometry (see Section 10.3.3.5), adapted from medical physics,[72] one can make the focal plane size much smaller than A_{c}, so relaxing the requirements on the detector, but leaving unchanged the fundamental properties of the *non-imaging* collimated instrument.

As we have seen, cosmic X-ray astronomy arose out of the first "failed" attempt at remote X-ray sensing of the Moon. Now at last, astronomers are repaying the favour, with the introduction into solar system studies of novel, low mass, *imaging* X-ray telescopes, based on the Wolter Type 1 geometry and realized using square-pore microchannel plate (MCP) technology.[70] Beginning with the MIXS-T instrument on *BepiColombo* (Section 10.3.3.5), this "technology transfer" brings new capabilities to orbital XRF.

10.3.2.2 Telescope Principles

In a Wolter Type 1 grazing-incidence telescope, X-rays from a distant source are brought to a true focus by reflection first from the internal surface of a parabola of rotation, then from the surface of a confocal hyperbola of rotation. Conventional Wolter Type 1 telescopes, made, as in the case of the *Chandra* High Resolution Mirror Array, from thick Zerodur (a zero-expansion glass ceramic) or, as in the case of the European Space Agency (ESA)'s *XMM-Newton* observatory, from thin replicated nickel shells, have effective areas A_{eff} of $\sim$1000 cm^2 coupled to masses M_{tel} of $\sim$100–1000 kg. Such single-experiment masses are completely impractical in the context of planetary exploration, where there are always many distinct measurements to be made by any science package. The required improvement in mass efficiency of one or two orders-of-magnitude (*i.e.* $A_{\mathrm{eff}} \sim$ 100 cm^2 for M_{tel} $\sim$1 kg) has been realized by the development of microchannel plate (MCP) or micropore (MPO) optics[62,70] where the length of the reflecting surfaces is of order of $\sim$mm, rather than $\sim$metres and the lateral separation of the reflecting surfaces is $\sim$10 μm. In contrast with the collimated instrument of the previous section, the angular resolution of the X-ray telescope ($\sim$arcminutes) may be much less than the field of view ($\sim$1°), making possible the mapping from a given orbit of much smaller surface features, if counting statistics allow.

Consider now the response of a miniature Wolter Type 1 telescope to the same source and background fluxes expressed above in Equations (10.5) and (10.6). The counterpart of Equation (10.5) is:

$$N_{\mathrm{t}} = FT_{\mathrm{f}}QA_{\mathrm{t}}(\theta = 0°)\left[\int_0^{\theta\,\max}\int_0^{2\pi}\theta V(\theta)\mathrm{d}\theta\;\mathrm{d}\phi\right] \qquad (10.10)$$

where T_{f} and Q have their previous meanings, A_{t} $(\theta = 0°)$ is the on-axis geometric area of the telescope and $0 < V(\theta) < 1$ is the off-axis vignetting function, describing the decrease in telescope throughput away from the optical axis. The vignetting function is properly computed by Monte Carlo raytrace methods, but assuming a linear fall-off in area with increasing angle leads to the simplification:

$$N_{\mathrm{t}} = FT_{\mathrm{f}}QA_{\mathrm{t}}(\theta = 0°)\frac{\pi}{3}\theta^2_{\mathrm{max}} = FA_{\mathrm{eff}}(\theta = 0°)\frac{\pi}{3}\theta^2_{\mathrm{max}} = FG_{\mathrm{t}} \qquad (10.11)$$

The background count rate for the 2D position-sensitive detector in the focal plane of the telescope, finally, is given by Equation (10.12):

$$B_{\mathrm{t}} = b[1 - r_{\mathrm{t}}]A_{\mathrm{fp}}\Delta E \qquad (10.12)$$

Note that in the case of the focusing system:

$$A_{\mathrm{fp}} < A_{\mathrm{t}}(\theta = 0°) \quad (10.13)$$

While it is not straightforward to calculate the absolute, mission-specific line and continuum fluxes for a given instrument, orbit and planetary body, estimating the *relative* performance of an imaging and non-imaging instrument is simple. Let the observing (F, t), filter (T_{f}) and detector (Q,ΔE) parameters be common to both systems. Then the count rate ratio is simply the ratio of the instrument grasps, or:

$$\left[\frac{N_{\mathrm{c}}}{N_{\mathrm{t}}}\right] = \left[\frac{A_{\mathrm{c}}T_{\mathrm{o}}}{A_{\mathrm{t}}}\right]\left[\frac{3\theta_{\mathrm{w}}^2}{\pi\theta_{\mathrm{max}}^2}\right] \quad (10.14)$$

Owens *et al.*[70] have estimated, using raytrace methods in the context of Mercury, the count rates in (a) a collimated instrument with $\theta_w = 2.5°$ and $A_c T_o = 12.5\,\mathrm{cm}^2$ and (b) a $100\,\mathrm{cm}^2$ Wolter Type 1 microchannel plate optic with an aperture of $\theta_{\mathrm{max}} = 1°$ for X-ray energies below 2 keV. Their numerical model gives a count rate ratio $3300/4400 = 0.75$. Direct substitution into Equation (10.14) gives a count rate ratio of 0.746, in rather good agreement.

In these calculations of performance, we have treated the telescope as a *flux collector* (or, more colloquially, *light bucket*), whose most powerful property is the decoupling of the effective area for photons (signal) from that for charged particles (background). The focusing action of the Wolter Type 1 geometry makes possible the use of a focal plane detector that is much smaller than the frontal area of the telescope. The parameter I in Equation (10.15) is the *Imaging Advantage* of the telescope system:

$$I = \sqrt{[A_{\mathrm{t}}(\theta = 0°)/A_{\mathrm{fp}}]} \quad (10.15)$$

In the MIXS-T design (Section 10.3.3.5), I is approximately the square root of (100/4), or 5 times, meaning that if, pessimistically, the particle rejection efficiency in the telescope focal plane was the same as in the case of planar collimator of identical grasp, the telescope would be more sensitive by a factor five as a means of collecting X-rays. The real advantage of the telescope, however, is that in elevated solar flare states and for the major rock-forming elements Mg, Al and Si, its sub-aperture angular resolution makes possible the geochemical mapping of planetary surfaces on linear scales of (<10 km), which no practical collimator can match from any realistic orbital altitude.

10.3.3 Missions

The following sections briefly describe in turn the missions to the Moon (Sections 10.3.3.1 and 10.3.3.2), the asteroids (Sections 10.3.3.3 and 10.3.3.4) and to the planet Mercury (Section 10.3.3.5) which have carried, or will deploy, X-ray spectrometers.

10.3.3.1 Apollo 15 and 16[4,73–75]

Little publicized at the time, the Apollo Command-Service Modules provided a platform for remote sensing of the Moon from low, near-equatorial orbit, while the Lunar Module famously carried its astronaut cargo down to the surface and back again. The fifth and sixth lunar landing missions in 1971–2 carried X-ray spectrometers based on simple collimators and gas proportional counters, which produced the first (and, to date, still the most extensive) maps of the element ratios Mg : Si and Al : Si from any planetary surface.

The Apollo spectrometers consisted of three lunar-pointing gas proportional counters with a total area of $75\,cm^2$ and a single, much smaller ($3\,mm^2$) solar monitor gas counter with a quasi-hemispheric ($\sim 106°$) field-of-view. The counting gas was argon−9.5% carbon dioxide−0.5% helium at one bar pressure and the common detector entrance window was 25 μm beryllium. The single anode wire was surrounded by guard wires to discriminate against charged particle tracks and so reduce background. Because of the low ($h < 100$ km) orbital height, a coarse collimator (60° FWZM) provided acceptable linear resolution on the lunar surface. A shortcoming of any XRF spectrometer design based on a gas proportional counter is that the energy resolution is insufficient to separate the major rock-forming elements Mg, Al and Si by pulse height alone. A set of *balanced filters* must be added to the common, gas-tight counter window to differentiate the responses of the individual detectors. The detector windows:

1. 25-μm Be
2. 25-μm Be + 5-μm Mg
3. 25-μm Be + 7.5-μm Al.

provide sensitivity to fluorescence from (1) all elements above sodium, (2) the element magnesium and (3) the elements magnesium and aluminium, respectively. This balanced filter approach is still being employed, some 40 years after Apollo (see Sections 10.3.3.3 and 10.3.3.5).

The Apollo X-ray spectrometer was co-mounted with an α-particle spectrometer in a common package weighing ~26 kg. Together, Apollos 15 and 16 accumulated over 180 hours of X-ray data covering about 20% of the lunar surface. Despite being acquired at solar minimum, the X-ray spectra showed significant differences in composition between the lunar maria and highland regions. Some of the X-ray coverage was of the "hidden" far side of the Moon, including the large crater Tsiolkovsky. On their return coast trajectories from the Moon to the Earth, the spectrometers were also used to observe the bright galactic X-ray sources Sco X-1 and Cyg X-1. Observing the variability of point-like astronomical sources was informative in resolving the anomalous departure of the honeycomb collimator response function from the ideal triangular profile expressed in Equation (10.3). Note that astronomical measurements with a large collimated aperture must become *confused* for all but the brightest cosmic sources – that is, at any given brightness level, more than one source is in the collimator field-of-view at any given time. The second use of a planetary XRF spectrometer for cosmic X-ray astronomy during transfer to the target planetary body is yet to be fully exploited.

10.3.3.2 SMART-1 and Other Lunar Missions[76–80]

At the time of writing (November 2007), the world's space agencies are embarked upon "A Return to the Moon". Since the emphasis of this new phase of lunar exploration is on exploitation as much as investigation, it is not surprising that X-ray spectrometers are key payload elements of almost all the new generation of orbiters and landers. Of course, there still remain important unanswered scientific questions regarding the composition of the Moon: Did a global melting event give rise to the Al-rich crust? Does the South Pole –Aitken basin – an immense impact feature – contain material excavated from deep within the mantle?

Following on from the detailed study by the European Space Agency (ESA) of a large lunar observatory – *MORO*[76] – Europe has led "The Return" with its *SMART-1* technology mission, launched on September 27th 2003, primarily as a test-bed for ion propulsion, but carrying in addition a set of demonstrator experiments, including the *Demonstration Compact Imaging X-ray Spectrometer* (D-CIXS).[77] D-CIXS is a low mass (4.5 kg) instrument whose key elements are (a) a dual microstructured collimator, formed from UV sensitive photoresist, (b) an array of 24 "swept charge" detectors (SCDs)[77] arranged in three parallel facets of eight detectors each and (c) a Peltier-cooled, 500 μm Si PIN diode-based X-ray Solar Monitor (XSM).[79]

The SCD is a form of charge coupled device (CCD) in which charge is transferred to the single low-capacitance, low noise output node not in cartesian coordinates but by a single set of electrodes lying orthogonal to the main diagonal of the $1 \times 1\,cm^2$ device. This geometry has been shown to combine good spectroscopic performance ($\sim$250 eV FWHM) with – for a silicon-based X-ray detector – large area, good resistance to particle-induced radiation damage and relatively high temperature operation. As in terrestrial applications, the ability to operate a semiconductor detector in space without active (*e.g.* thermoelectric) or passive (*i.e.* radiator-based) cooling leads to a much simpler spectrometer design.

In *SMART-1*'s 300 km mapping orbit, reached in March, 2005, the 8° field-of-view central facet (facet 2) of D-CIXS pointed towards nadir while the remaining offset facets (1 and 3) with their 12° collimation pointed back 10° and forward 10° along the spacecraft's ground track. Facet 3 incorporated a 6 μm thick magnesium filter to augment the ability of the SCDs to separate the Mg K lines from those of Al and Si. Taking the collimator and Al-polyimide UV filter transmissions into account, the effective area of D-CIXS was $14\,cm^2$, about one-fifth that of the *Apollo 15* and *16* X-ray instruments. Published results from D-CIXS[78] to date include only one lunar spectrum, obtained during an M-class solar flare while *SMART-1* was flying over the Mare Crisium region on the Moon's eastern limb. The observed energy resolution (~380 eV FWHM) is degraded compared to pre-launch calibration and the detector response to Ca K X-rays appears distinctly non-Gaussian. Since D-CIXS is composed of 24 independent SCD detectors, each possibly with its own particular radiation dose accumulated during the 15 month long transfer of *SMART-1* from Earth to lunar orbit by ion propulsion, it is not surprising that the instrument or facet responses are poorer than expected for a single, undamaged SCD.

Observations of the strong cosmic X-ray source Sco X-1, published online by ESA as part of a 2004 *SMART-1* mission status report and analysed by the present author, gave a source spectrum that was plausible in terms of normalization and slope, but accompanying offset pointings gave a higher-than-expected background. The XSM solar monitor channel of D-CIXS, however, did successfully return long observations of the Sun in the 1–20 keV band; the cross-calibration of the XSM with GOES is described by Vaananen *et al.*[79] The radiation resistance of the XSM detector is described by Laukkanen *et al.*,[79] who also define a thermal annealing strategy to help restore energy resolution lost with accumulated proton dose.

The *SMART-1* mission ended with a planned impact on the Moon in September, 2006. An upgraded version of D-CIXS, including the XSM, is being prepared for launch on the Indian Chandrayan-1 lunar orbiter in April, 2008. The Chinese *Chang 'E-1* (launched October 24th 2007) and the Japanese *SELENE* (Selenological and Engineering Explorer), renamed *Kaguya* after its launch on September 14th 2007, are the other entrants in the new Asian-led Moon race. While little is known of the *Chang'E-1* payload, an X-ray spectrometer is certainly an impressive part of the *SELENE* payload.[80] The lunar XRF detector (XRF-A) consists of a large (100 cm^2) mosaic of 16 separate X-ray CCDs preceded by 5 μm thick beryllium windows and passively cooled to 230 K by means of a very large (800 cm^2) radiator. The field-of-view of XRF-A is 12°, producing 20 km surface resolution from the spacecraft's 100 km near-polar mapping orbit. The solar monitor function is performed by a four-fold array of filtered Si PIN diodes, sensitive to 1, 2, 4 and 8 keV X-rays (SOL-B) and by a single CCD viewing a standard sample plate (SOL-C). The standard sample plate concept is described in more detail in Section 10.3.3.4. The total mass of the *SELENE* X-ray Spectrometer is 23.5 kg and the power budget is 35.5 W, both resource figures that are much higher than their equivalents for the miniaturized D-CIXS on *SMART-1*. Overall, *SELENE* carries 14 scientific instruments with a total mass of 270 kg.

10.3.3.3 NEAR Shoemaker[81–86]

Some key problems in asteroid research are:

1. The relationship of the observed optical/IR spectral class (*e.g.* S – stony; C – carbonaceous) to the actual composition of the surface.
2. The relationship between (assumed) asteroid parent bodies and the daughter meteoritic materials that arrive at the surface of the Earth.

More than for any other class of planetary body, therefore, XRF is an essential technique for understanding the asteroids. In the case of a spacecraft in orbit about such a small body as a typical asteroid, a non-imaging (collimated) instrument can formally only observe X-ray fluorescence from the surface if the body fills its entire field of view. If not, the observation is confused by the cosmic X-ray background radiation and (possibly) by point cosmic X-ray sources. A collimated instrument therefore has an associated critical orbital height h_{max} above which surface observations may be compromised:

$$h_{\text{max}} = a \tan \theta_{\text{w}} \qquad (10.16)$$

where $2a$ is the characteristic size of the planetary body. For a 1 km asteroid and a 5° FWHM collimator, h_{max} is about 12 km, implying highly accurate interplanetary navigation to place the spacecraft platform in a close orbit around an object to which it can barely be gravitationally bound.

Because there are large, diverse populations of both "main belt" and "near earth" objects, the future number of dedicated asteroid missions is nevertheless potentially huge. The first such undertaking – the eventful *Near Earth Asteroid Rendezvous (NEAR)* mission – was launched by NASA towards the large S-class asteroid 433 Eros in February 1996. Before its final soft landing on Eros in February 2001, *NEAR* was temporarily lost in space and acquired the name of the late Gene Shoemaker, the celebrated American planetary geologist. *NEAR* was the first spacecraft to orbit an asteroid and carried as part of its instrument suite a combined X-ray/gamma-ray spectrometer, the XRGS. The main detectors for the X-ray channel were updated copies of the gas counters carried on *Apollo 15* and *16*, 25 years before. The intrinsic energy resolution of these counters was only ~850 eV FWHM at ~6 keV so that discrimination of Mg, Al and Si required the use of the "balanced filter" technique (Section 10.3.3.1). The collimators were made of Cu-Be alloy and had a 5° FWHM field-of-view. One of the two solar monitor counters was based on an Amptek[28] silicon PIN diode, the other was a small gas counter. The gamma-ray channel consisted of a NaI(Tl) scintillator with a bismuth germanate (BGO) shield. The somewhat conservative nature of the XRGS design is traceable to the fact that the entire instrument programme – from project kick-off to delivery of the XRGS to the spacecraft – took less than 22 months.[84]

Analysis of the X-ray signal received from an irregular asteroid body involves the unravelling of complex relationships between source, surface and detector orientations. In its 50 or 35 km circular mapping orbits, *NEAR* orbited Eros above the terminator – the line between the night and day "hemispheres" – that gave rise to unfavourable viewing geometries for the XRGS. The most important X-ray data set was obtained on May 4th 2000, when a 40-minute long M-class flare led to the spatially-resolved determination of Mg/Si. Al/Si and Fe/Si abundance ratios and, hence, to the conclusion that the surface of Eros had never been differentiated by melting.[85,86] Overall, the number of solar flares that occurred during optimal viewing conditions was rather small.

10.3.3.4 *MUSES-C*/*Hayabusa*[87–91]

Slips in launch date can cause the target for a given asteroid mission to change several times. The target object for *MUSES-C* changed, during mission planning, from 4660 Nereus to 1989ML to, finally, the Apollo-class asteroid 25143 Itokawa. The third engineering test mission in the programme of the (then) Japanese space agency ISAS, this highly ambitious spacecraft was launched on May 9th 2003 and renamed *Hayabusa* ("Falcon") after successful insertion into its interplanetary transfer orbit. Like *SMART-1* (Section 10.3.3.2), *MUSES-C* used an ion engine to reduce the travel time to its target, but its mission was not limited to simple remote study of a planetary surface. Its further aims were to land on that surface and to then return a small soil sample to the Earth for detailed analysis. Landings on the 548-m long Itokawa (formerly designated 1998 SF36) with its curious "head and body" shape took place on November 20th and 26th 2005; spacecraft problems have delayed the return to Earth until June 2010. On its way to the asteroid, *Hayabusa* obtained, in May 2004, a gravity assist from an Earth–Moon swingby, which permitted X-ray observations of the far side of the Moon.

The *Hayabusa* X-ray Spectrometer (XRS) was innovative in several regards: it was the first flight instrument to employ cooled silicon Charge Coupled Devices (CCDs) in the detection plane and also the first (along with D-CIXS on *SMART-1*) to attempt dual use of an X-ray spectrometer – for astronomy (observing the cosmic X-ray background) during its long cruise to target and for remote geochemical surveying on arrival. As noted in Section 10.2.1, silicon CCDs convey a significant advantage in energy resolution over the proportional counter instruments previously flown on *Apollo* and *NEAR*. The 140 eV FWHM delivered by a CCD at the energy of Al Kα is sufficient to separate Al and Mg K-shell emission without the need for bandpass filtering. A 5 μm thick Be filter excluded optical and UV light and provided 20% X-ray transmission at 0.7 keV, the energy of Fe L X-rays (recall the conclusions of Table 10.1) The *Hayabusa* XRS is described in Table 10.2. The 3.5°

Table 10.2 Characteristics of the *MUSES-C*/*Hayabusa* X-ray Spectrometer (XRS).

Parameter	*Value*
Mass	4.5 kg (1.7 kg detector + 2.8 kg electronics)
Power	12 W
Detector	4 1024 × 1024 pixel Hamamatsu front side-illuminated CCDs, with 24 μm pixels
Collecting area	25 cm^2
Collimator field-of-view	3.5° FWHM
Count rate	$<1000\ s^{-1}$, dependent on solar activity
Bandpass	0.7–10 keV
Operating temperature	<230 K
Solar monitor	CCD of the same type as main detector, viewing the standard sample plate

field-of-view of the phosphor bronze collimator matches the expected size of the asteroid from the 10 km nominal survey orbit [see Equation (10.16), above]. The standard sample plate (SSP) is a glassy material of artificial composition intermediate between geological chondritic and basaltic materials. The SSP provides a reference signal that directly folds the variable solar X-ray spectrum through a detection channel physically and electronically identical to that of the main spectroscopic detector.

The main contribution of the XRS to our understanding of asteroid Itokawa was to show that, despite its irregular shape, the body showed no differences in elemental composition between its "eastern" and "western" sides. The major element ratios measured during *Hayabusa's* descent to the surface were, according to Okada *et al.*:[90]

$$Mg/Si = 0.78 \pm 0.09 \text{ and } Al/Si = 0.07 \pm 0.03;$$

the uncertainties being too large to definitively identify the sub-class (H or L) of chondrite. Most XRS observations, unfortunately, took place in conditions of extreme solar quiet. Lunar observations were limited to only 30 minutes duration, with the Moon occupying only about 3% of the XRS field-of-view. After subtracting the known cosmic X-ray background spectrum, significant peaks of Al and Si were observed in a ratio "consistent with a typical anorthosite".

10.3.3.5 Messenger and BepiColombo[92–95]

The importance of X-ray remote sensing for the exploration of the planet Mercury is now well-established. Geochemically, Mercury is the least known of the terrestrial planets, with only Mariner 10 colour difference data obtained during three flybys in 1974–5 to indicate relative depletion of surface Fe and Ti relative to the Moon and predominantly Earth-based radar and optical albedo measurements to hint at the presence of water ice and/or sulfur deposits at the poles. High quality compositional data for Mercury are crucial to our understanding of the evolution of the inner solar system. At present, there are no clear associations drawn between the terrains (inter-crater plains, smooth plains, highlands) mapped by *Mariner 10* over less than half the surface of Mercury with any geochemical boundaries. Discussion of the planet's tectonic and volcanic histories has therefore proceeded unconstrained.

Explanation of the planet's anomalously high uncompressed density – implying a large core and thin mantle – also requires detailed compositional data, to distinguish between competing histories of Mercury's formation out of the solar nebula. Did a massive impact remove the outermost layers of the protoplanet? Mercury's crust will not reflect the planet's bulk composition but will have been modified by vulcanism, space weathering and impacts. XRF measurements of surface composition will need to be spatially correlated with surface terrains and landforms; crater ejecta and central peaks give access to material excavated from depth, but demand surface resolutions on scales of 10 km or less.

Understanding Mercury requires both global XRF measurements at relatively coarse surface resolution and very high resolution measurements over a limited (but representative) fraction of the planet. The sputtered alkali metal rich "exosphere" of Mercury may provide an interesting foreground X-ray signal, while *Mariner 10*'s observations indicate that Mercury's magnetosphere may be the most dynamic in the solar system,[95] riven by magnetic substorms and potentially a variable source of intense X-ray emission. All these science questions will hopefully be answered by two very different X-ray instruments operating in the very difficult thermal and radiation environment of the innermost solar system.

The NASA *Messenger* spacecraft, launched in 2004, is due to make its first flyby of Mercury in January 2008, before entering its highly elliptical (200 by 15 000 km) near-polar (inclined 80° with respect to the planet's rotation axis) mapping orbit in April, 2009. During the nominal one-year mission, the *Messenger* X-ray Spectrometer – effectively an improved rebuild of the *NEAR* XRS with a 12° FWHM collimator – will be capable of 40 km surface resolution in the northern hemisphere, but only 3000 km in the south. The corresponding exposure times are ~100 and 2000 s, respectively. *Messenger* XRS[96] will therefore essentially address only those questions associated with *global* composition. On *Messenger*, the gamma-ray channel is combined with the neutron detector, rather than with the X-rays sensors. A Si PIN diode monitor counter with an active area of only 0.12 mm^2 faces the Sun. Silicon was considered for the main detection plane, but was ultimately rejected on the grounds that the attainable area was much less than could be achieved (30 cm^2 in total) by gas counter technology.

The ESA/JAXA *BepiColombo* mission to Mercury, due for launch in September 2013, consists of two spacecraft, the Mercury Magnetospheric Orbiter (MMO), built by Japan, and the European Mercury Planetary Orbiter (MPO). The latter spacecraft includes in its payload the Mercury Imaging X-ray Spectrometer (MIXS) and its sister solar monitor the Solar Intensity X-ray Spectrometer (SIXS). The complementary MIXS-C (collimator) and MIXS-T (telescope) channels are designed to address *both* the local and global aspects of Mercury's geochemistry. Most significantly, MIXS-T will be the first imaging telescope to be deployed in planetary XRF, based on the novel technologies of low-mass, square-pore microchannel plate (MCP) optics and silicon macropixel DEPFET arrays. The imaging MIXS-T and non-imaging MIXS-C are described in detail in Table 10.3 and in the schematic Figure 10.5. The surface resolution achievable with MIXS-T depends on the solar flare state and the element in question. Despite *BepiColombo*'s arrival at Mercury in September 2019 closely coinciding with solar minimum, this resolution could be of the of order only a few kilometres in those swathes of the planet imaged by MIXS-T while illuminated by M- or X-flare solar X-ray intensities.

Table 10.3 Key parameters of the *BepiColombo* MIXS instrument. The mass budget for both channels, their common digital processor unit and common power supply is 7.2 kg. The MPO mapping orbit is a 400 × 1500 km, 2 hour period, near-polar ellipse.

Parameter	*Value*
MIXS-T	
Geometry	Two-stage Wolter Type 1 approximation
Aperture	21 cm
Focal length	1.0 m, limited by height of spacecraft bus
Microchannel plates	Radially-packed arrays of square cross-section channels, radii of curvature: 4.0 and 1.33 m; 36 paired sectors arranged in three annuli
Detector	32 × 32 array of 300 μm silicon active pixels; 450 μm depletion depth
Angular resolution	Pixel limit: 2 arcminutes
Surface resolution	~1 km (pixel limit at apoherm)
Field of view	1.1°
Grasp	$0.006\ cm^2$ sr at 1 keV
Exposure (dwell) times	22 s at apoherm; 22 s at periherm
MIXS-C	
Geometry	Radial collimator
Aperture	8 × 8 cm
Detector-optic distance	55 cm
Microchannel plates	2 × 2 array of square-packed, square pore channel plates, spherically slumped to a radius of 5 cm
Detector	Identical to MIXS-T
Field of view	6.5° (core); 10.4° total
Grasp	$0.035\ cm^2$ sr
Exposure (dwell) times	210 s at apoherm; 28 s at periherm

Figure 10.5 Schematic view of the MIXS experiment; the larger circular aperture is the Wolter Type 1 telescope.

Acknowledgements

GWF wishes to thank his colleagues on the *Beagle 2* XRS and *BepiColombo* MIXS teams for many interesting discussions on planetary X-ray fluorescence. Thanks especially to James Carpenter (University of Leicester) for providing Figure 10.4 and to Phil Potts (Open University) who commissioned this chapter and then patiently bore the consequences.

References

1. D.F. Blake, *Adv. X-ray Anal.*, 2000, **43**, 487.
2. R. Rieder, H. Wanke, T. Economou and A. Turkevich, *J. Geophys. Res.*, 1997, **102**(E2), 4027.
3. Y. Surkov, *Exploration of Terrestrial Planets from Spacecraft*, Wiley/Praxis, Chichester, 1997, pp. 245–311.
4. I. Adler and J.L. Trombka, *Geochemical Exploration of the Moon and Planets*, Springer Verlag, Berlin, 1970.
5. K. Das Gupta, H.W. Schnopper, A.E. Metzger and R.A. Shields, *Adv. X-ray Anal.*, 1966, **9**, 221.
6. H. Ida and J. Kawai, *Spectrochim. Acta.*, 2005, **B60**, 89.
7. J.Z. Wilcox, E. Urgiles, R. Toda, S. Douglas, T. George and J. Crisp, *Geophys. Res. Abstr.*, 2005, **7**, 00311.
8. J. Bruckner, R.Rieder, G. Klingelhofer and R. Gellert, The Rosetta alpha X-ray spectrometer, preprint available at: http://www1.mpch-mainz.mpg.de.
9. V. Radchenko and seven co-authors, *Appl. Radioactivity Isotopes*, 2000, **53**, 521.
10. M. Omand, J.L. Campbell and J.A. Maxwell, *Nucl. Instrum. Methods*, 2005, **B229**, 123.
11. D. Wegrzynrk, A. Markowicz, E. Chinea-Cano, S. Bamford and P.J. Potts, *Adv. X-ray Anal.*, 2003, **46**, 388.
12. A.C. Huber and J.A. Pantazis, *Nucl. Instrum. Methods*, 1995, **B99**, 665.
13. Ketek GmBH, Gustav Heinemann Ring, D-18739 Munich, Germany.
14. P. Kump, M. Necemer and J. Drinovic, *X-Ray Spectrom.*, 2004, **33**, 107.
15. R. Gellert, ten co-authors, *J. Geophys. Res.*, 2005, **111**, E02S05.
16. R.D. Cockfield and T.S. Chan, in *IEEE Proc. Energy Conversion Conf. 2002*, 2004, p. 134.
17. J.-P. Fleurial and six co-authors, in *Proceedings International Forum on Space Technology and Applications*, Albuquerque NM, 2000.
18. C. Ponnamperuma, A. Shiyoyama, M. Yamada, T. Hobo and R. Pal, *Science*, 1977, **197**, 455.
19. P. Toulmin III, A.K. Baird, B.C. Clark, K. Keil and H.J. Rose Jr., *Icarus*, 1973, **20**, 153.
20. B.C. Clark and A.K. Baird, *Earth Planetary Sci. Lett.*, 1973, **19**, 359.
21. B.C. Clark and nine co-authors, *J. Geophys. Res.*, 1977, **82**, 4577.

22. B.C. Clark, P. Toulmin III, A.K. Baird, K. Keil and H.J. Rose Jr., *Science*, 1976, **193**, 804.
23. K. Larsen, R.E. Arvidson, B.L. Joliff and B.C. Clark, *Lunar and Planetary Science Conference XXX*, 1999, available online at www.lpi.usra.edu.
24. N.T. Bridges and J.A. Crisp, *Lunar and Planetary Science Conference XXX*, 1999, available online at www.lpi.usra.edu.
25. R. Rieder, H. Wänke, T. Economou and A. Turkevitch, *J. Geophys. Res.*, 1997, **102**, 4027.
26. R. Reider and seven co-authors, *Science*, 1997, **278**, 1771.
27. H. Wänke, J. Brückner, G. Drebus, R. Reider and I. Ryabchikov, *Space Sci. Rev.*, 2001, **96**, 317.
28. Amptek Inc., 14 De Angelo Drive, Bedford MA 01730, USA.
29. V. Linkin and forty-eight co-authors, *Planet. Space. Sci.*, 1998, **46**, 717.
30. C.N. Foley, T.E. Economou, W. Dietrich and R.N. Clayton, *Lunar Planetary Science Conference XXXI*, 2000, available online at www.lpi.usra.edu.
31. C.N. Foley, T.E. Economou and R.N. Clayton, *Lunar Planetary Science Conference XXXIII*, 2002, available online at www.lpi.usra.edu.
32. J.L.C. Stewart and eight co-authors, *Geophys. Res. Abstr.*, 2003, **5**, 09869.
33. D.L. Talboys and seven co-authors, Investigation of the feasibility of in situ dating on Mars using the Beagle 2 gas analysis package and X-ray spectrometer, in *Proceedings of 36th COSPAR Assembly, Beijing*, 2006.
34. G. Klingelhőfer and ten co-authors, *Hyperfine Interact.*, 2002, **144–145**, 371.
35. G.I. Butcher and five co-authors, *Nucl. Instrum. Methods*, 2006, **A564**, 585.
36. C.L. Mallett, J.M. O'Meara, J.A. Maxwell and J.L. Campbell, *X-Ray Spectrom.*, 2006, **35**, 329.
37. R. Rieder, R. Gellert, J. Brückner, G. Klingelhöfer, G. Dreibus, A. Yen and S.W. Squyres, *J. Geophys. Res.*, 2003, **108**, E12, 8066.
38. R. Gellert and the APXS Science Team, Remarks on the calibration of the alpha-particle X-ray spectrometers on board the MER Rovers, NASA Planetary Data System user note, 2006.
39. R. Gellert and nine co-authors, Notes on Alpha Particle X-ray Spectrometer (APXS) data reduction, NASA Planetary Data System user note, 2004.
40. T. Economou, R. Pierrehumbert, D. Banfield and G.A. Landis, *Proceedings Seventh International Conference on Mars*, Pasadena CA, 2007, available online at www.lpi.usra.edu.
41. J. Jensen and eight co-authors, *X-Ray Spectrom.*, 2006, **34**, 359.
42. The Spirit and Opportunity APXS data sets are archived within the public NASA Planetary Data System, accessible via http://pds.nasa.gov.
43. H.Y. McSween and thirty-four co-authors, *Science*, 2004, **305**, 842.
44. S.W. Squyres and eighteen co-authors, *Science*, 2004, **306**, 1709.
45. R. Gellert and eleven co-authors, *Lunar Planetary Sci. Conf. XXXVI*, 2005, available online at www.lpi.usra.edu.

46. A.R. Vasavada and the MSL Science Team, *Proceedings Seventh International Conference on Mars*, Pasadena, CA 2007, available online at www.lpi.usra.edu.
47. D.F. Blake and six co-authors, *Geophys. Res. Abstr.*, 2005, **7**, 02160.
48. D.F. Blake and seven co-authors, CHEMIN: a definitive mineralogy instrument on the Mars Science Laboratory (MSL '09) Rover, *Proceedings Seventh International Conference on Mars*, Pasadena CA, 2007, available online at www.lpi.usra.edu.
49. L. Marinangeli and the MARS-XRD team, *Lunar and Planetary Science Conference XXXVIII*, 2007, available online at www.lpi.usra.edu.
50. L. Struder, *Nucl. Instrum. Methods*, 2000, **A454**, 73.
51. G. Bertuccio and four co-authors, *Nucl. Instrum. Methods*, 2004, **A518**, 433.
52. J.E. Lees and six co-authors, *Nucl. Instrum. Methods*, 2007, **A578**, 226.
53. M. Collon and seven co-authors, *Acta Astronaut.*, 2006, **59**, 1052.
54. K. Dennerl, *Astron. Astrophys.*, 2002, **394**, 1119.
55. A. Bhasdwaj and thirteen co-authors, *Planetary Space Sci.*, 2007, **55**. 135. See also the *Chandra* Science Center website: http://www.cxc.harvard.edu/newsletters/news_11/solar.html.
56. J.H.M.M. Schmitt, S.L. Snowden, B. Aschenbach, G. Hasinger, E. Pfeffermann, P. Predehl and J. Trumper, *Nature*, 1991, **349**, 583.
57. B.C. Edwards, W.C. Priedhorsky and B.W. Smith, *Geophys. Res. Lett.*, 1991, **18**, 2161.
58. R. Giacconi, H. Gursky, F.R. Paolini and B. Rossi, *Phys. Rev. Lett.*, 1962, 9, 439. See also W. Tucker and R. Giacconi, *The X-ray Universe*, Harvard University Press, Cambridge, MA, 1985, ch. 5.
59. Y. Kamata, T. Takeshima, T. Okada and K. Terada, *Adv. Space. Res.*, 1999, **23**, 1829.
60. B.J. Wargelin and five co-authors, *Astrophys. J.*, 2004, **607**, 596.
61. R.P. Lin and R. Gopalan, *Rev. Sci. Instrum.*, 1991, **62**, 660.
62. A.P. Martin, A.N. Brunton and G.W. Fraser, *Nucl. Instrum. Methods*, 1999, **A422**, 567.
63. GOES X-ray data is available from the NOAA Space Weather Prediction Center (http://www.swpc.noaa.gov).
64. P.E. Clark and J.I. Trombka, *J. Geophys. Res.*, 1997, **102**, 16361.
65. P. Truscott, C. Dyer and C. Peerless, Basalt X-ray fluorescence study for ESA, report no.DERA/CIS/CIS2, 2000.
66. M. Grande, S.K. Dunkin and B. Kellett, *Planetary Space Sci.*, 2001, **49**, 1553.
67. J. Trombka, L.G. Evans, R. Starr, P.E. Clark and S.R. Floyd, *Nucl. Instrum. Methods*, 1999, **A428**, 199.
68. I.M. Govill, *Curr. Sci.*, 2001, **80**, 1542.
69. R. Giacconi, H. Gursky and L.P. van Speybroeck, *Annu. Rev. Astron. Astrophys.*, 1968, **6**, 373.
70. R. Willingale, G.W. Fraser, A.N. Brunton and A.P. Martin, *Exp. Astron.*, 1998 **8**, 281; A. Owens and eight co-authors, *Proc. SPIE*, 2001, **4506**, 136.

71. T. Okada, *Proceedings Lunar and Planetary Science Conference, 2004, XXXV*, ed, J. Näränen, H. Parviainen and K. Muinonen, Proc. IAU Symposium No. 236, 2006, p. 243, available online at www.lpi.usra.edu.
72. S.R. Cherry, J.A. Sorensen and M.E. Phelps, *Physics in Nuclear Medicine*, 3rd edn., Grune and Stratton, 2003.
73. I. Adler, R. Schmadebeck, J.I. Trombka, P. Gorenstein and P. Bjorkholm, *Space Sci. Instrum.*, 1975, **1**, 305.
74. I. Adler, J. Trombka, L. Yin, P. Gorenstein, P. Bjorkholm and J. Gerard, *Naturwissenschaften*, 1973, **60**, 231.
75. J.L. Whitford-Stark and B.R. Burke, *Lunar and Planetary Science Conference*, 1973, XIII, 861–862, available online at www.lpi.usra.edu.
76. B.H. Foing, G. Racca and the MORO Science Team, *Adv. Space. Res.*, 1996, **18**, 85.
77. M. Grande and forty-seven co-authors, *Planetary Space Sci.*, 2003, **51**, 427.
78. M. Grande and twenty-three co-authors, Lunar elemental composition and investigation with D-CIXS X-ray mapping spectrometer on SMART-1, *Proceedings Lunar and Planetary Science XXXV*, 2004; M. Grande and thirty co-authors, The D-CIXS X-ray spectrometer on the SMART-1 mission to the Moon – first results, *Planetary Space Sci.*, 2007, **55**, 494, available online at www.lpi.usra.edu.
79. J. Huovelin and nineteen co-authors, *Planetary Space Sci.*, 2002, 50, 1345; J.Laukkanen and nine co-authors, *Nucl. Instrum. Methods*, 2005, **A538**, 496; M. Vaananen and seven co-authors, "X-ray solar monitor (XSM) – a stellar X-ray spectroscope of the Sun onboard SMART-1", University of Helsinki preprint, 2006.
80. T. Okada and seven co-authors, *Adv. Space Res.*, 2002, **30**, 1909.
81. J.O. Goldsten and fifteen co-authors, *Space Sci. Rev.*, 1997, **82**, 169.
82. S.R. Floyd, J.I. Trombka, H.W. Leidecker, P.E. Clark, R. Starr, J.O. Goldsten and D.R. Roth, *Nucl. Instrum. Methods*, 1999, **A422**, 577.
83. T.P. McClanahan and thirteen co-authors, *Nucl. Instrum. Methods* **A422**, 1999, 582; J.I. Trombka and eleven co-authors, *Nucl. Instrum. Methods*, 1999, **A422**, 572; R. Starr and nine co-authors, *Nucl. Instrum. Methods*, 1999, **A428**, 209.
84. J.O. Goldsten, R.L. McNutt, R.E. Gold, S.A. Gary, E. Fiore, S.E. Schneider, J.R. Hayes and J.I. Trombka, *Acta Astronaut.*, 1997, **39**, 281.
85. T.P. McClanahan and twelve co-authors, *Nucl. Instrum. Methods*, 2001, **A471**, 179; C. Foley and six co-authors, *Icarus*, 2006, **184**, 338.
86. L.R. Nittler and fourteen co-authors, *Meteoritics Planetary Sci.*, 2001, **36**, 1673.
87. T. Okada, M. Kato, A. Fujimura, H. Tsunemi and S. Kitamoto, *Adv. Space Res.*, 2000, **25**, 345.
88. T. Okada, M. Kato, K. Shirai, Y. Yamamoto, T. Matsuda, H. Tsunemi and S. Kitamoto, *Adv. Space Res.*, 2002, **29**, 1237.

89. T. Okada and six co-authors, First X-ray observation of lunar farside from Hayabusa X-ray spectrometer, *Proceedings Conference Lunar and Planetary Science XXXVI*, 2005, available online at www.lpi.usra.edu.
90. T. Okada and six co-authors, X-ray fluorescence spectrometry of asteroid Itokawa by Hayabusa, *Science*, 2006, **312**, 1338.
91. A. Fujiwara and twenty co-authors, Global properties of 25143 Itokawa observed by Hayabusa, *Proceedings Conference Lunar and Planetary Science XXXVII*, 2006, available online at www.lpi.usra.edu.
92. A. Owens, M. Bavdaz, M. Beiersbergen, A.N. Brunton, G.W. Fraser, P. Nieminen, A. Peacock and M.G. Pia, *Proc. SPIE*, 2000, **4506**, 136.
93. BepiColombo, *An Interdisciplinary Cornerstone Mission to the Planet Mercury, – System and Technology Study Report ESA-SCI*, 2001, p. 1.
94. J. Brückner and J. Masarik, *Planet. Space Sci.* 1997, **45**, 39; Y. Langevin, *Planet. Space Sci.*, 1997, **45**, 31; P.E. Clark and J.L. Trombka, *Planet. Space Sci.*, 1997, **45**, 57; V.V. Shevchenko, *Adv. Space Res.*, 2004, **33**, 2147.
95. J.A. Slavin, *Adv. Space. Res.*, 2004 **33**, 1859; L.G. Blomberg and J.A. Cumnock, *Adv. Space Res.*, 2004, **33**, 2161.
96. R.D. Starr and six co-authors, The X-ray spectrometer for Mercury Messenger, *Proceedings Conference Mercury: Space Environment, Surface and Interior*, 2001; R.E. Gold and twenty-nine co-authors, *Planetary Space Sci.*, 2001, **49**, 1467.

Subject Index

www.ingramcontent.com/pod-product-compliance
Ingram Content Group UK Ltd.
Pitfield, Milton Keynes, MK11 3LW, UK
UKHW062338170926
8624IPUK00029B/415